Advanced MEMS Sensors and Actuators

ADVANCED MEMS SENSORS AND ACTUATORS

Prof. Dr. R.K. Singh

ARCLER
PRESS

www.arclerpress.com

Advanced MEMS Sensors and Actuators

Prof. Dr. R.K. Singh

Arcler Press
4164 Lakeshore Road
Burlington, ON L7L 1A4
Canada
www.arclerpress.com
Email: orders@arclereducation.com

ISBN: 978-1-77956-982-0 (Hardcover)

ABOUT THE AUTHOR

Dr.R. K. Singh, born on 11th June, 1969, is a Professor and worked as Vice Chancellor, Patliputra University, Patna, Bihar, a state university of Bihar. Having throughout first division career, he obtained his Ph.D. from Allahabad University and M. E. from BITS, Pilani. He is recipient of numerous awards and honours. He is engaged in teaching, research, administration and consultancy for the last 28 years. Apart from this, he has worked as Director, B.T. Kumaon Institute of Technology, Dwarahat, Almora, Uttarakhand an autonomous Institution of Government of Uttarakhand and appointed as Director since 01/10/2015 at B.T. Kumaon Institute of Technology, Dwarahat, Almora,

Dr. Singh started his esteemed career as teaching assistant at Birla Institute of Technology and Science (BITS), Pilani (one and half years, Oct. 4, 1990 - July 1993) and as RA & Scientist(P) at CEERI, Pilani (more than one and half years, July 1993 – March 20, 1995), as an Assistant Professor March 21, 1995 – July 30, 1998) at BT Kumaon Institute of Technology (BTKIT), Dwarahat, formerly known as Kumaon Engineering College (KEC), Dwarahat, Almora (more than 3 years) and Associate Professor July 31, 1998 – July 30, 2006) at BT Kumaon Institute of Technology, Dwarahat (8 years). He became professor (July 31, 2006 – till date) at the age of 37 years, at BT Kumaon Institute of Technology, Dwarahat. He held the positions of HOD ECE, Principal, BT Kumaon Institute of Technology, Dwarahat, Officer on Special Duty, Registrar and Controller of Examination, at Uttarakhand Technical University, Dehradun.

Prof. Singh has received felicitations and awards by professional bodies such as Rastriya Samman Purskar- presented by Indian Society for Intellectual Development in presence of His Excellency, ShriSatya Narayan Reddy, Ex. Governor, U.P. and His Excellency, Shri Bhism Narayan Singh, Ex. Governor, Bihar, Jewel of India award- presented by International Institute of Education and Management in presence of His Excellency, Shri Bhism Narayan Singh, Ex. Governor, Bihar and Hon. Minister, Health, Govt. of Delhi, Shri Yoga Nand, Rastriya Ekta Award- presented by Global Economic Council, New Delhi, in presence of His Excellency, Shri Madan Lal Khurana, Governor, Rajasthan on 28th April, 2004 and Arch of Excellence Award-2004- presented in National Seminar-All India Achievers Conference in presence of His Excellency, Shri Satya Narayan Reddy, Ex. Governor, and U.P. He is member of institution of Engineers (India). He has worked as Programme Officer of National Service Scheme at BTKIT, Dwarahat.

Prof. Singh has held positions of responsibility such as - Member, various committees of UTU University and Govt. of Uttarakhand, First Appellate Authority, BTKIT, Dwarahat (for Right to Information Act, 2005); Dealing World Bank assisted Technical Education Quality Improvement Program for BTKIT; Expert Member, of various committees of several Institutions and organizations of national and international repute. Technical Expert for AICTE, NBA, and UGC of Government of India

Prof. Singh has guided 22 Ph.Ds. and about 15 M.Tech.thesis. He has participated in various National and International conferences, and published more than 120 research papers in national and international journals of repute. He has also authored eight books. Besides these, he has undertaken numerous consultancies for state government and various PSUs. He has provided technical assistance as member to groups of expert set up for various tasks.

Prof. Singh is member of the various prestigious bodies like Institution of Engineers (India), Kolkata, IEEE, etc.

Prof. Singh has visited various International countries as key note speaker and session chair-man. Some of visited countries are Peris France, Hong Kong, Zurich, Switzerland, Kuala lumpur, Malaysia and Bangkok, Thailand where he Delivered Key note speech and chaired the session of International Conference.

TABLE OF CONTENTS

LIST OF FIGURES

PREFACE

The Internet of Things is no longer something far off; it is a living tapestry of daily existence. Houses raise and lower thermostats on their own, factories broadcast sensor data around the clock, and hospitals link up monitors and pumps to clinical systems in real time. These advantages are built on untold millions of small computers—many low-power, resource-limited, and deployed in the millions. The same attributes that make them revolutionary also expose them: contentious processors which cannot run heavy defenses, longer lifetimes lasting beyond vendor support, and physically exposed systems which will be mounted on a wall, in the field, or inside a vehicle. Security for such systems can't be a legacy IT controls copy-and-paste. It must support bursty short messages, loss networks, periodic power, and heterogeneous devices cobbled together by diverse vendors. Most important, it must be respectful of safety: a faulty security control on an infusion pump or grid sensor is not merely a service outage—it is a killer.

It is not just an issue of securing individual devices but securing the interactions among them. IoT systems combine sensing, actuation, communication, and cloud analytics in extended chains of trust. A compromised link—a default password on a camera, an unpatched gateway protocol, or poorly designed API—is enough to defeat a botnet or a data breach. Attackers have even gotten good at weaponizing scale, taking over dozens of thousands of nodes to spam services or mine sensitive data, but defenders must work within constrained bandwidth and power budgets. Effective IoT security therefore integrates hardware root of trust, performance-aware cryptography, robust identity, carefully bounded network exposure, continuous monitoring, and responsible update techniques. It also requires transparent management—procurement practices, compliance frameworks, and operation runbooks—so that security persists from design and supply chain through deployment and end-of-life.

Cybersecurity in the IoT Era is engineered to meet these realities. The book explains the fundamentals of IoT systems and how attackers might exploit them, and then equips you with real-world defenses that respect device constraints and operational requirements. It finds a middle ground between conceptual simplicity and implementation detail, showing how to design for minimal exposure, select protocols that maintain overhead low, implement cryptography that tolerates limited hardware, and implement detection appropriate to high-volume, low-signal telemetry. Each chapter combines plain-English explanations, structured guidance, and practitioner-experienced methods.

Chapter 1 sets out the basic pieces of IoT and resultant security goals. It defines device classes, gateways, edge computing tasks, and cloud back ends; defines constraints in terms of limited CPU, memory, and power; and maps these facts onto confidentiality, integrity, availability, and safety goals.

Chapter 2 is focused on IoT architecture and the many places an attacker is able to get a foothold. You will see how sensors, actuators, firmware, bootloaders, radios, local networks, gateways, message brokers, APIs, and cloud services create a wide attack surface. The chapter illustrates minimizing exposure using segmented designs, least-privilege connectivity, secure boot chains, and update paths that are rollback and tamper-resistant. Design patterns and anti-patterns are compared to help you make trade-offs early, when they matter most.

Chapter 3 discusses common threats and attacks in IoT, from credential stuffing and device cloning to side-channel abuse, radio jamming, replay, and supply-chain injection. Real-world attack patterns—botnet recruitment, command hijacking, data exfiltration from sensors, and physical tampering—are mapped to kill-chains that identify detection and prevention potential. You will see how careless errors, like leaving dev interfaces open or shipping shared keys, snowball into systemic risk.

Chapter 4 discusses IoT device network and communication protocols and their security implications. It contrasts short-range technologies (BLE, Zigbee, Z-Wave, Thread) with long-range ones (Wi-Fi, cellular IoT, LoRaWAN), and messaging layers such as MQTT, CoAP, AMQP, and HTTP/2. For all of them, the chapter explains authentication modes, session protection, reliability properties, and common misconfigurations. It further discusses segmentation, broker hardening, certificate handling, rate limiting, and gateway policies that preserve function while robustly limiting exposure.

Chapter 5 outlines cryptography for constrained IoT devices with a focus on robustness and feasibility. You will learn when to use symmetric versus asymmetric methods, how to choose curves and ciphers best suited for constrained hardware, and how to manage keys via secure elements, TPM-like hardware, or software vaulting. Secure boot, measured boot, firmware signing, sealed storage, and protocols like DTLS and OSCORE are all covered in this chapter and bring strong protections to constrained stacks. Real-world guidance is given regarding entropy sources, nonce handling, key rotation, and avoiding unsafe shortcuts that make it into embedded code.

Chapter 6 treats privacy and data protection in IoT, where operational and personal data streams flow continuously. It translates principles such as data minimization, purpose limitation, and storage limitation into device and cloud design choices: on-device pre-processing, anonymization at the edge, tokenization, and controlled aggregation. The chapter covers consent models, legal bases for processing, subject data rights, and cross-border transfer implications, translating legal obligations into practices that scale.

Chapter 7 defines access control and identity management with IoT realities. It illustrates how to issue and manage device identities during manufacturing, onboarding, and field replacement; how to bind identities with cryptographic material and attestation evidence; and how to apply authorization models such as RBAC and ABAC to devices, services, and operators. You will find templates for just-in-time provisioning, ephemeral credentials, scoped tokens, and command authorization that resists abuse even if a single node is compromised.

Chapter 8 covers intrusion detection and anomaly detection in systems that are capable of producing sparse, noisy, or high-rate telemetry. It compares signature-based detection, model-based baselining, and physics-informed monitoring, and then shows how to combine them up in resource-aware ways. Topics addressed are selecting signals that warn of device abuse, filtering to reduce false positives, placing sensors on gateways and brokers, and using device attestation and heartbeat checks to spot tampering or cloning. The chapter also addresses how to feed detections into incident response playbooks suitable for fleets of devices across large geographic expanses.

Chapter 9 covers security standards and frameworks that drive procurement, certification, and assurance. It compares large-scale efforts by international organizations and industry consortia, and highlights common controls such as unique identities, secure update, vulnerability disclosure, and baseline cryptography. You will learn how to use these frameworks in your environment, generate compliance evidence during design, and use standardized labels or profiles to encourage vendor responsibility without constraining innovation.

Chapter 10 discusses next-generation IoT security technologies as becoming feasible. Topics include edge confidential computing, post-quantum cryptography preparation for long-lived devices, verifiable supply-chain artifacts (SBOMs with signing and provenance), zero-trust patterns applied to constrained endpoints, and hardware-backed attestation with policy enforcement in the cloud. Chapter 10 rounds out with guidance on evaluating new tools: how to pilot, estimate overhead, and decide when the security benefit is worth the cost and complexity of adoption.

Cybersecurity in the Age of IoT is targeted at professionals tasked with designing, developing, deploying, or assessing connected devices and the systems beneath them. Security practitioners will utilize it to bring enterprise practices to resource-constrained environments. Embedded and firmware developers will benefit from insightful cryptography, secure boot, and update guidance that they can implement on real hardware. Network and cloud architects will inherit the patterns for gateways, brokers, and APIs that govern exposure but not performance. Compliance and risk teams will benefit from the clear linkage to standards and the direction on obtaining credible evidence without slowing down projects. Product owners, technical leaders, and students will get a formalized view that maps device realities to business risk so that they can make informed decisions through the product life cycle.

CHAPTER 1

INTRODUCTION TO ADVANCED MEMS

Contents

INTRODUCTION

MEMS is a virtuous technology that entails the construction of tiny devices that are mechanically, optically, and electrically functional. It not only encompasses the design and engineering of small actuators and sensors but also constitutes a vital element in applications such as automotive systems, biomedical devices, and consumer electronics. The key draw of MEMS is its ability to miniaturize complex systems without sacrificing performance and reducing the consumption of power. With growing need for low-cost, high-performance microsystems, there is an even stronger need for simplicity in the field. This involves understanding the goals of microsystems technology (MST) in providing a framework for comprehending its multi-faceted character, demonstrating how diverse applications and standardization efforts converge to push the frontiers of MEMS technology (Fluitman et al., 1996). A closer look at MEMS reveals its applicability towards addressing mundane issues, particularly sensing and control systems. For instance, MEMS sensor development has enabled driving better applications in environmental monitoring, healthcare, and automobile safety. MEMS sensors provide mission-critical information with unparalleled accuracy and reliability, making them invaluable for uses ranging from pollution detection to life signs monitoring. Moreover, the sensors typically apply sophisticated signal processing algorithms, which enhance their capabilities in hostile environments. The applicability of MEMS is further brought to light by ongoing research into new materials and fabrication methods that can enhance the strength and flexibility of MEMS devices. Therefore, the use of new methods is focused on addressing common problems, such as noise interference and fluctuating environmental conditions, and their implementation in numerous industries (Ahmad et al., 2018). The direction of the future of MEMS will be characterized by greater design, integration, and application innovations. New interdisciplinary collaborations are expected to foster innovation, leading to a new generation of intelligent microsystems that can make decisions autonomously and in real-time. The integration of MEMS with border technologies such as IoT and AI will also be a key contributor in this revolution, making smarter, more efficient devices possible. As the MEMS environment evolves and gets ever more complex, the overall conceptual roadmap set out by MST will serve as a guide for research scientists and engineers. This approach not only provides the means for identifying potential challenges but also presents avenues for interoperability and standardization in the realm of cutting-edge MEMS applications (Fluitman et al., 1996).

Definition and Overview of MEMS

Micro-Electro-Mechanical Systems or MEMS are a groundbreaking integration of mechanical engineering and microelectronics that create devices that perform mechanical functions at a microscopic scale. MEMS employ microfabrication techniques to enable the integration of electronics, sensors, actuators, and mechanical components on a single chip. The integration enables more functionality with reduced power and size compared to traditional systems. The versatility of MEMS technology has, nonetheless, found itself used in a range of applications, from car sensors such as accelerometers and gyroscopes to consumer products and medical devices. With the evolution of the technology, it is vital to grasp the underlying principles of MEMS to branch out to more complex technologies and how this effect innovation in business (Fluitman et al., 1996). The applications of MEMS technology transcend mere component functionality. The field encompasses a wide range of subject areas, necessitating a pragmatic approach to its definition and application. A more in-depth study of generic issues and particular applications in MEMS can give better avenues for standardization and further research. By focusing on practical objectives, the MEMS community aims to create common approaches to microsystem design, fabrication, and integration.

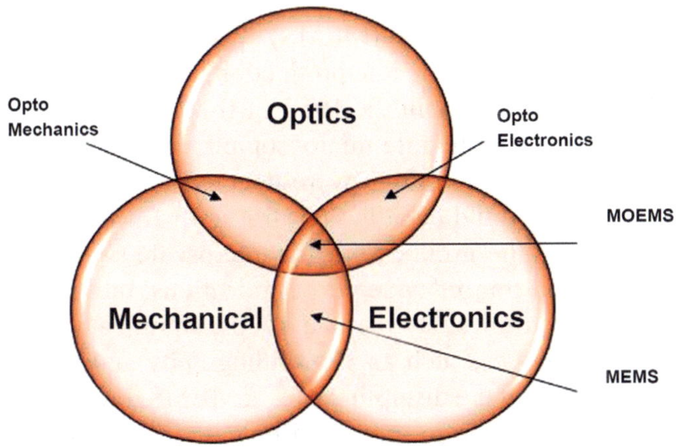

Figure 1.1. Micro-electro-mechanical systems (MEMS) represent a groundbreaking convergence of microelectronics and mechanical engineering, creating devices that perform mechanical functions at the microscale.

Source: https://www.engineersgarage.com/wp-content/uploads/2019/07/A-Diagram-Illustrating-Micro-Electro-Mechanical-System-Opto-Electro-System-Micro-Opto-Electro-Mechanical-System.jpg.

This project seeks to apply MEMS in various industries, thereby encouraging innovation and even more increasing the applicability of MEMS to modern technology. The dispersed nature of the MEMS field calls for collaboration and concurrence among researchers in boundary marking of the essential areas of the field (Fluitman et al., 1996). Particularly dominant within the MEMS approach is the development of RF devices that are capable of addressing the growing needs of wireless communication. As the industry concentrates on reconfigurable functions in devices such as filters, oscillators, and antennas, MEMS components seem like choices due to their efficiency and size. Among them, MEMS electromechanical resonators stand out for their capacity to replace older technologies such as surface acoustic wave (SAW) filters with lower power consumption and reduced integrated functions. The frequency response of these resonators is extremely sensitive to both material and structural parameters, providing the opportunity for new innovation in frequency tuning and device tuning. There remain challenges, such as environmental effects on performance, but the advance of MEMS technology is making way for more complex device function and applications across industries from telecommunications through consumer electronics (Figure 1.1).

Historical Development of MEMS Technology

Micro-electro-mechanical systems (MEMS) technology development is inextricably linked with semiconductor production, materials science research, and high-precision engineering. In the 1950s, MEMS began when researchers began looking for ways to integrate microscopic mechanical components into electronic systems. The initial efforts were devoted to etching and patterning silicon—a material that would be at the forefront of MEMS technology development. This book laid the groundwork for microscale sensors and actuators that would allow engineers to devise new applications, most notably automotive and consumer electronics. By the late 1980s, significant developments in microfabrication techniques, such as photolithography and surface micromachining, further progressed the discipline so that MEMS devices could be mass-produced, which would go on to transform numerous industries. Entering the 1990s, MEMS technology broke out of its traditional sensor applications into more advanced fields, driven by the growing demand for electronics miniaturization and the advent of wireless technology. The decade saw the marriage of MEMS and wireless technology, leading to innovations such as micro-mirrors for optical switching and microfluidics for lab-on-a-chip technology. MEMS's flexibility helped them infiltrate diversified industries like healthcare, telecommunication, and the aerospace industry. During these years, some key industry-academic partnerships emerged. Organizations recognized the potential of MEMS to foster economic growth and moved towards establishing research programs and funding programs—which were instrumental in capitalizing on the base technological achievements. Such collaborations emphasize responding to new issues in technology as well as market trends as part of the evolution path of MEMS (Kokshagina et al., 2015). Looking ahead to the crucial advancements in the 21st century, the MEMS industry has continued to expand, with increased focus on increasing functionality and integration between systems. Improvements in materials, such as the application of polymers and metals to make MEMS, have revealed new ranges of applications. Currently,

MEMS devices permeate the technology scene, showing up in phones, automotive control systems, and even in environmental sensors. The history of MEMS is a quintessential example of trends toward developing generic technology that can be applied to solve dual unknowns of performance solution development. As technological domains grow in complexity, strategic portfolio management becomes a requirement in MEMS based on the results that generic technologies can value in numerous applications (Gad-El-Hak et al., 2017). Not only is the history of MEMS technology development a chronicle, but it also serves as a blueprint for future engineering and integration of technology development.

Importance of MEMS in Modern Applications

The development of Micro-Electro-Mechanical Systems (MEMS) has revolutionized a wide range of technologies, particularly sensor design and packaging. MEMS sensors are essential in numerous applications, enabling precise measurements and control systems in real-time. Their application in active flow control systems, for instance, has been found to be highly beneficial, resulting in efficiency and safety enhancements in aerospace and automotive applications. As specified by the findings of a bachelor thesis, selection among sensors is made simple by stringent testing and verification processes, emphasizing the utmost significance of operation reliability in actual use. This verification builds trust in MEMS technology, demonstrating how it serves as an infrastructure for new applications, which demand high performance and precision, hence cementing its position in current technological advancement (Kopylov et al., 2015). Also, the integration of MEMS with novel materials further enhances their functional capability, creating new applications in various industries. Integration of MEMS with adaptive and responsive materials, such as electroactive polymers and bio-responsive hydrogels, is one such future-oriented integration. These materials enable MEMS devices to dynamically react to external stimuli, expanding their applications from medicine to intelligent clothing. Through the assistance of the unique properties of the materials, designers are able to design high-tech systems with human-centric applications, thus establishing a more personalized relationship with technology. The potential with this synergy is endless, from developing intelligent materials that enhance user experience and functionality in everyday products (Oliver et al., 2009). Overall, the utility of MEMS in modern applications is brought out by their integrability and flexibility, and this has played a key role in bringing about significant advancements across a range of industries. From biomedical devices tracking patient health in real time to consumer electronics providing a comfortable experience, MEMS technology continues to push innovation boundaries. The consistent development of MEMS, enabled by rigorous testing and the application of new materials, ensures that they can serve the increasing demands of modern society. With their development, these technologies provide the opportunity for smarter solutions not only to improve efficiency but also to improve life, pointing to the key role that MEMS have in shaping the future of technology use (Figure 1.2).

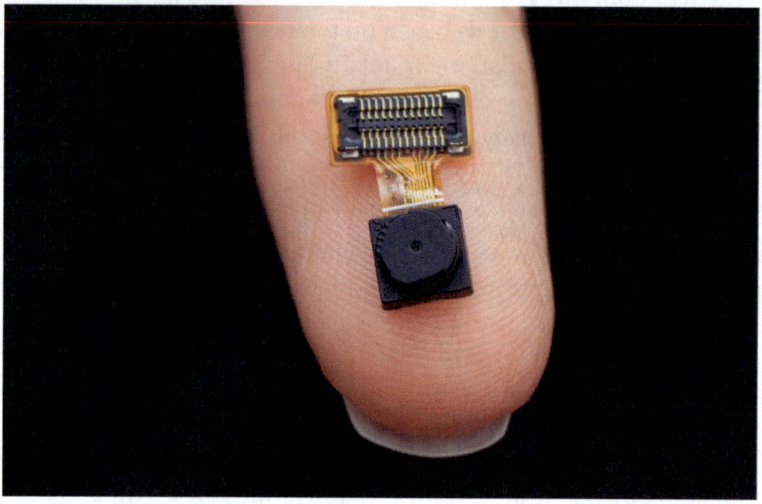

Figure 1.2. MEMS sensors are pivotal in numerous applications, allowing for precise measurements and control mechanisms in real-time.

Source: https://wevolver-project-images.s3.us-west-1.amazonaws.com/0.h36l-f1o9zp6Camerafromamobilephoneonafingercloseuponblackbackgroundstockphoto.jpeg.

1.1. FUNDAMENTAL PRINCIPLES OF MEMS

The fundamental technology of Microelectromechanical Systems (MEMS) is the integration of mechanical components, sensors, actuators, and electronics on a shared microscopic scale. These very small devices use sophisticated manufacturing processes to provide a significant amount of functionality within compact packages. The heart of MEMS operation is the conversion of electrical signals to mechanical motion, which can be observed in devices like accelerometers or gyroscopes. This conversion is generally enabled through mechanisms such as piezoelectricity and electrostatic actuation. Over time and technological advancements, the evolution of these principles gives rise to applications in healthcare, automotive, and consumer electronics. It is important to grasp these basic principles in a bid to achieve trajectory enhancements in MEMS because it provides mechanisms for innovation that result in enhancement of performance and reliability in markets. A significant aspect of MEMS technology is researching fabrication methods that ensure the precision and integrity of microstructures. Among the most popular methods is Deep Reactive Ion Etching (DRIE), which has transformed the landscape of MEMS manufacturing. It involves applying high-density plasmas at cryogenic temperatures to control the silicon structure etching rates with great precision. The relative etching and deposition rates, particularly in the case of passivation with silicon oxide-fluorine layers, directly influence the profile and functionality of MEMS devices. The effect of parameter interaction, such as gas composition and electrode temperature, is included in being aware of the details required for optimizing design. Effective utilization of these methods proves their foundation nature in achieving the desired properties of MEMS devices, which underscores the significance of studying these basics for further developments (Boer et al., 2002). In addition, MEMS devices become more susceptible to advanced operating conditions, such as conditions of noise and interference, which can affect performance to a significant degree. For instance, the efficiency of systems such as iris recognition is compromised by intractable environments that introduce noise and degrade image data so that segmentation becomes challenging. Current studies have investigated several methods to enhance image processing and enhance the accuracy of recognition, with an emphasis on methods that adapt better to environmental noise (Ahmad et al., 2018). These challenges emphasize the need for innovation in MEMS design to withstand harsh conditions but still be operational effectiveness. These major challenges, therefore, need to be met through advanced strategies so that MEMS technology can be made robust and able to address the demands of existing applications (Figure 1.3).

Figure 1.3. Fundamental to MEMS operation is the conversion of electrical signals into mechanical movement, which can be observed in devices like accelerometers or gyroscopes.

Source: https://www.i-pex.com/ips/sites/default/files/inline-images/Structure_MEMS.jpg.

1.1.1. Microfabrication Techniques

Microfabrication as a discipline gives rise to the premise of Micro-Electro-Mechanical Systems (MEMS) innovations, which are enabled by a variety of methods that play pivotal roles in matter manipulation at the microscale. Among the key techniques that are involved are photolithography, etching, and deposition, which enable the construction of intricate structures required in functional devices. Photolithography, for example, employs light to replicate geometric designs from a photomask to a light-sensitive chemical and subsequently followed by etching processes to strip off the unwanted material in order to create the desired microstructures. Such precise layering provides a platform for subsequent operations, and thus the significance of microfabrication in preparing devices that operate on a broad spectrum of applications, from medical instruments to aerospace. With researchers constantly exploring beyond the limits of what is possible, advancement and growth of these microfabrication techniques take prime importance in keeping pace with the rapidly evolving technology landscape (Cowan et al., 2002). Microfabrication processes have also found place in the development of high-level MEMS and have positively influenced other fields, like medicine. For instance, the creation of intelligent stent devices embodies the linchpin role of microfabrication in the treatment of cardiovascular disease, which remains a leading cause of mortality in current society. The devices capitalize on enhanced electric properties through microfabrication techniques that enable better disease diagnostics and treatment, such as central line access failure and atherosclerosis. The mechanized process of fabricating these medical implants mirrors the priority on accuracy; the specific fabrication challenges stemming from their geometry necessitate a solid understanding of microfabrication concepts. These developments not only benefit patient care but also recognize the significance of continued research in the field of microfabrication techniques (Bussooa et al., 2018). With the increased demand for smaller, more efficient devices, the evolution of microfabrication techniques becomes increasingly relevant. Future work in this area will be aimed at providing new technologies that will enhance functionality and MEMS integration. As an example, laser-driven accelerators are a cutting-edge use of microfabrication, as laid out in work that discusses their construction with sophisticated processes. By exploring these trial processes, implications for future structures can be observed, pointing towards the potential breakthroughs in other applications beyond mainstream fields. This study shows the importance of interdisciplinarity, as future advancements in microfabrication technologies can drive MEMS into new territory, so that they will remain major drivers of technological innovation and benefit to society in terms of functionality and efficiency (Cowan et al., 2002) (Figure 1.4).

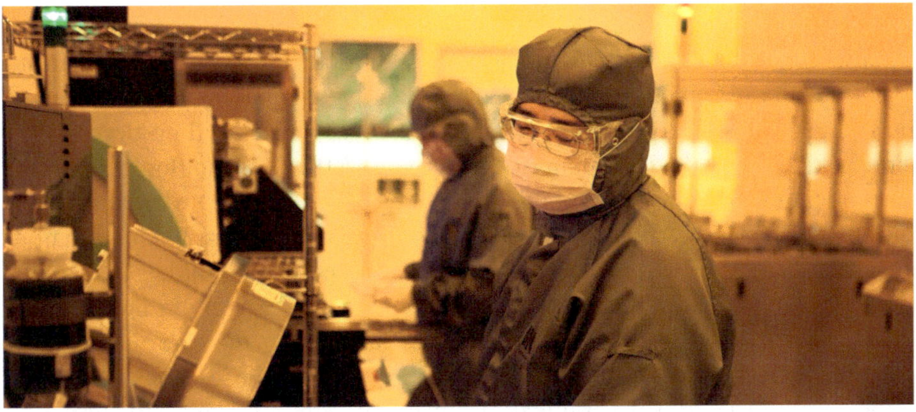

Figure 1.4. Photolithography, for example, utilizes light to transfer geometric patterns from a photomask to a light-sensitive chemical, followed by etching processes that remove unwanted material to yield the desired microstructures.

Source: https://nanomelbourne.com/app/uploads/2015/04/Lithography_Banner.jpg.

1.1.2. Mechanical and Electrical Properties

The mechanical and electrical properties of microelectromechanical systems (MEMS) are inherent in their functionality, influencing design and reliability across applications. Mechanical properties such as stress and strain, for instance, are fundamental in the functionality of MEMS devices, particularly where switching mechanism is involved. For instance, contact closure timing in MEMS switches is critical, with both mechanical and electrical performance affected. Interpreting the contact force, in the range 10–900 μN, is important for the prediction of contact behaviors like contact resistance and reliability in various materials and geometries. The present research enables engineers to design optimal devices, leading to improved-performing products for applications in fields like radio frequency (RF) communication and actuator technologies (Anwar et al., 2019). Electrical characteristics are also crucial because they govern the interaction of MEMS devices with electronic circuits to which they are integrated. The resistance of the contact and the speed of electrical contact closure dictate the efficiency of power transfer in the system. As an example, the electrical contact closure time is measured to be about 0.2 ms under some conditions, and this has profound practical significance for high-speed applications. Increased understanding of the electrical properties provides proper measurement and optimizations and guarantees that MEMS devices transition between states efficiently, thereby enhancing performance.

Besides, the shift towards RF communications applications highlights the importance of dealing with such properties, as poor electrical properties may cancel out the efficiency of MEMS in antenna systems (Charalampidis et al., 2008). Lastly, the interplay between mechanical and electrical properties plays a crucial role in the development of cutting-edge MEMS technologies. Through the application of expertise in both fields, engineers can create systems which not only perform perfectly but also perform perfectly in their functional capacities. The more advanced MEMS devices become, the higher the demand for good mechanical and electric performance, especially given the increased demands for miniaturization and high functionality. When they are able to handle mechanical stresses and still provide fast electrical responses, they create new opportunities in a variety of applications, from telecommunication to biomedical sensors. The use of strong test methods, such as the new microcontact support structures for post-mortem testing, also proves to be important in understanding these properties in real-life situations (Anwar et al., 2019) (Figure 1.5).

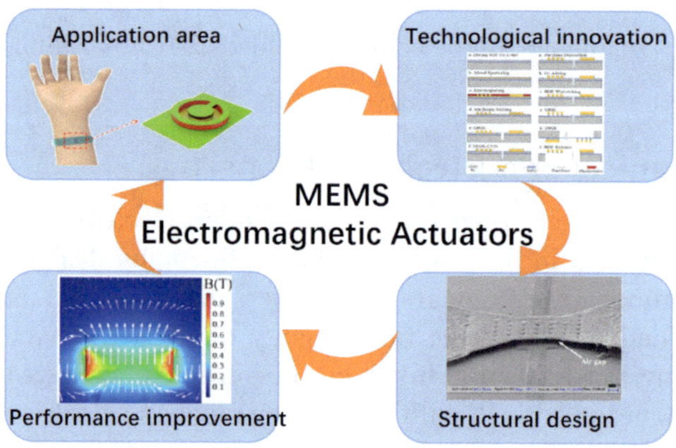

Figure 1.5. Electrical properties are equally significant, as they govern the interaction between MEMS devices and the electronic circuits they integrate with.

Source: https://www.mdpi.com/micromachines/micromachines-16-00670/article_deploy/html/images/micromachines-16-00670-g001.png.

1.1.3. Scaling Effects in MEMS Devices

Scaling effects in Microelectromechanical Systems (MEMS) devices are significant aspects of both the operation and design of these high-level components. As MEMS technology advances miniaturization levels, traditional theories of mechanics and dynamics increasingly become more complex. Device miniaturization leads to drastic changes in material properties, which affect performance characteristics such as sensitivity, response time, and operating stability. For instance, when components reduce in size from macro to micro scales, surface force effects become proportionally more dominant over bulk forces, necessitating the re-evaluation of design parameters. Understanding such scaling laws allows engineers to design MEMS devices that utilize their size in a way that provides enhanced performance while overcoming potential limitations due to miniaturization, ultimately leading to breakthroughs in areas of application including sensors, actuators, and communications systems. MEMS technology also finds echo in the luxury applications of wireless sensors and resonators with tunability, where scaling effects are employed to achieve improved functionality. According to recent studies, the use of such phenomena as isospectral party time reciprocal scaling (PTX) symmetry can significantly enhance the performance characteristics of such devices. This approach enables it to create record-breaking sensitive and ultrahigh-resolution wireless microsensors, much more than can be achieved by traditional passive sensors. The PTX symmetry in electronic circuits not only gives rise to equivalent eigenfrequencies with improved circuit profiles, but also a route to further enhance the extrinsic quality factor (Q factor) of sensors. This success demonstrates the advancement potential of scaling effects in MEMS technology that could result in innovations in a range of fields including medical diagnosis and environmental monitoring (Alù et al., 2018). Besides, embedding MEMS devices on new platforms by, for instance, the MISAT research cluster is another demonstration of applicability of scaling effects. By converting micro-systems technology to space use, MEMS architectures aim to leverage miniaturization and enhance system performance. Current plans involve the development of plug-and-play wireless sensor networks, which indicates the necessity for scaling effects in building effective systems that answer current mission requirements. These successes pave the way towards future missions requiring miniature size, light mass, and low power, hence solidifying MEMSs position on ground-and space-based missions. As such, the present research into scaling effects will significantly contribute to the direction of MEMS technologies in the future, rendering them viable to a broad spectrum of scientific and industrial applications (Gill et al., 2007).

1.2. TYPES OF MEMS DEVICES

Microelectromechanical systems (MEMS) are a general class of devices that combine mechanical and electrical components at the microscopic scale. Devices can be classified into sensors, actuators, and microstructures at large. Sensors are crucial for the sensing of physical phenomena such as pressure, temperature, and motion. For instance, MEMS accelerometers are widely employed in consumer electronics, automotive applications, and aerospace to measure acceleration and changes in orientation. Their miniaturization ensures great sensitivity and low power consumption, crucial in handheld devices. The development in MEMS technology has ensured the integration of advanced features such as Wireless Sensor Networks (WSNs) that has expanded their applications in various fields, further propelling them into the modern-day technological world. Actuators are also another critical class of MEMS devices, supporting precise control of motion and positions. Examples include micro mirrors and micro motors, which are part of applications ranging from optical switching to biomedical devices. Optical MEMS, which utilize tiny mirrors to manipulate light, are a key innovation, enhancing data transmission in telecommunication. Technologies such as the Silex Through Silicon Via (TSV) process make it possible for intricate interconnect features in such devices, allowing the goals of reduced die size and improved wafer-level packaging (Bauer et al., 2008). Additionally, with increasingly sophisticated actuators, their integration with CMOS processing makes them compatible with diverse applications in consumer markets, improving their functional ability and industrial use.

Figure 1.6. Microstructures, which form the foundational components of MEMS devices, are essential in enabling miniature systems to perform complex tasks.

Source: https://www.bisinfotech.com/wp-content/uploads/2024/08/Microelectro-mechanical-Systems-MEMS-Miniaturization-and-Precision-in-Industrial-Sensors.jpg.

Microstructures, which form the building blocks of MEMS devices, are responsible for making miniature systems capable of performing sophisticated functions. Various fabrication methods make it possible to create intricate designs, permitting the production of devices used in catalysis, microfluidics, and energy harvesting. Microfluidic MEMS devices, in particular those used in controlling small amounts of fluids to be used in biochemical analysis, have emerged as highly prominent in the medical sector. Such systems usually run into challenges when maintaining performance under changing conditions, like noise interference during data collection in sensitive applications like iris recognition. Research shows that using proper segmentation methods, like the Circular Hough transform, can enhance output data quality, ensuring stable performance even in hostile environments (Ahmad et al., 2018). Therefore, sustained innovation and modification of the microstructures are essential to the development of MEMS technologies (Figure 1.6).

1.2.1. Sensors: Types and Applications

In Micro-Electro-Mechanical Systems (MEMS), sensors are the stars in enabling the possibility of having a variety of applications in many different industries. These sensors, which convert physical phenomena into measurable electrical signals, are generally categorized into two: analog and digital sensors. Analog sensors produce continuous signals that vary as a function of variation in the physical quantity to be measured, while digital sensors produce discrete signal, hence more ideal for integration into modern digital systems. Digital sensors find applications from environmental monitoring through to biomedical diagnostics, illustrative of their versatility. With the industries demanding increased data processing intensity, innovation in sensor technology has been essential, particularly in applications where real-time monitoring and control are required, such as developments towards active flow control applications (Kopylov et al., 2015). The application of sensors in MEMS is not restricted to traditional fields with automobile technology and healthcare recording significant impacts. In automotive systems, sensors are integral for vehicle navigation, stability control, and safety mechanisms, including airbags and anti-lock braking systems. Similarly, within healthcare, MEMS sensors enable precise monitoring of physiological parameters, leading to advancements in telemedicine and personalized health solutions. For instance, wearable devices equipped with sensors can continuously track heart rate, temperature, and activity levels, providing valuable data for both patients and healthcare professionals. The growing emphasis on miniaturization and energy efficiency in sensor development further enables these applications, directly fostering innovation that contributes value to user experiences and improves outcomes in various areas. Furthermore, the evolution of sensor technology poses challenges to researchers intended to be overcome through innovative solutions. For example, based on research comparing image processing segmentation methods, the data quality garnered can reduce significantly when conditions are adverse, affecting applications such as iris recognition (Ahmad et al., 2018). This is towards the direction of the importance of continually enhancing not only the sensors themselves but also algorithms used to interpret what they output. By tackling such issues, researchers can create more resilient systems that can work with high accuracy under different conditions. The marriage of sensor technology with high-end MEMS is poised to overcome these barriers and facilitate greater performance and reliability in upcoming applications (Figure 1.7).

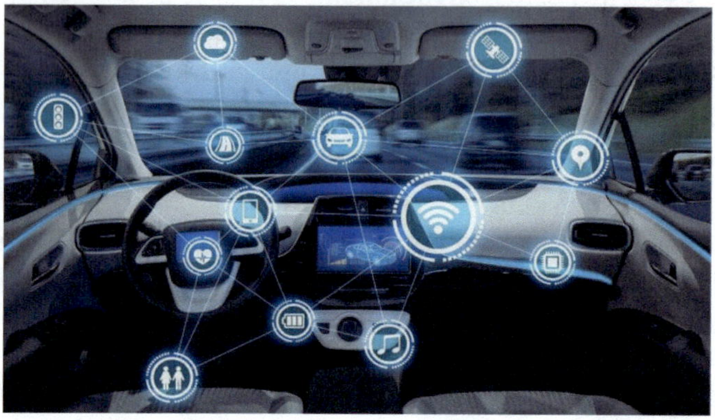

Figure 1.7. In automotive systems, sensors are integral for vehicle navigation, stability control, and safety mechanisms, including airbags and anti-lock braking systems.

Source: https://b3h9h6w4.delivery.rocketcdn.me/wp-content/uploads/2022/02/1-Security-Safety-Auto-768x450.jpg.

1.2.2. Actuators: Mechanisms and Uses

Actuators play a crucial role in the realm of micro-electro-mechanical systems (MEMS) as the principal appliances for converting electrical energy to mechanical motion. This is realized through various principles of operation, such as piezoelectric, thermal, and electromagnetic effects. Among these, MEMS electromagnetic actuators have garnered much attention due to their high accuracy and rapid response rates, which are essential for applications ranging from micro-optical devices to microrobotics. Such an actuator's operation efficiency, based on magnetic attraction/repulsion and Lorentz force phenomena, dictates the need to continually progress actuator fabrication and design. Advances in microfabrication process and materials have further enhanced the performance of MEMS actuators, making them able to meet the demanding requirements of modern applications while remaining compact and lightweight (Cheng et al., 2025). The use of actuators in many technology areas proves their extensive versatility and functionality. Rehabilitation technology is one area that clearly illustrates this usage, such as through the development of rehabilitation lower limb exoskeletons (RLLEs). New devices like these employ actuators to interact with ease with the user's limbs, thereby improving the efficacy of physical therapy and rehabilitation. The selection of appropriate actuators and power transmission systems is an important factor for ensuring proper human-machine interaction, which has a direct impact on therapy efficiency and patient comfort. New technology development leads to increasingly significant aspects of design in regard to actuators, thus providing a huge number of technical solutions adaptable to different application needs (Aliman et al., 2024).

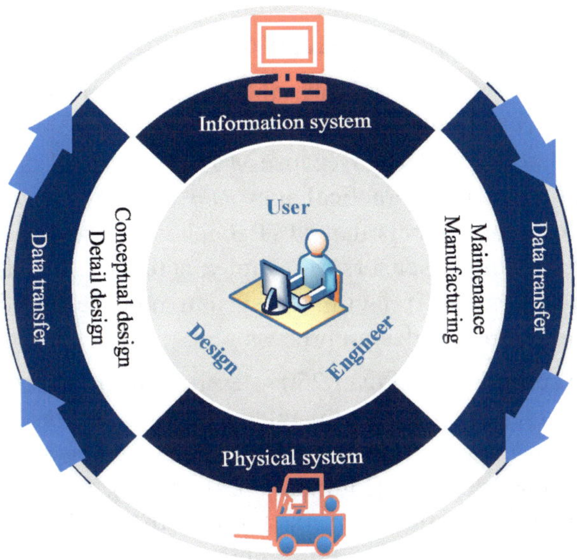

Figure 1.8. The selection of appropriate actuators and power transmission mechanisms is critical to ensuring smooth human-machine interactions, which significantly influence patient comfort and therapy effectiveness.

Source: https://www.mdpi.com/sustainability/sustainability-16-04158/article_deploy/html/images/sustainability-16-04158-g006.png.

This evolution not only challenges researchers to enhance existing technologies but also provokes novel approaches to developing actuators. Furthermore, the trajectory of actuator technology gives one hope for what is ahead as far as new applications and advancements in the MEMS paradigm. While the majority of MEMS actuators remain in their experimental stages, their potential for revolutionary impacts in numerous industries cannot be questioned. New disciplines such as microfluidics and precision micromanipulation will benefit considerably from advancements in actuator performance, including reliability and energy efficiency. Challenges still remain, particularly in issues related to miniaturization and integration, so a concerted effort in the direction of research and development is necessary. The future is characterized by the promise of breakthroughs in adaptive actuators and intelligent control systems, which will allow for more functionality at the microscale. With present constraints being resolved and interdisciplinary collaboration being embraced, the future of MEMS actuators is not only to extend their mechanical functionality but also to expand their range of applications across several emerging areas (Cheng et al., 2025) (Figure 1.8).

1.2.3. Microstructures: Design and Functionality

The microstructure design within the new Micro-Electro-Mechanical Systems (MEMS) is naturally governed by the diverse objectives included in microsystems technology (MST). Because MST is described as complicated and diverse, what is required is a practical approach that categorizes its constituent elements. The different subjects that MST touches on create a situation where the use ranges from simple sensors to complex actuators. It is crucial to define the sub-territories because it facilitates system module and interconnection standardization and thus the design process.

Through the compilation of over 250 references from literature, there is now a comprehensive map of the objectives and generic issues concerning MST, which serves as an educative map for developers and researchers willing to venture into this intricate topic (Fluitman et al., 1996). Such an organizational strategy is invaluable, particularly for future MEMS designers that would want to be innovative while operating within prevailing frameworks. Within microstructure design, among the most convincing innovations is in the aspect of stretchable electronics (SE), most particularly in relation to the development of in-body and medical diagnostic devices. The promise of SE technology is matched by the difficulty of fabricating high-density interconnects that can sustain a significant deformation. As microfabrication techniques continue to advance, integration of ubiquitous CMOS-type processes is especially important in the development of highly stretchable interconnect structures. Innovations such as Flex-to-Rigid (F2R) post-processing technology allow free-standing structures to minimize their three-dimensional kinematic confinement through complex deformation mechanisms like buckling and bending. These advances not only render such microstructures more functional but also enhance their potential for high-volume, reproducible manufacture for medical application (Dekker et al., 2017). This complementarity of stretchability and function opens up new horizons for MEMS technology in life-critical health monitoring and therapy. Microstructure functionality is intrinsically bound to its design parameters, which take on very central roles in the case of new applications. Microstructures optimized for responsive environments must possess particular mechanical and electrical properties to provide the diverse operating needs of various MEMS devices. In one case, an equilibrium between strength and flexibility enables uses such as wearable sensors that must endure repeated movement without loss of performance. Furthermore, the optimization of fabrication makes it possible to tailor microstructure properties, whose commensurability with specified functionalities can lead to drastic enhancements in the effectiveness of devices. As technology evolves, it will become essential to understand how design complexity and functional performance relate to each other in order to drive future microstructure innovation so that it not only keeps pace with but exceeds next-generation MEMS requirements (Dekker et al., 2017; Fluitman et al., 1996) (Figure 1.9).

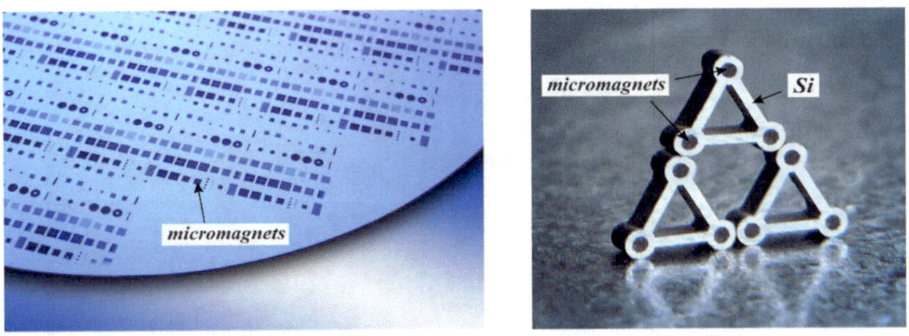

(a) (b)

Figure 1.9. The design of microstructures within advanced micro-electro-mechanical systems (MEMS) is fundamentally influenced by the diverse objectives encapsulated in microsystems technology (MST).

Source: https://encyclopedia.pub/media/item_content/202204/625913514881amicromachines-13-00398-g012.png.

1.3. MATERIALS USED IN MEMS

The design of Microelectromechanical systems (MEMS) relies heavily on the selection of sufficient materials that meet the stringent demands of performance, reliability, and integration. Silicon has been the most popular material throughout due to its high mechanical properties, semiconductor quality, and incompatibility with prevalent microfabrication techniques. The ability to use wet chemical bulk micromachining can enable the creation of complex microstructures with tilted sidewalls and unsupported geometries critical for uses like sensors to actuators. Furthermore, the enormous amount of R&D in the wet chemical etching processes makes silicon even more acceptable to use in applications, thereby becoming the workhorse material in MEMS technology (Pal et al., 2017). As the technology advances, scientists are also investigating new materials to increase the functionalities and applications of MEMS devices. In addition to silicon, other materials such as polymers and metals are being used in MEMS too. Polymers such as SU-8 enjoy cost-effective and high-throughput processing; their light weight and flexibility characteristics enable new design liberties, particularly for bio-MEMS and disposable devices. Conversely, metal-based structures like gold, silver, and aluminum are utilized to fabricate electrical interconnects and mechanical support for RF MEMS switches as well as other devices. The application of these materials in MEMS technology makes it possible to integrate hybrid systems that take advantage of the unique characteristics of each material to improve the overall performance, strength, and reliability of the device (Charalampidis et al., 2008).

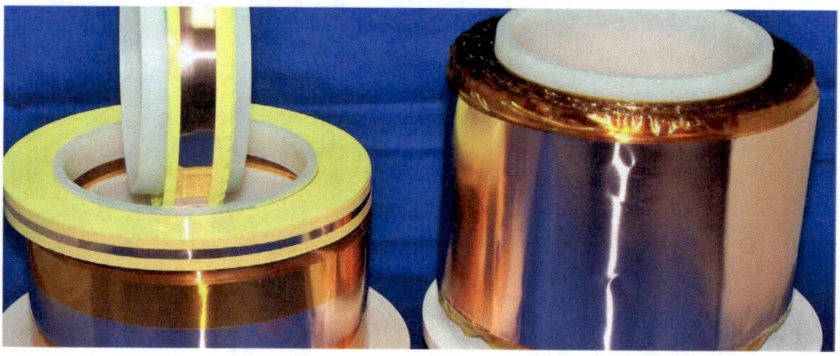

Figure 1.10. Metal-based materials, including gold, silver, and aluminum, are essential for developing electrical interconnects and providing mechanical support in RF MEMS switches and other devices.

Source: https://www.raypcb.com/wp-content/uploads/2023/02/RF-Shielding-Material.jpg.

Research of these disparate materials not only creates new opportunities for MEMS applications but also meets the growing demands for miniaturization and multi-functionality in many technology areas. Material choice interacting with MEMS design plays an important role in broadening the potential of what these systems can do. Integration of recently emerging materials such as piezoelectric materials and shape memory alloys, for example, has the potential to significantly influence device function. Piezoelectric materials enable MEMS devices to convert mechanical energy into electrical signals, thus enabling advanced sensing and actuation applications. Similarly, shape memory alloys introduce unique capabilities for designing microvalves and microactuators that can adapt to dynamic conditions. With the materials being researched and developed on a continuous basis, field professionals need to understand their unique properties and potential. The incorporation of such new materials into MEMS design not only optimizes function but puts MEMS technology at the forefront of innovation in a host of fields from consumer electronics to medical device and beyond (Figure 1.10).

1.3.1. Silicon and Its Alloys

Silicon is used most in MEMS due to its favorable electronic and mechanical characteristics. Silicon's semiconductor nature provides for the precise control employed in most MEMS devices, including sensors, actuators, and switches. Silicon's compatibility with standard silicon-based fabrication techniques, such as photolithography and etching, also makes it more useful for fabricating miniature devices. The use of silicon alloys further broadens its applicability, allowing the mechanical stiffness, conductivity, and thermal properties to be optimized. Alloying the silicon with other elements such as germanium or carbon allows designers to tailor the material properties to suit particular performance requirements, broadening the scope of possible MEMS design. The manufacture of MEMS typically involves the deposition of various alloy films to enhance the electrical contact characteristics of switches and sensors. The interdependence of hardness and resistivity of the materials is crucial to the optimization of their functionality. Research has established that the incorporation of noble metal alloys, such as gold alloyed with platinum, rhodium, and ruthenium, has a significant impact on the contact resistance and wear characteristics. For instance, as these metals form alloys with gold, their hardness would reduce and resistivity rise, impacts that could directly affect the operation and lifespan of MEMS devices. Through such understanding, it becomes easier to design alloy compositions that strike a balance between electrical and mechanical strength in order to ensure maximum operation in high-end MEMS applications (Coutu et al., 2006). Furthermore, material composition and microstructure-dependent process-specific properties, such as curvature in MEMS cantilevers, play a crucial role. Back-end processing and stacking of multiple materials to form layers, in particular, play major roles in the mechanical properties and performance of these microstructures. Recent studies have demonstrated the importance of the number of layers and tungsten via design on cantilever bending behavior. Additionally, processing conditions after release can induce substantial changes in curvature, making the choice of appropriate materials and post-processing processes important in realizing optimal MEMS design (Barrachina et al., 2017). This sensitive interrelation between silicon alloys and MEMS processing processes gives an example of how state-of-the-art materials science is inextricably linked to the development of MEMS technology (Figure 1.11).

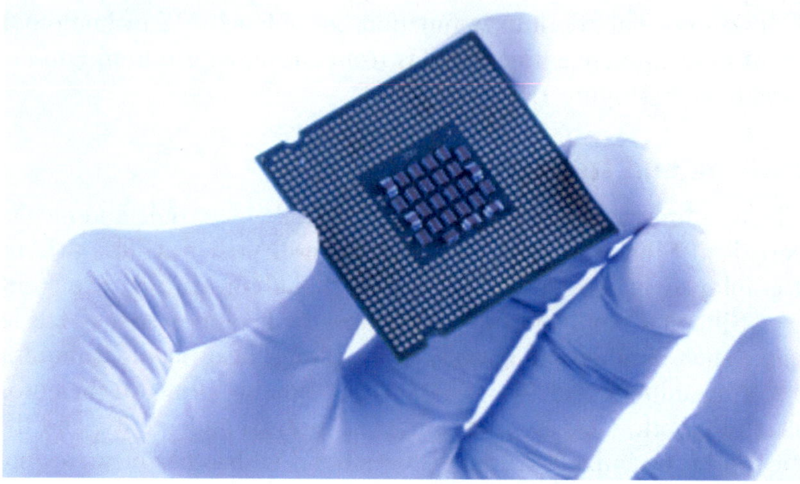

Figure 1.11. Silicon's semiconductor characteristics enable the precise control needed for various MEMS applications, including sensors, actuators, and switches.

Source: https://www.elprocus.com/wp-content/uploads/2014/02/17.jpg.

1.3.2. Polymers and Their Role in MEMS

The use of polymers within micro-electromechanical systems (MEMS) has evolved into a revolutionary leap that has significantly enhanced their performance and feasibility across various industries. Polymers, particularly electroactive and photoactive polymers, have proven to be ideal candidates for the engineering of responsive and adaptive systems to intelligently interact with their environment. Organic thin film electronic devices (OFEDs) and bio-responsive hydrogels show this capability, as these are easily integrated into MEMS and nano-electromechanical systems (NEMS). By studying the unique characteristics of these polymers, researchers can create new applications that adapt to human-focused needs, such as intelligent fabrics, biomedical equipment, and smart building material, thereby setting the stage for an era of ambient living platforms and interactive consumer products (Oliver et al., 2009). The applicability of polymers extends even beyond their intrinsic properties because they make possible innovations in micro-fabrication techniques essential to MEMS. Existing initiatives on non-MEMS-based micro-manufacturing show the increasing demand for micro-components that need RTD in order to transform conventional production methods to micro-scale applications. Creation of polymer materials on this aspect facilitates scalability and versatility required for successful manufacturing methodologies, whether MEMS or other forms of micro-manufacturing. This shift towards manufacturing process optimization points to the ability polymers bring to enhancing structural and functional characteristics of MEMS devices and thus its feasibility in various industries (Albers et al., 2010). Continued studies on polymers in MEMS inspire innovation as well as sustainable development through the production of products that are recyclable and biodegradable. The region is witnessing increased interest in the application of sustainable polymers in MEMS design, with a driving force for the development of environmental footprint-reducing devices with enhanced performance. Furthermore, the polymers' versatility facilitates multifunctional properties to be incorporated into MEMS, thus the expansion of their applications in areas such as renewable energy and biomedical engineering. By concentrating on the synergy of progress in polymer science and MEMS fabrication methods, researchers are opening the door to revolutionary applications that solve today's problems and foster an intelligent built environment (Albers et al., 2010; Oliver et al., 2009).

1.3.3. Emerging Materials for Advanced MEMS

MEMS development is increasingly propelled by the innovation of new materials that enhance functionality and performance. Smart materials such as electro-active polymers and bio-responsive hydrogels are spearheading this revolution. These smart materials have the capability of real-time adaptation and interaction with their environment, allowing MEMS applications to be extremely sensitive and efficient. The integration of these kinds of materials into MEMS/NEMS devices greatly expand the range of potential applications, from intelligent clothing to interactive medical products. Through the tunable nature of such materials, companies can create revolutionary products responding to various human-centered needs, hence revealing revolutionary improvement possibilities in design and technology, as seen in the potential of smart materials in existing uses (Oliver et al., 2009). Biomedical use in the MEMS field is very promising since future materials reveal revolutionary solutions in health monitoring and therapy. Bio-responsive hydrogels, for instance, may be employed in soft robots that interact with biological systems in a soft and adaptive fashion. These hydrogels can sense change in their environment and respond accordingly and are therefore the best materials to employ in smart bandages or drug delivery systems releasing drugs in response to physiological signals. This flexibility not only enhances the utility of MEMS devices in medicine but also results in improved patient outcomes and user satisfaction. In addition, with evolving medical devices, the application of advanced materials translates into new therapeutic approaches that are responsive to the dynamic status of patients and hence constitute a critical turning point towards personalized medicine for the domain of MEMS (Oliver et al., 2009). Besides, academic and industrial context around MEMS is being upgraded by massive research efforts in the field of new materials, which are evident in institutions like the Boston University Photonics Center. The centers' focus on multidisciplinary research into optoelectronic materials, and their applications reflects how various fields of science are interrelated and necessary for MEMS technologies to be enhanced. With significant investment in photonics and biophotonics research and innovation, institutions have fostered an environment that is innovative in the development of MEMS. Focus on applications in real-world settings, such as cancer diagnosis and novel imaging modalities, illustrates the real-world applicability of utilizing new materials in MEMS. By developing standards for interdisciplinary research, researchers are in a good position to keep exploring the intersection of material science and MEMS and thus bring about a new age of technological advancement that will have its advantages in various industries (Figure 1.12).

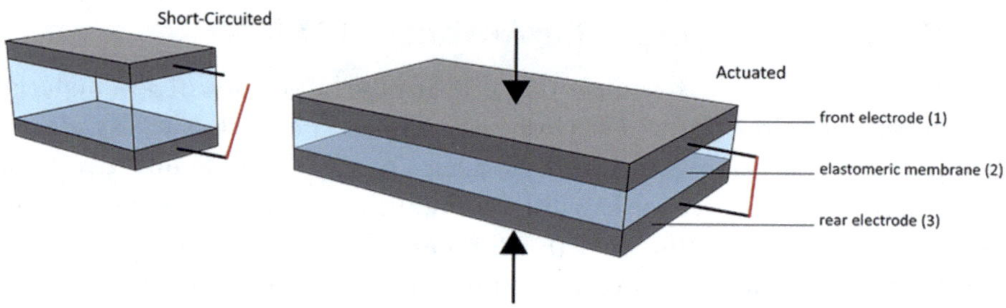

Figure 1.12. Electro-active polymers.

Source: https://materiability.com/wp-content/uploads/2013/03/m_eap-principle.jpg.

1.4. APPLICATIONS OF MEMS TECHNOLOGY

MEMS technology versatility spans a wide array of applications across various industries, significantly enhancing equipment performance and efficiency. In telecommunication, MEMS resonators feature prominently in precise frequency control, which is integral to modern-day communication systems. The advent of phononic frequency combs (PFCs) is a significant advance that makes it possible to integrate improved signal processing capability. These resonators apply nonlinear mode coupling, generating higher harmonics that can be utilized in the case of sophisticated measurement techniques. This characteristic of more control over telecommunication is the way MEMS technology is not merely a device but a revolutionary force in itself, enabling progress in data transmission as well as in the overall architecture of communication networks, justifying its essential part in the evolution of modern electronics (Bhosale et al., 2024). Aside from telecommunications, MEMS technology is making significant strides in Internet of Things (IoT) applications, particularly in structural health monitoring (SHM). Low-power sensors based on MEMS allow for continuous monitoring of vital structures such as bridges and skyscrapers, which would help avoid disaster failures. Since the sensors are meant to be self-sustaining, stable power sources are needed to run them for extended periods. Scientists are examining novel solutions such as piezoelectric vibration energy harvesters (PVEH) to address this issue. These energy harvesters, by converting ambient mechanical power into electrical power, provide a maintenance-free supply of energy pivotal to the functionality and longevity of IoT systems. The innovation reflects the broader trend of utilizing MEMS technology to boost monitoring functionality, ultimately enabling safer and more resilient structural frames (Raghavan et al., 2023). Furthermore, the scope of application of MEMS technology extends to medical and biomedic industries where it serves as a catalyst towards novel diagnostic tools and therapeutic devices. MEMS sensors have revolutionized medical diagnosis by enabling the possibility of miniaturized but very sensitive devices that can detect a variety of biomolecules with high precision. With the ability to make such sensors at microscale levels, they can be incorporated into wearable devices where patients' health can be continuously monitored. This capability is of the highest value in chronic disease management, where early discovery and ongoing observation can dramatically change the patient's outcomes. As increasing demand for tailored medicine continues to grow, MEMS technology will certainly play a decisive role in the development of next-generation diagnostic approaches that are not only effective but also cost-effective and affordable, thereby heralding a new era in healthcare delivery.

1.4.1. Consumer Electronics: Smartphones and Wearables

The greater convergence of microelectromechanical systems (MEMS) into consumer devices, notably smartphones and wearables, has revolutionized user interaction with technology. MEMS technology adds more capabilities in the form of miniature sensors tracking movement, orientation, and environmental conditions. For instance, smartphones nowadays contain accelerometers and gyroscopes to enhance gaming and add fitness tracking features. In wearable technology, the application of MEMS not only facilitates basic health monitoring functionalities but also progresses into complex functions, such as heart rate and oxygen saturation sensing. With the increasing applications, the need for strong, small MEMS components continues to rise, marking the central role that MEMS has in enhancing consumer electronics to multifunctional devices that suit different customers. The ongoing innovation in MEMS technology highlights the importance of these devices in enhancing user experience and functionality in today's consumer electronics. MEMS technology evolution has revolutionized smartphone and wearables design and functionality, particularly navigation and positioning. With the use of location-based services becoming increasingly vital, inclusion of MEMS-based sensors is essential. Bosch Sensortec's latest developments illustrate how such sensors can enhance navigation not only outdoors but also indoors, where traditional GNSS systems are ineffective.

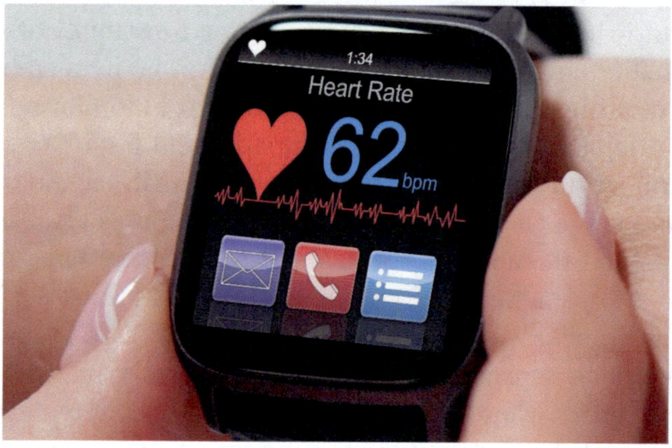

Figure 1.13. In wearable technology, the implementation of MEMS not only facilitates basic health monitoring functions but also extends to sophisticated features, such as heart rate and oxygen saturation detection.

Source: https://www.azosensors.com/images/Article_Images/ImageForArticle_3063_17193513053861706.jpg.

Vertical positioning is supported by MEMS technology, enabling possible applications such as precise location tracking for emergency responders, as mandated by legislation like e911. Also, the embedded navigation capabilities offered via MEMS sensors in such settings as airports or meeting rooms validate their role in elevating user experiences on more advanced environments. These advancements validate that MEMS are not merely supplements; instead, they are indispensable components driving the evolution of consumer electronics. On top of this, the arrival of the Internet of Things (IoT) and consumers' demands for intelligent consumer electronics are further broadening the applicability horizons of MEMS in wearables and smartphones. MEMS technology can potentially transcend functionality, such as intricate design elements specifically tailored to provide improved performance. Polymeric material selection in MEMS sensor packages is crucial as it influences the behavior of the sensors under varied conditions, such as temperature and humidity. An advanced simulation approach, aimed at conditions such as moisture and temperature, accentuates MEMS packaging intricacies. Focusing on the moisture concentration in interfaces of materials, researchers are moving forward to improve the reliability and precision of MEMS devices. Therefore, the ongoing investment in high-tech MEMS technologies not only improves today's consumer electronics but also leads the way towards even more innovative products which are likely to emerge in the near future (Figure 1.13).

1.4.2. Automotive Industry: Sensors and Safety Systems

The auto industry has witnessed enormous development in sensor technology, which has significantly enhanced car safety systems. Integration of Micro-Electro-Mechanical Systems (MEMS) has significantly changed how cars monitor and respond to various road conditions. Microscopic sensors implanted in the car can detect acceleration, temperature, and pressure, enabling automakers to deploy advanced safety features such as collision prevention and adaptive cruise control. MEMS technology application allows miniaturization, reduced cost, and improved performance, which is used to fulfill increasing requirements for reliability and safety in modern vehicles. Therefore, the automotive sector is moving towards more responsive and intelligent systems, which not only safeguard drivers and occupants but also contribute to the global reduction of road accidents and fatalities. Reviewing the crucial role of MEMS in car safety, one is required to explore implications of various sensor technologies developed for real-world applications. Proper review of such sensors shows their position as core to active safety systems, especially in operations such as automatic emergency braking and lane-keeping assistance. The sensors used in such systems must meet high-performance levels and operate under harsh conditions, according to relevant literature. This calls for balance between sensitivity and robustness, whereby the protection systems unfold exactly and in a timely manner in emergency situations. The ongoing research in MEMS technology will keep on leading to increasingly sophisticated applications aimed at driver and passenger protection and towards the next generation of autonomous driving models. Hence, it is essential to grasp the intricacies of sensor technologies to realize their groundbreaking impact on the automotive sector (Kopylov et al., 2015). Besides, the new field of microsystems technology (MST) is critical in shaping the destiny of future automotive safety systems. As MST encompasses a broad variety of applications, its overall principles are the foundation for standardized sensor modules essential for seamless vehicle integration. Through the creation of a platform that accommodates modular systems and interfaces, engineers and scientists can make safety features easy to understand for different models of vehicles. Standardization of this nature encourages innovation while addressing the diversity of MST by separating it into its many applications and technologies. Moreover, finding these sub-territories of MST enables manufacturers to maximize their design, ultimately leading to better vehicular safety. As sensors become increasingly sophisticated and multi-dimensional, there must be an effective approach towards their integration to provide the integrity and security required in current cars (Fluitman et al., 1996).

1.4.3. Medical Devices: Innovations and Impact

The advent of advanced Micro-Electro-Mechanical Systems (MEMS) has deeply revolutionized the field of medical devices, enabling innovations that enhance patient care and operational efficiencies. Micro-scale devices, which straddle the biological and digital domains, enable the development of advanced diagnostic solutions and treatment solutions. With the addition of MEMS, devices such as glucose sensors and implantable drug delivery systems have become a possibility, which can provide real-time health information for immediate interventions. The move toward such intelligent systems is the way to the future, not just in terms of engineering progress but also a shift toward individualized medicine, which will revolutionize the life of patients suffering from chronic diseases very fundamentally. As the functionality of MEMS continues to advance, it will only result in more innovative and effective medical treatments. Furthermore, the complexity of integrating such technologies is suggestive of several issues that will need to be overcome by organizations to achieve market adoption. Concerns regarding the achievement of acceptable sensing capability in real-world applications are of utmost concern, as indicated by concerns regarding compatible sensor-biology interfaces. Also, challenges to the commercialization of sophisticated MEMS must be addressed, such as technical and non-technical ones, for instance, regulatory, cost-effectiveness, and environmental considerations. Analysis of the roadblocks is crucial in the direction of future innovation in medical devices so that innovations may progress with ease from the laboratory to the clinical environment. By proposing a participative development agenda and deployment plans, the stakeholders are then able to direct the utmost effort in primary areas, thereby guaranteeing a harmonious coexistence of technological potential and application in medicine. The potential contribution of IoT-supported sensor networks to the healthcare sector further demonstrates the capability of MEMS to revolutionize. These frameworks enable the push towards smart cities, where converged technology solutions aim to address problems of urban life such as access to medicine and resource management. Use of IoT-based sensors not only improves the effectiveness of healthcare delivery but also enables socio-economic inclusion within the city. As per recent research, a machine learning solution to leveraging this data is imperative, making masses of information actionable and manageable. Focus on sustainable, innovative healthcare solutions positions MEMS at the forefront of next-generation medical devices. This convergence of technological innovations with societal needs opens the door for a new generation of healthcare innovation that converges with the vision of creating smarter, healthier communities (Luo et al., 2023; Taher et al., 2021) (Figure 1.14).

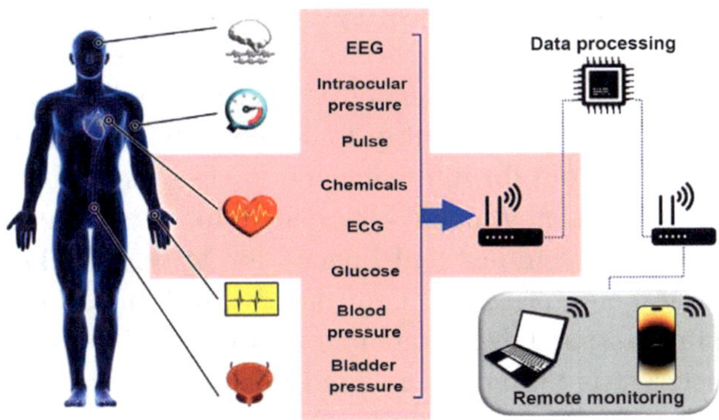

Figure 1.14. With the integration of MEMS, devices such as glucose monitoring sensors and implantable drug delivery systems have emerged, providing real-time health data that allow for timely medical interventions.

Source: https://ars.els-cdn.com/content/image/1-s2.0-S2590006423002442-gr2.jpg.

1.5. CHALLENGES IN MEMS DEVELOPMENT

The production of Microelectromechanical Systems (MEMS) is intrinsically marred by a variety of problems that could potentially restrain their use and deployment in state-of-the-art technologies. One of the dominant issues is the high level of sophistication of the fabrication processes, which entail tight control over size and material properties at the microlevel. The in-built complexity of these processes leads to variability of performance and reliability across devices, for which tight testing and quality control protocols have to be adopted. Additionally, the use of standardized platforms can help make production simpler; however, maturity level differences in technology can also further exacerbate such problems and make the process of development even more complex, making MEMS devices less competitive in markets (Bogaerts et al., 2020). Overcoming such fabrication problems is essential to enhance the dependability and magnitude of MEMS technologies in a wide range of applications. Another key challenge in developing MEMS is packaging the devices into established systems and platforms. With more sophisticated MEMS components, their integration with traditional electronic components or newer photonic components remains a significant challenge. Packaging MEMS-enabled components into silicon photonic platforms has proven potential to realize maximum functionality at minimal power consumption (Bogaerts et al., 2020). However, standardization methods for delivering unproblematic integration between MEMS and other systems remain a challenge. Secondly, developing hardware and software interfaces that can sustain the native character of MEMS is crucial to the development of MEMS in telecommunication and sensor networks applications. Inability to address integration challenges may rule MEMS technology out for future general application and implementation. Finally, a sense of urgency must be created to overcome the handicaps imposed by resource constraints and finance for MEMS R&D. The majority of emerging MEMS enterprises are confronted with limited budgets that suck out innovative thinking and decelerate them. Collaboration initiatives, such as the MISAT cluster, strive to create a sound microsystems foundation through fostering cooperation among public institutions and industry stakeholders. These collaborations are essential to bringing resources and talent together, ultimately supporting the development of MEMS technologies for space to Earth applications. Even so, the inherent challenge of having stable funding and resources continues to hold back the rapid development of MEMS. As such, the overcoming of these funding challenges will be vital to unlocking the full potential of MEMS in future technology development.

1.5.1. Manufacturing and Scalability Issues

The manufacturing and scalability issues in Micro-Electro-Mechanical Systems (MEMS) present massive challenges that may hinder their widespread application and efficacy in numerous uses. The initial major challenge that accompanies the production of MEMS is in the intricacy that characterizes their fabrication and design processes. Miniaturization to realize MEMS devices commonly leads to intricate architectures necessitating advanced manufacturing techniques. Traditional macro-fabrication processes are not easily scalable to the micro-range, resulting in inefficiency and increased cost. Some of the issues that require new manufacturing approaches—capable of guaranteeing both accuracy and scalability—include the following. These issues must be overcome for MEMS to achieve their potential in applications such as sensors, actuators, and other high-reliability, high-performance multifunctional systems; hence, the need for enhancing manufacturing methods becomes pivotal (Studor et al., 2012). The advent of new micro-manufacturing technologies, some of which are not based on MEMS, exhibits the effort being made to bypass these manufacturing issues. Technological innovation and study in this field have speeded up with an increased demand for micro-components in various industries. EU-funded initiatives have been focused especially on enhancing manufacturing techniques, such as hybrid-processing methods that combine more than one method to achieve desired outcomes at the micro-scale.

Figure 1.15. The miniaturization required for MEMS devices often leads to intricate architectures that necessitate advanced manufacturing techniques.

Source: https://product.minebeamitsumi.com/en/tech/mems/about/img/technology_img_001.jpg.

The innovation is targeted towards making production easier, thereby enhancing throughput, and reducing costs associated with MEMS fabrication. By promoting the synergy between industry and academia, these research projects hold the potential to improve current manufacturing processes alongside the investigation of newer, scalable approaches capable of meeting the increasing demand for complex micro-scale devices (Albers et al., 2010). In conclusion, the road map to overcoming manufacturing and scalability problems in MEMS relies on ongoing innovation and cooperation within the research community. As methodologies keep evolving, it becomes increasingly urgent to study and implement advancements in micro-manufacturing that can satisfactorily respond to the particular issues that MEMS technologies pose. Success in this regard not only renders MEMS devices more dependable and effective but also simplifies their deployment into higher-volume applications, further solidifying their niche in the technological landscape. Lastly, addressing such manufacturing challenges is crucial to ensure sustainable development of the MEMS sector, which has the capability to transform a number of industries, including healthcare, aerospace, and consumer electronics, among others (Studor et al., 2012) (Figure 1.15).

1.5.2. Reliability and Longevity Concerns

In Microelectromechanical Systems (MEMS), reliability and durability are the major factors influencing the application of these technologies in a variety of applications, particularly in biomedical and aeronautical engineering. As MEMS devices continue to be more integrated into sophisticated systems, ensuring their consistent functioning under a wide range of operating conditions is extremely critical. Material degradation, mechanical fatigue, and environmental stressors are generally major concerns. For instance, implantable medical devices must be operational without failure for a long duration of time; thus, advancement in biocompatibility and material durability is critical. The key to making stable MEMS technologies a reality is intrinsically intertwined with solving these problems and ensuring that devices are capable of performing stably under dynamic environments, as is attested through recent advancements in micro/nanotechnology and biomaterials (Amar et al., 2015). The electrical and mechanical characteristics of MEMS sensors and switches are responsible for their performance and durability. The most recent technology developments, such as WNx nano-ribbon-based NEM switches, illustrate the potential for high reliability under extreme environments. They are powered at lower voltages but have high switching speeds and long operation cycles, up to 8 trillion cycles without apparent degradation (Alhashim et al., 2024). The advancements made in material science, especially the elasticity and mechanical strength of the materials employed, are fundamental to the improvement of the durability of these systems. By reducing stiction, mechanical welding, and electrical short-circuit-related problems, scientists are opening the doors to MEMS devices with greater capabilities for withstanding a wider range of working challenges, thereby greatly enhancing their reliability in applications where they are deployed. Secondly, power management remains a critical factor in reducing the reliability and lifetime problems of MEMS devices, especially in implantable devices where continuous operation is required without any periodic power supply interruptions. The lifespan of power sources has a direct influence on the overall performance and lifetime of the devices. As identified by evolving research, the integration of energy harvesting methods and power management system design that is efficient are becoming increasingly critical to sustaining performance over a period of time (Amar et al., 2015). With the increasing need for wearable and implantable devices in the healthcare sector, the focus on new power solutions is sure to increase. By enhancing power supply mechanisms' reliability along with mechanical integrity, the MEMS field approaches the development of robust systems that can offer revolutionary healthcare solutions and beyond, facilitating individualized and continuous monitoring of physiological parameters.

1.5.3. Integration with Other Technologies

Blending newer micro-electro-mechanical systems (MEMS) with newer state-of-the-art technologies is the key to transforming most industries. At the center of such technological advancements is Through Silicon Via (TSV) technology, where down to 30 micrometers pitch through-wafer connections in substrates above 600 micrometers thickness are enabled. Such a feature greatly reduces die size and enables true "Wafer Level Packaging," and these are critical specifications for application in the consumer market. The ability to connect MEMS devices to advanced electronic systems allows for higher performance and processing capabilities, thereby promoting the application of MEMS in sectors such as telecommunications and automotive systems. As companies rapidly adopt TSV technologies, the process is ever more becoming the standard in the MEMS industry, setting the stage for further integration with complementary technologies (Bauer et al., 2008). In addition to the electric connections, incorporation of adaptive and responsive materials into MEMS technologies is also revolutionizing applications in various fields. Electro- and photo-active polymers, and bio-responsive hydrogels, bring new capabilities for MEMS/NEMS devices. These new materials allow for the development of intelligent systems that are able to adapt to environmental changes, thereby enhancing the function of conventional MEMS applications. For instance, smart fabrics integrated with MEMS sensors can be programmed to adapt to a wearer's state, leading to new medical innovations, such as smart bandages that offer real-time vital signs monitoring. Conjoining these new materials with MEMS design is part of an overall movement towards creating multimodal technologies to improve user experience and operational efficiency in markets like consumer electronics and medicine (Oliver et al., 2009). Apart from this, the combination of MEMS with other premium technologies holds immense potential for developing smart environments facilitating everyday life. Ambient living platforms that combine MEMS technologies with interactive materials offer seamless interaction between people and their surroundings. For instance, the integration of MEMS sensors with organic thin film electronic devices may lead to the creation of environments responsive to the needs of the user, further advancing concepts such as intelligent homes and buildings. The systems can efficiently track environmental parameters and adapt to user needs, facilitating an improved pattern of energy consumption. It is through the networking of MEMS with other emerging technologies that firms are able to come up with innovative product designs that meet the evolving needs of modern society as well as foster general user experience (Oliver et al., 2009) (Figure 1.16).

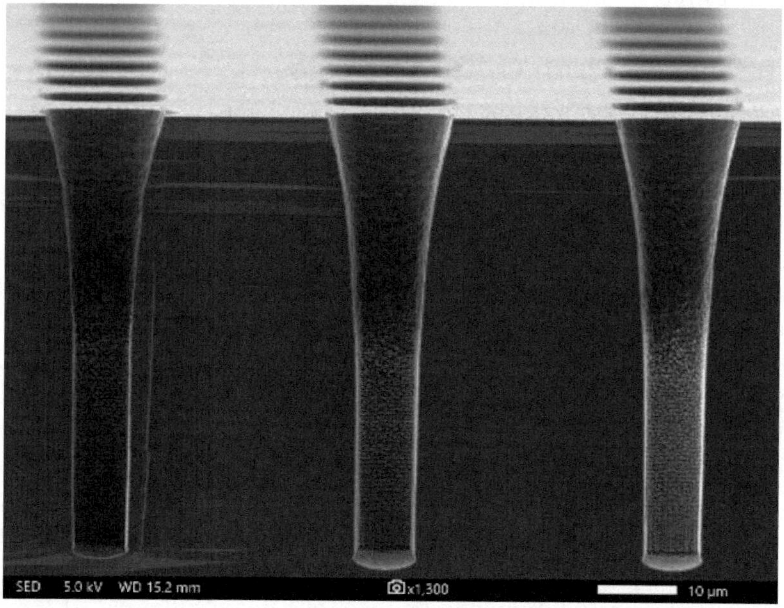

Figure 1.16. Through-silicon via (TSV) technology stands at the forefront of these innovations, as it facilitates pitches down to 30 micrometers for through-wafer connections in substrates that can exceed 600 micrometers in thickness.

Source: https://www.okmetic.com/wp-content/uploads/2021/11/TSV-wafers-Okmetic.jpg.

1.6. FUTURE TRENDS IN MEMS

The fate of Micro-Electro-Mechanical Systems (MEMS) technology rests at the intersection of integration and innovation with greater functionality and miniaturization. As the demand for smaller but efficient devices grows, MEMS technologies are moving into various fields such as biomedical engineering, the automotive industry, and IoT (Internet of Things). Sophisticated systems employ sophisticated microsensors and actuators to enhance precision and responsiveness in usage. The combination of MEMS with technologies like Artificial Intelligence (AI) and machine learning is especially groundbreaking, with the ability to process data in real time and with autonomy. This phenomenon not only makes MEMS devices more accurate and efficient but also gives rise to a new generation of smart products. These advancements are most likely to pave the way for a clearer roadmap towards standardization in microsystems technology, as sub-territories emerge and require coordinated effort for inter-operability and integration (Fluitman et al., 1996). In addition, the incorporation of smart materials into MEMS is a milestone shift bound to re-engineer applications in the coming days. With electroactive and photoactive polymers, scientists are working on exploring their capabilities to create responsive and adaptive products. This innovative solution has applications in the entire domain of assistive healthcare to interactive consumer devices. More broadly, the combination of organic thin film electronic devices (OFEDs) with MEMS technology makes it possible to produce smart bandages to monitor wounds and digital treatments tailored to the needs of patients. This can revolutionize patient care and optimize the effectiveness of treatments. Such convergence also leads to new product configurations along with satisfying the growing demands for tailored solutions in healthcare and consumer markets (Oliver et al., 2009). With continuous advancements in such intelligent materials, it is a surety that these will continue to make MEMS systems more applicable and capable of reacting to evolving human needs. In the coming years, the convergence of MEMS technology with clean energy solutions will continue to be of paramount significance. The integration of MEMS in energy systems, such as organic photovoltaics and solar fabric, highlights the possibility of creating energy-efficient devices that are functional and sustainable. As the use of renewable energy resources continues to increase, MEMS can potentially play a significant role in the creation of high-performance energy harvesting devices that are more universal and applicable. Moreover, this integration makes possible a transition towards smarter built environments, both enhancing resource efficiency and user experience. The ability to track and manage energy consumption in real-time with MEMS-based systems promises to initiate unprecedented advances in smart grid systems. In the long run, the future MEMS trends will not only be innovation drivers for various diverse industries but also an integral component of the broader sustainability and energy efficiency discussion (Fluitman et al., 1996).

1.6.1. Advances in Nanotechnology and MEMS

Latest advances in nanotechnology have positioned Micro-Electro-Mechanical Systems (MEMS) at the forefront of technological innovation. The integration of nanoscale materials into MEMS devices has resulted in enhanced performance, miniaturization, and functionality, making possible multiple applications. Special attention has been given to the integration of "smart" materials such as electroactive polymers and bio-responsive hydrogels, which have enabled the fabrication of adaptive and responsive MEMS components. Such materials not only respond actively to environmental inputs but also enhance the operating ranges of MEMS devices to allow greater versatility in applications ranging from healthcare to environmental monitoring and consumer electronics. Thus, this interface of nanotechnology and MEMS is a revolutionary shift in device functionality to facilitate the creation of smart systems to meet diverse user needs and operations. Further, advanced MEMS technologies leverage advances from cross-disciplinary research, according to the 2014–2015 annual report of the Boston University Photonics Center.

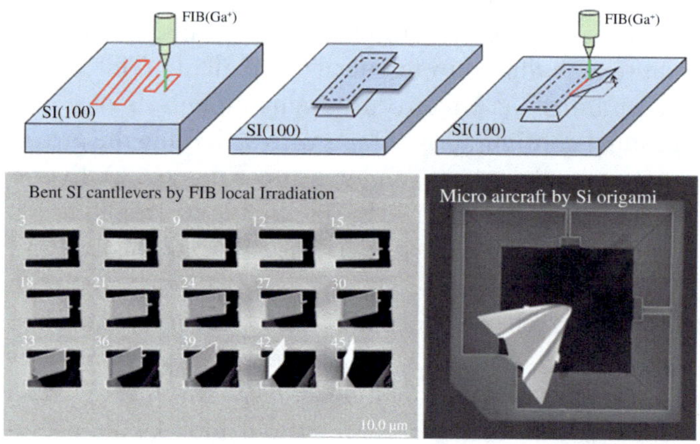

Figure 1.17. The integration of nanoscale materials into MEMS devices has led to enhanced performance, miniaturization, and functionality, facilitating a wide array of applications.

Source: https://onlinelibrary.wiley.com/cms/asset/04374ba5-3ee2-40cf-aaf1-a1067ad99b88/tee23747-fig-0011-m.jpg.

The center's focus on materials science and multi-disciplinary research has resulted in breathtaking advancements in photonics and MEMS, particularly the development of multi-scale modeling methods for electronic materials. These developments not only optimize the design and manufacturing process of MEMS devices but also optimize their operation in many different applications, including personalized medicine and environmental sensing. The centers work emphasizes the need to converge several areas of science and engineering to push the frontiers of what MEMS technologies can achieve. With substantial funding and collaboration, photonics continues to propel innovations that overlap with MEMS, positioning them firmly to push technology frontiers. Second, the application of innovative design methods in MEMS devices incorporates cutting-edge technologies to address practical problems in the real world. This is implemented in the creation of organic thin-film electronic devices (OFEDs), which can be implemented in MEMS systems to create multifunctional applications. Research indicates that through the use of these technologies, engineers can design intelligent fabrics and wearables with real-time feedback for human-centered applications. The developments have widespread implications, ranging from assistive health products like intelligent bandages to enhanced consumer products that combine form and function. As the technology of MEMS advances in cutting-edge fields, it is necessary to evolve in accordance with these new technologies that enable us to develop more sophisticated and holistic systems that meet the requirements of a constantly changing world, in closing the gap between technology and everyday life (N/A, 2015; Oliver et al., 2009) (Figure 1.17).

1.6.2. Integration with IoT and Smart Systems

The merging of Microelectromechanical Systems (MEMS) with Internet of Things (IoT) and smart systems is a monumental technology leap, particularly in data processing and acquisition. MEMS sensors form the backbone of IoT application through the offer of real-time monitoring and control in various environments. These small sensors, which can measure physical events such as temperature, pressure, and motion, assist in enriching the capabilities of smart systems by providing critical information that can be analyzed for decision-making needs at the time. The increasing application of data-driven algorithms in handling sensor data marks the defining factor in integrating MEMS with IoT because the integrity of damage detection and structural health monitoring systems directly depends upon the data these sensors collect. Ultimately, this convergence facilitates a synergy bridge between the physical and digital worlds, leading to more intelligent and efficient operational systems (Cherta et al., 2020).

Besides this, the growing demand for autonomous as well as low-energy consuming devices has amplified the role of energy harvesting technologies as a means to propel MEMS in IoT systems. Battery-powered systems in the conventional sense typically lack maintenance and sustainability issues, particularly when thousands of devices are networked. Energy harvesting — the process of collecting ambient energy from sources such as solar, thermal, or mechanical means — offers a realistic solution to such a situation. This technology not only gives MEMS devices in embedded smart systems endless operation but also enhances the deployment of IoT applications where traditional power sources are unavailable. By utilizing power from the environment, MEMS do not necessarily require constantly draining or having to recharge batteries, thereby increasing their utility and suitability in real projects (Dhananjaya et al., 2017). Furthermore, the interaction between MEMS, IoT, and smart systems will transform various industries with enhanced automation and enhanced efficiency. With the growth of MEMS technology, it will be easy to develop networks of devices that have the ability to talk to one another and share information, which can help improve sectors such as healthcare, transport, and environmental monitoring. Combining MEMS sensors with advanced IoT platforms makes it easy to develop predictive algorithms for monitoring and maintenance of infrastructure. This high-level integration among the technologies facilitates responsive interventions as well as anticipatory maintenance processes, playing a significant role in reducing downtime and operational cost. The impact of this holistic approach extends to delivering safer and more resilient systems, reflecting the revolutionary role of MEMS in the broader scheme of IoT and smart systems (Cherta et al., 2020).

1.6.3. Potential for New Applications in Various Fields

The integration of Micro-Electro-Mechanical Systems (MEMS) technology into various industries demonstrates its immense potential for new applications. One of the significant ones is the Through Silicon Via (TSV) process, which has been applied in MEMS design to pack more components in and reduce die sizes. Miniaturization is significant, especially in consumer electronics where space and efficiency are crucial. As stressed out by Bauer et al. (2008), besides improving interconnectivity within MEMS devices, the TSV technology also enables the integration of sophisticated optical interconnects, sensors, and microfluidic systems. In addition, CMOS integration takes the functionality and variety of MEMS products to a different level, paving the way to revolutionary innovations across different applications from telecommunications to biomedical devices. Besides consumer goods, MEMS technology is also opening new frontiers in the field of health and biomedicine. Boston University Photonics Center is a typical example of the inter-synergistic possibilities of MEMS technology and photonics, particularly in minimally invasive surgery and point-of-care diagnostics. As illustrated by N/A (2013), advances in the field of biophotonics herald a convergence of a number of disciplines that accelerate innovations such as optogenetics and new diagnostic modalities for oncology treatment. By incorporating MEMS in these uses, researchers are creating devices that can analyze and track biological samples in real time, thus making clinical interventions effective and timely. With the healthcare industry demanding more sophisticated and less resource-intensive solutions, MEMS technology emerges as the prime enabler of next-gen healthcare devices with the potential for improved patient outcomes and streamlined medical processes. The uses of sophisticated MEMS extend to new fields such as environmental monitoring and intelligent infrastructure. The ability to create sensor networks by using MEMS technology enables the deployment of small, cost-effective, and sensitive devices to monitor various environmental parameters such as air quality, water quality, and temperature fluctuations. This can be employed to enable smart cities and smart environments that are data-driven and decision-making in nature. Besides, MEMS' capacity for the development of miniaturized sensors that can withstand extreme conditions opens up new possibilities for industrial automation and energy management applications. With the support of MEMS technologies' high-end capability, industries gain the ability to drive operational effectiveness while addressing the critical challenges of urbanization and climate change. This is a witness to the tremendous potential for innovation that MEMS technology continues to unleash across various industries (Figure 1.18).

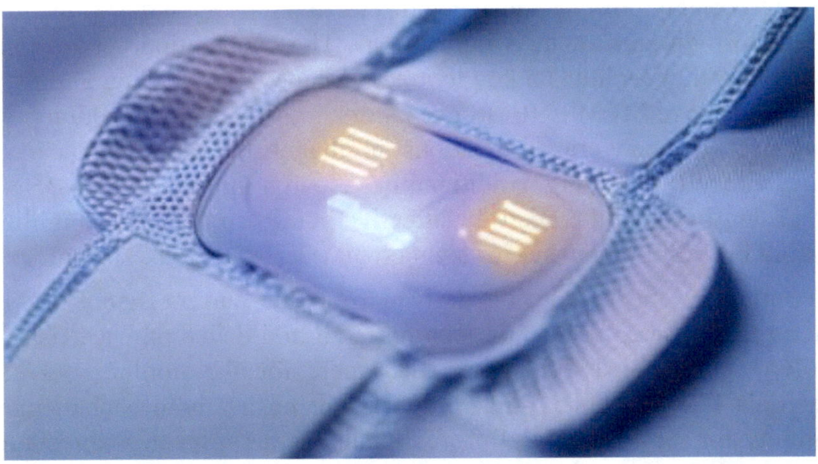

Figure 1.18. As the healthcare industry demands more sophisticated and resource-efficient solutions, MEMS technology emerges as a critical enabler of next-generation healthcare devices that promise improved patient outcomes and streamlined medical processes.

Source: https://res.cloudinary.com/tbmg/ar_16:9,c_thumb,g_auto:subject,f_auto,q_auto/c_limit,w_400/v1737160370/sites/mdb/econa-16x9/52396/400/011725-mdb-podcast-adobe.jpeg.

1.7. CONCLUSION

The introduction of Micro-Electro-Mechanical Systems (MEMS) into most fields is a revolutionary change in technology, even more so in medicine and the monitoring of pollution. With continued developments, the application of new materials and highly sensitive sensors is crucial. Among these are semiconductors and nanomaterials like Boron Nitride Nanotubes (BNNTs), which are leading to improved diagnostics and therapy, especially in personalized medicine. The potential of such technologies to identify volatile compounds and enable early diagnosis of disease is a major development in clinical practice. This is notable evidence of a need for interprofessional collaboration between clinicians, researchers, and technologists to further develop these advances and significantly improve patient outcomes, as evident from ongoing interest in the management and diagnosis of lung cancer in recent literature (Stella et al., 2023). In latest MEMS, the generation of sensor technologies plays a crucial role in enhancing the sensitivity, stability, and overall performance of the devices.

Utilizing the unique properties of nanomaterials, great progress has been achieved to design more responsive and selective sensors that are able to meet the needs of high-performance applications. The cooperation between micro- and nanotechnologies holds a key to creating innovative sensing abilities. For instance, the use of MEMS/NEMS resonators in sensing and detection presents the prospect for new-era applications in diagnostics and healthcare non-invasive, as well as sensing. Additionally, research into the use of BNNTs in mass sensing presents the need for effective yet green materials, highlights a significant breakthrough in sensor technology (Deshwal et al., 2021).

In summary, the conclusion derived from the current status of MEMS is that an urgent need for continuous research and development exists. With technology continuously changing, the quest for the nonlinear dynamics of materials like BNNTs is underemphasized, posing both a challenge and opportunity for researchers and professionals in this respect. Appreciation of the complex behaviors of these materials is required for the creation of more effective mass sensors that could revolutionize medical diagnostics and many other applications. Thus, building increased collaboration between the scientific community, industry collaborators, and clinical practitioners is imperative to effectively transfer innovative MEMS technology from laboratory to applications. The extensive incorporation of current advancements will not only enhance current technology, making it more efficient, but also push upcoming innovations that can make significant improvements in patient outcomes and care (Deshwal et al., 2021).

1.7.1. Summary of Key Points

Within the Micro-Electro-Mechanical Systems (MEMS) field, knowledge of the working principles and technological innovations is crucial for realizing their applications. An important element within this domain involves the creation and utilization of sensors, which are instrumental in active flow control systems. As elucidated within recent studies, the list of sensors created for these purposes includes new prototypes developed and existing ones effectively utilized in real-life circumstances. Proper selection of these sensors relies on actual flight test data, so they are effective in practice. Such ground insight takes into account not only the technical potential of MEMS sensors but also their ability to meet some requirements for operations, which are most important for further research and development in the area of active flow control (Kopylov et al., 2015). In addition, MEMS ventures into the region of indoor positioning and navigation, where empirical evaluation is an essential component in demonstrating various localization systems.

The Indoor Positioning and Indoor Navigation (IPIN) conference is a perfect example of this through its structured competition intended to assess the performance of indoor localization systems globally. The competition, performed in a realistic environment of a large shopping mall, provides a close-to-reality scenario to evaluate system performance. The achieved accuracies reflect the highest possible available indoors navigation, and the competing teams demonstrated the capability to achieve high accuracy without having to deploy additional infrastructure like beacons. This project not only demonstrates advancements in MEMS technologies but also emphasizes the importance of systematic evaluation methods while developing robust indoor navigation solutions (Ali et al., 2019). The integration of MEMS technology in real-world applications emphasizes the importance of continuous innovation and evaluation. As the examples of active flow control and indoor positioning demonstrate, the effectiveness of these systems relies on a complete knowledge of their surroundings and sensors. Further advances will likely focus on improving the accuracy and reliability of MEMS technology and creating new applications across a wide range of industries, from aerospace to consumer electronics. Sustained research activities, as enabled at such conferences as IPIN, serve to develop the discipline, stimulate collaboration among researchers, and stimulate technological advancements. Ultimately, an understanding of these key considerations in the development and practical applications of MEMS will pave the way for further development and greater operational efficiency in the dynamic discipline (Ali et al., 2019; Kopylov et al., 2015).

1.7.2. The Future of MEMS Technology

The MEMS technology promise heralds a revolutionary era in many disciplines, most particularly in energy management. As renewable energy sources gain dominance, the incorporation of advanced MEMS devices is paramount for reducing the inherent volatility brought by these energies. The studies of Battery Energy Storage Systems (BESSs) indicate that MEMS can contribute to energy stability and resilience enhancement. By facilitating efficient storage systems, MEMS technology can be the driving force behind the transition to sustainable energy systems. Furthermore, the applicability of MEMS from micro-scale systems up to the scales of large utilities makes them the right choice to transform and redefine energy storage solutions, thereby providing an energy future with reduced dependence on fossil fuels and increased sustainability for the environment (Dong et al., 2025). Wearable technology is also a field in line to be significantly developed using MEMS technology, particularly in the healthcare sector. Development of miniaturized sensors and flexible electronics allows for precise monitoring of vital health parameters, thus enhancing the quality of patient care. Existing trends indicate that MEMS technology is starting to emerge as a significant component in wearable devices for the management of diabetes, cardiovascular disease monitoring, and sleep tracking.

The integration of MEMS not only maximizes the accuracy of health monitoring but also supports real-time data collection, which is invaluable for clinical decision-making. Moreover, upcoming research focuses on bypassing market and technical challenges associated with these technologies, eventually guiding the path of wearable MEMS applications in healthcare that can enable personalized medicine and improved patient outcomes (Wang, 2025). The future of MEMS technology is outside its conventional applications, with hope for convergence with artificial intelligence and machine learning to drive system performance. With the utilization of high-level algorithms, including reinforcement learning, MEMS devices will be more autonomous and capable of making independent decisions in real-time from massive sets of data. These capabilities will significantly enhance the performance of energy storage systems and medical monitoring devices to make them more efficient and reduce operational risks. As researchers continue to search for ways to couple MEMS with emerging technologies, innovative applications continue to expand limitlessly. This collaborative stretch across borders not only redesigns established paradigms but also lays the ground for game-changing innovation in energy management and healthcare, thus placing MEMS technology as an indispensable resource in spearheading sustainable development across various fields.

1.7.3. Final Thoughts on the Impact of MEMS in Society

The pervasive effect of Micro-Electro-Mechanical Systems (MEMS) in contemporary society stems from their remarkable versatility and effectiveness across various sectors. MEMS technologies have transformed industries through the ability to enable efficient and inexpensive solutions for enhancing performance, accuracy, and functionality. MEMS innovations cross health, automotive, consumer electronics, and telecom sectors. For instance, MEMS sensors are critical to medical diagnosis and treatment and showcase their potential to deliver intelligent healthcare solutions that promote better patient outcomes and more healthcare access. By integrating MEMS with novel materials like organic thin film electronic devices and bio-responsive hydrogels, researchers are creating the possibility of a new continuum of intelligent applications that can react to the sophistication of users in both public and private domains (Oliver et al., 2009).

In addition to the concerns of developing healthcare, MEMS technology holds tremendous potential for sustainable development since it specifically translates to renewable energy solutions. Devices such as organic photovoltaics and solar cloth demonstrate the capabilities of MEMS to increase energy efficiency and harvesting for the demands of today's environment. The use of MEMS in energy devices not only has the potential to boost the rate of energy conversion but also highlights the importance of actively responsive smart materials in responding to environmental stimuli. This harmony with environmental sustainability also emphasizes the role of building technology that enables ambient living environments, hence creating a more sustainable world (Oliver et al., 2009). With continuous struggles against climate change, these MEMS-based technologies will be a key factor in ensuring a greener route. Moreover, the applications of MEMS extend beyond traditional manufacturing and engineering disciplines, offering new paradigms for interactive consumer products. Intelligent, interactive consumer products designed for daily use exhibit a basic shift in consumer demand, emphasizing interactivity and individuality. This tendency is most obviously evident within the domain of personal care and cosmetics, where wearable tech is being merged with MEMS increasingly. These uses can optimize user experience and deliver customized sensory requirements, speaking to the adaptability of MEMS within the consumer space. Moreover, the all-metal RF MEMS switches demonstrate how robust designs can render devices dependable and more versatile, once again significantly addressing the increasing demand for integrated devices with minimal size supplemented by robust abilities. MEMS technologies ultimately advance technological frontiers not only but also create a more individualized, smart, and connected society.

REFERENCES

1. Agis Cherta, D., Gomez Vargas, R. C., Pedraza, C., Pozo Montero, F., et al. (2020). Damage identification in structural health monitoring: A brief review from its implementation to the use of data-driven applications. https://core.ac.uk/download/288625955.pdf (accessed on 14th October 2025).
2. Ahmad, A. S., Ahmar, A. S., Asmuni, H., et al. (2018). Iris segmentation. https://core.ac.uk/download/19162806.pdf (accessed on 14th October 2025).
3. Albers, A., Gillner, A., Razali, A., Yabe, A., Brockett, A., Razali, A. J., Sanchez-Salmeron, A., et al. (2010). Micro-manufacturing: Research, technology outcomes and development issues. https://core.ac.uk/download/9023391.pdf (accessed on 14th October 2025).
4. Alhashim, H. H., Eftekhari-Zadeh, E., Kurdahi, F., Mayet, A., et al. (2024). Sub-0.3 volt amorphous metal WNx based NEMS switch with 8 trillion cycles. https://core.ac.uk/download/642837841.pdf (accessed on 14th October 2025).
5. Ali, M. U., Ben-Moshe, B., Chien, Y. R., Cho, Y., et al. (2019). Evaluating indoor positioning systems in a shopping mall: The lessons learned from the IPIN 2018 competition. https://core.ac.uk/download/287941266.pdf (accessed on 14th October 2025).
6. Aliman, N., Ramli, R., & Amiri, M. S. (2024). Actuators and transmission mechanisms in rehabilitation lower limb exoskeletons: A review. *Biomedical Engineering/Biomedizinische Technik, 69*, 327–345. https://www.semanticscholar.org/paper/98ddfbd83a3f2f8b26af9ceb2c26fc6c95cf537e (accessed on 14th October 2025).
7. Alù, A., Chen, P. Y., Cheng, M., Cui, T., et al. (2018). Generalized Parity-Time Symmetry condition for enhanced sensor telemetry. http://arxiv.org/abs/1812.08296 (accessed on 14th October 2025).
8. Anwar, F., Coutu, R. A., & Mahanta, P. (2019). Novel test fixture for characterizing MEMS switch microcontact reliability and performance. https://core.ac.uk/download/228028443.pdf (accessed on 14th October 2025).
9. Barrachina, L., Fernández Martínez, D., Madrenas Boadas, J., Valle Fraga, R., et al. (2017). Curvature of BEOL cantilevers in CMOS-MEMS processes. https://core.ac.uk/download/157810913.pdf (accessed on 14th October 2025).
10. Bauer, T., & Rimskog, M. (2008). High density through silicon via (TSV).

https://core.ac.uk/download/15497651.pdf (accessed on 14th October 2025).

11. Ben Amar, A., Cao, H., & Kouki, A. B. (2015). Power approaches for implantable medical devices. https://core.ac.uk/download/237711608.pdf (accessed on 14th October 2025).

12. Bhosale, K., & Li, S. S. (2024). Multi-harmonic phononic frequency comb generation in capacitive CMOS-MEMS resonators. *Applied Physics Letters*. https://www.semanticscholar.org/paper/7609a626188a0a1f1cd8ca4f05f7fa6ec4ecad3d (accessed on 14th October 2025).

13. Bogaerts, W., Edinger, P., Errando-Herranz, C., Gylfason, K. B., et al. (2020). MEMS-enabled silicon photonic integrated devices and circuits. https://core.ac.uk/download/287939692.pdf (accessed on 14th October 2025).

14. Bussooa, A., Mercer, J. R., & Neale, S. (2018). Future of smart cardiovascular implants. https://core.ac.uk/download/161513601.pdf (accessed on 14th October 2025).

15. Charalampidis, N., Kampitaki, D., Mallios, N., Nilavalan, R., et al. (2008). An ohmic RF MEMS switch for reconfigurable microstrip array antennas built on PCB. https://core.ac.uk/download/336424.pdf (accessed on 14th October 2025).

16. Charalampidis, N., Nilavalan, R., Spasos, M., Tsiakmakis, K., et al. (2010). Analysis and design of an all metal in line series ohmic RF MEMS switch for microwave applications. https://core.ac.uk/download/336972.pdf (accessed on 14th October 2025).

17. Cheng, J., Xue, N., Qiu, B., Qin, B., Zhao, Q., Fang, G., Yao, Z., et al. (2025). Recent design and application advances in micro-electro-mechanical system (MEMS) electromagnetic actuators. *Micromachines, 16*. https://www.semanticscholar.org/paper/f12c30913941508784cfe33c0a2d949e423887b3 (accessed on 14th October 2025).

18. Coutu, R. A., Lee, H., Leedy, K. D., et al. (2006). Characterization of metal and metal alloy films as contact materials in MEMS switches. https://core.ac.uk/download/213087972.pdf (accessed on 14th October 2025).

19. Cowan, B. (2002). Microfabrication of laser-driven accelerator structures. http://arxiv.org/abs/physics/0304097 (accessed on 14th October 2025).

20. de Boer, M. J., Elwenspoek, M., Gardeniers, J. G. E. (Han), Gilde, M., et al. (2002). Guidelines for etching silicon MEMS structures using fluorine high-density plasmas at cryogenic temperatures. https://core.ac.uk/download/pdf/11457799.pdf (accessed on 14th October 2025).

21. Dekker, R., Geers, M. G. D., Hoefnagels, J. P. M., Joshi, S., et al. (2017). Ultra-stretchable interconnects for high-density stretchable electronics. http://arxiv.org/abs/1709.04434 (accessed on 14th October 2025).
22. Deshwal, D., & Narwal, A. K. (2021). An extensive review of nanotubes-based mass sensors. *Journal of Micro and Nano-Manufacturing, 9*. https://doi.org/10.1115/1.4051261 (accessed on 14th October 2025).
23. Dhananjaya, M. S. (2017). A survey of energy harvesting sources for IoT device. https://media.neliti.com/media/publications/239746-a-survey-of-energy-harvesting-sources-fo-27529cf9.pdf (accessed on 14th October 2025).
24. Dong, Z., Tao, Y., Lai, S., Wang, T., & Zhang, Z. (2025). Powering future advancements and applications of battery energy storage systems across different scales. *Energy Storage and Applications*. https://www.semanticscholar.org/paper/caf19441964464aebae27d67c29124739f16082c (accessed on 14th October 2025).
25. Fluitman, J. (1996). Microsystems technology: Objectives. https://core.ac.uk/download/pdf/11453403.pdf (accessed on 14th October 2025).
26. Gad-el-Hak, M. (2017). In defense of science—What would John do? https://core.ac.uk/download/79821948.pdf (accessed on 14th October 2025).
27. Gemelli, M., Sakauchi, R., Block, T., & Scheiermann, S. (2020). Low-cost MEMS based systems augmenting location and navigation performance in consumer and industrial electronics applications. *2020 IEEE/ION Position, Location and Navigation Symposium (PLANS)*. https://www.semanticscholar.org/paper/700896bceae611e6717b86a983807d553c33931d (accessed on 14th October 2025).
28. Ghazal, T. M., Hasan, M. K., Alshurideh, M. T., Alzoubi, H. M., Ahmad, M., Akbar, S. S., Al Kurdi, B., et al. (2021). IoT for smart cities: Machine learning approaches in smart healthcare—A review. *Future Internet, 13*, 218–218. https://doi.org/10.3390/fi13080218 (accessed on 14th October 2025).
29. Gill, E., Monna, G. L. E., Scherpen, J., et al. (2007). MISAT: Designing a series of powerful small satellites based upon micro systems technology. https://core.ac.uk/download/pdf/12941826.pdf (accessed on 14th October 2025).
30. Kokshagina, O., Le Masson, P., & Weil, B. (2015). Portfolio management in double unknown situations: Technological platforms and the role of cross-application managers.

31. Kopylov, G. (2015). Overview of sensors suitable for active flow control methods. https://core.ac.uk/download/30294554.pdf (accessed on 14th October 2025).

32. Luo, Y., Abidian, M. R., Ahn, J. H., Akinwande, D., Andrews, A. M., Antonietti, M., Bao, Z., et al. (2023). Technology roadmap for flexible sensors. *ACS Nano, 17*, 5211–5295. https://doi.org/10.1021/acsnano.2c12606 (accessed on 14th October 2025).

33. N/A. (2013). The Boston University Photonics Center annual report 2013–2014. https://open.bu.edu/bitstream/2144/29752/1/Photonics-Center-Final-Annual-Report-2013-2014-v2.pdf (accessed on 14th October 2025).

34. N/A. (2015). The Boston University Photonics Center annual report 2014–2015. https://open.bu.edu/bitstream/2144/29753/1/Photonics-Center-Annual-Report-2015-final.pdf (accessed on 14th October 2025).

35. Oliver, R., Tillotson, J., & Toomey, A. (2009). Bioengineered textiles and nonwovens – The convergence of bio-miniaturization and electroactive conductive polymers for assistive healthcare, portable power and design-led wearable technology. https://core.ac.uk/download/9587153.pdf (accessed on 14th October 2025).

36. Pal, P., & Sato, K. (2017). Silicon wet bulk micromachining for MEMS. https://core.ac.uk/download/80948524.pdf (accessed on 14th October 2025).

37. Raghavan, S., Gupta, R., & Sharma, L. (2023). Applications of a novel tunable piezoelectric vibration energy harvester. *Micromachines, 14*. https://www.semanticscholar.org/paper/a980a858f5a29568f2def630f276aa3ade547fc0 (accessed on 14th October 2025).

38. Stella, G. M., Lettieri, S., Piloni, D., Ferrarotti, I., Perrotta, F. M., Corsico, A. G., & Bortolotto, C. (2023). Smart sensors and micro technologies in the precision medicine approach against lung cancer. *Pharmaceuticals, 16*, 1042–1042. https://doi.org/10.3390/ph16071042 (accessed on 14th October 2025).

39. Studor, G. F. (2012). 2012 PWST workshop summary. https://core.ac.uk/download/pdf/10571173.pdf (accessed on 14th October 2025).

40. Talbot, A., Florin, B., Durand, C., Renaud, D., Judong, F., Casset, F., Bannon, F. D., et al. (2007). Silicon on Nothing MEMS electromechanical resonator. https://core.ac.uk/download/52700979.pdf (accessed on 14th October 2025).

41. Wang, S. (2025). Wearable devices in healthcare: Technological advancements, clinical applications, and future horizons. *Applied and*

Computational Engineering. https://www.semanticscholar.org/paper/8cebca447f825cebba24e7a7b81bf8be1f356408 (accessed on 14th October 2025).

42. Yalagach, M., Fuchs, P. F., Wolfberger, A., Gschwandl, M., Antretter, T., Feuchter, M., Tak, C., et al. (2019). Numerical analysis of the influence of polymeric materials on a MEMS package performance under humidity and temperature loads. *2019 IEEE 69th Electronic Components and Technology Conference (ECTC)*. https://www.semanticscholar.org/paper/2da2e247f806887ad1dc33cbfc7261b05f2b3cf8 (accessed on 14th October 2025).

CHAPTER 2

MECHANICS OF MICROSTRUCTURES

Contents

INTRODUCTION

An appreciation of the fundamental principles involving microstructures is central to any investigation based on the mechanics of materials. Microstructures refer to the internal materials' structures that play a significant role in determining their properties and behaviors under varied conditions. They can supply vital information regarding the behavior of a material under mechanical stress, thermal fluctuation, or chemical reaction and thus influence design in various fields of engineering. Microstructural analysis becomes increasingly relevant as modern materials science continues to move towards the optimization of composite and novel materials. The integration of microstructural mechanics into engineering curricula can prepare students with a unified grasp of microstructure-property relationships, positioning them for high-end applications and novel design breakthroughs in materials science and engineering (Sanjeev et al., 2020). Microstructure study not only enhances our understanding of traditional materials but also underscores their role in the achievement of advanced materials. In particular, mechanics of microstructures is interested in a meticulous study of how grain, phase, and defect structure influences material behavior. For instance, processes of deformation and fracture are intrinsically linked with the microstructural characteristics of materials. Estimation of these relationships is essential in designing load-carrying members or parts with dynamic loads. Improvements in how to instruct the mechanics of materials, as evident in endeavors in schools of learning, reveal that combining theoretical ideas with their practical applications in microstructural examination can further deepen students' comprehension and skills (Sanjeev et al., 2020). Hence, this approach promises to produce a generation of engineers capable of addressing the problems posed by new, high-tech materials. Moreover, the importance of dynamic events in microstructures is demonstrated with the recent introduction of smart materials. As is demonstrated by numerous research, the promise of altering microstructural characteristics in response to outside stimuli provides new avenues for applications from smart coatings to bendable electronics. The example of dynamic wrinkles, where the microstructure is influenced by environmental conditions, shows the creative aspect that comes with microstructural mechanics. By understanding the underlying mechanics and characteristics of microstructures, engineers can create materials that not only fulfill their respective functions but also transform based on their environment. This versatility is an important factor in the creation in the future of multifunctional materials and provokes continued research and study of the intricate relationship between microstructure and material performance in numerous applications.

Definition of Microstructures

Microstructure is basic to the study of mechanics of materials in many disciplines, ranging from civil and materials engineering. Microstructures are the complex configurations of phases, grains, and pores in a material at microscopic levels, usually between 1 nanometer and a few millimeters. These structures occur through the solidification, heat treatment, and other phase transformations materials go through. The specific microstructural characteristics, such as grain size, shape, and orientation, have a significant influence on the mechanical behavior of materials, i.e., strength, ductility, and fatigue resistance. Hence, critical examination of microstructures is necessary for material prediction under diverse conditions and loads and thus forms an integral part of materials science and engineering. A superb example of microstructural analysis can be found in the study of Portland cement paste, where porosity and phase distribution relate particularly to the materials performance during building. Simulations indicated that the microstructures of Portland cement paste, being dependent on water-to-cement ratio and curing time, offer various pore size distributions, which affect durability as well as the overall performance of the material. For instance, numerical simulations with the cement hydration model illustrates the development of nanoscale structures like calcium silicate hydrates into larger microstructural components over a curing time window of 1–28 days. Such computer simulations make use of advanced image segmentation methods to model the capillary and gel pore properties in the paste, with development of the microstructure being critical for achieving realistic performance requirements in engineering (Gao et al., 2015).

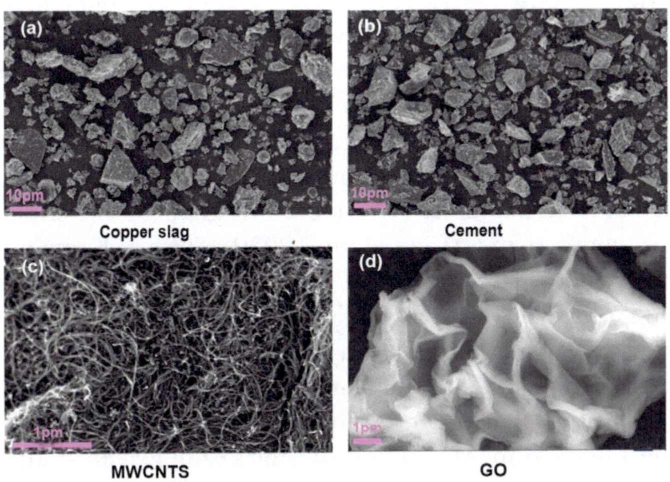

Figure 2.1. A notable example of microstructural analysis can be found in studies of Portland cement paste, where the porosity and arrangement of phases are directly linked to the material's performance in construction applications.

Source: https://media.springernature.com/full/springer-static/image/art%3A10.1038%2Fs41598-025-97227-5/MediaObjects/41598_2025_97227_Fig1_HTML.png.

More details regarding microstructures can be derived from the study of phase transformations in alloys, particularly martensitic transformations.

Mechanics of such transitions describe the complex process of microstructural development, as seen from the displacements between austenite and martensite phases in ultra-low hysteresis alloys. Applying a moving mask approximation, researchers have deduced a Hadamard jump condition describing the deformation characteristics relevant to martensitic microstructures. This high-level concept of phase interfaces allows for the description of energy-minimizing structures and identification of supercompatibility conditions that potentially influence the reversibility of such transitions. Microstructure investigation at this size thus shows how intricate behaviors on the micro-scale can influence material properties and processing capability (Porta et al., 2018) (Figure 2.1).

Importance in Material Science

The significance of material science to microstructure cannot be overstated as it is a background science that informs our understanding of how microscopic behavior influences macroscopic properties. By studying the atomic and molecular interactions, material scientists can account for how microstructures control material responses to various stresses and to environmental conditions. This encompasses all the way from metallic tensile strength to thermal and electrical characteristics of composites. For instance, the creation of curved structures demonstrates an intimate relationship between microstructure and mechanical performance. These microstructures possess greater strength-toughness pairing due to their unique grain size gradients, as evidenced by experimental data on crack resistance of gradient nickel compared to homogeneous counterparts (Cao et al., 2020). These findings reflect the significance of microstructure design in developing sophisticated materials with specified properties. Placing added emphasis on the significance of microstructures to materials science, composite material engineering has yielded new applications across numerous industries. The design of materials that combine attributes in order to attain optimal performance is based on understanding microgeometries involved. For example, the study of efficient elastic tensors of composites is directly related to their performance in terms of mechanics, showing how some combinations of different materials can lead to stiffness or flexibility when required by applications (Briane et al., 2016). The designating of these effective tensors allows the creation of materials with specified mechanical properties without necessitating the utilization of original, more expensive materials. This tailoring underlines the importance of accurate microstructural analysis, as it enables materials to be produced that can meet highly demanding performance requirements while remaining economical. Finally, the ongoing advancements in material science, particularly in the aspect of examining microstructures, have paradigm-breaking impacts on technology and engineering. The ability to process and engineer materials at a microstructural level has implications in high-technology applications such as aerospace, robotics, and bioengineering. Greater toughness of crack growth, for example, is critical to reliability in structural components subjected to harsh conditions. Studies on gradient materials enhance the probability of utilization in sectors where strength and toughness are a prerequisite (Cao et al., 2020). With advancements in the field, its emphasis on the interrelation between microstructure and mechanical properties will most likely yield revolutionary solutions that not only expand the limits of material but also enable sustainable operation. Thus, the importance of material science in terms of microstructure mechanics continues to expand, shaping the future of material development (Figure 2.2).

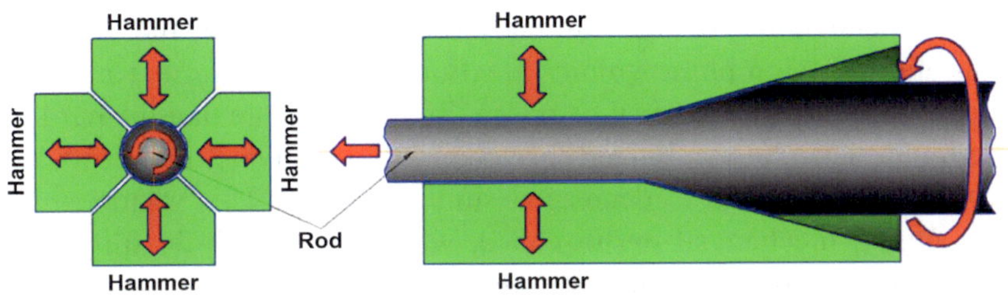

Figure 2.2. The emergence of gradient structures demonstrates a profound relationship between the microstructure and mechanical performance.

Source: https://www.mdpi.com/materials/materials-16-01706/article_deploy/html/images/materials-16-01706-g001.png.

Overview of Mechanical Properties Influenced by Microstructures

Microstructures have a significant influence on the mechanical properties of materials, and they dictate the material's response when forces are applied externally. Macroscopically, properties such as toughness, ductility, and strength tend to be controlled by the distribution and interaction between the different phases in the material. For instance, within two-phase materials that have a hard, brittle phase distributed throughout a soft, ductile matrix, the phase spatial distribution has the capacity to significantly affect failure modes. Sites of fracture initiation are prone to where hard-phase regions intersect the soft phase under arrangements that maximize stress concentration, demonstrating a complex interaction between microstructure and mechanical performance

(Appolaire et al., 2016). It is vital to understand these connections for the design and optimization of high-performance materials for extreme engineering uses. Besides, microstructural features also govern not only the macroscopic strength but even the ductility of materials. Multi-phase materials that under normal circumstances exhibit a balance of high strength and high ductility, on failure, like to happen by means of ductile mechanisms of fracture of the more compliant matrix phase. Research has identified some crucial geometric arrangements where the position of the hard particles in relation to the matrix can lead to high tensile hydrostatic stress and plastic strain, indicating higher susceptibility to damage (de Geus et al., 2015). Microstructural characteristics' dependence on mechanical behavior calls for meticulous examination of topological configurations since topological configurations dictate the material failure behavior under various load conditions, contributing towards an entire understanding of the resiliency of materials. Also, the variation of hardness and inclusion phase volume fraction can provide further insight into composite materials' mechanical response. Studies indicate that the hard phase volume fraction can be maximized for enhanced overall ductility, provided the hardness of the inclusion is decreased simultaneously. This relationship can lead to optimal mechanical performance, with the strength-ductility balance being important in most engineering applications. Through the utilization of advanced modeling techniques such as the Fast Fourier Transform, researchers can simulate and analyze a broad range of microstructures to predict their mechanical behavior (Appolaire et al., 2016). Therefore, further study into the influence of microstructure is required for the continued creation of materials that are intended to perform for specific applications, ultimately resulting in their increased reliability and performance in practical applications.

2.1. TYPES OF MICROSTRUCTURES

Microstructures are of main significance in determining the material properties of materials, and their classification is necessary in predicting material behavior under varying kinds of load. The two important types of microstructures are crystalline and amorphous. Crystalline microstructures are characterized by a well-ordered atomic structure, which is further subdivided into single crystals and polycrystals. Single crystals possess uniform properties in any direction and are required for situations where such special directional strength is required. However, polycrystalline materials consist of many grains with different orientations and exhibit higher-order mechanical behavior that is grain-boundary sensitive. Grain interaction may strengthen or weaken the material, depending on grain shape and size and stresses the importance of microstructural investigation in material design. Among metallic materials, some microstructural arrangements have a significant impact on performance characteristics. Face-centered cubic (fcc) structures, like those of copper, are very ductile and resistant to deformation. Body-centered cubic (bcc) structures, like those of tungsten, are more strength-efficient at the cost of lower room temperature ductility. Hexagonal close-packed (hcp) materials such as magnesium possess unique challenges and advantages due to their anisotropic deformation behavior. Each of these microstructural structures requires careful computational models to predict their mechanical responses properly. Referring to the new developments, crystal plasticity finite element method (CPFEM) models are widely used to simulate the intricate behavior of such microstructures, illustrating the deep relationship between material performance and microstructure (Tran et al., 2022). For further deepening insight, systematic approaches such as uncertainty quantification (UQ) may be used to determine the predictability of mechanical models in terms of microstructural data. These models account for variability in microstructure that may arise as a consequence of processing conditions and iterations during manufacturing, influencing resulting properties and measures of performance. Stochastic modeling techniques aid in determining the robustness of different constitutive models and sensitivity to underlying parameters. In polycrystalline materials, for instance, spectral formulations have progressed considerably, allowing researchers to study the mechanical behavior as a function of evolving microstructures (Ricardo et al., 2020). Through the use of UQ in microstructural analysis, engineers are able to better understand the reliability of material predictions and, in turn, optimize designs tailored to fulfill specific performance needs (Figure 2.3).

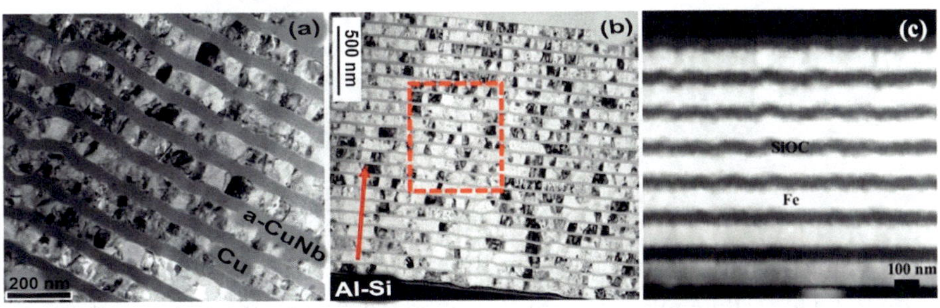

Figure 2.3. Crystalline microstructures are characterized by a highly ordered arrangement of atoms, which can be further subdivided into single crystals and polycrystals.

Source: https://pub.mdpi-res.com/materials/materials-16-02874/article_deploy/html/images/materials-16-02874-g001.png?1680603713.

2.1.1. Crystalline versus Amorphous Structures

The distinction between crystalline and amorphous forms is focal to the science of microstructure study, influencing material behavior and properties. Crystalline materials are characterized by a long-range ordered arrangement of atoms, leading to geometric form that is sharply defined and reproducible mechanical properties. This long-range order confers properties such as anisotropic behavior, where the properties vary direction. In contrast to them, amorphous materials lack such long-range order and, as a result, experience a disordered atomic arrangement that gives rise to isotropic properties. This structural property deviation has a very crucial influence on the mechanical behavior of materials, especially under conditions of stress, and consequently their use in semiconductor technology and structural engineering. The mechanical behavior of crystalline and amorphous structures can be described in terms of their respective plasticity mechanisms.

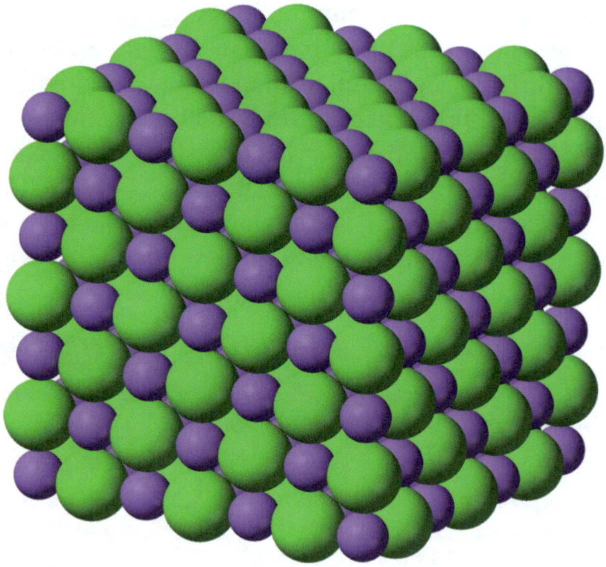

Figure 2.4. Crystalline materials are characterized by a regularly ordered arrangement of atoms, leading to well-defined geometric shapes and predictable mechanical properties.

Source: https://upload.wikimedia.org/wikipedia/commons/e/e9/Sodium-chloride-3D-ionic.png.

Crystalline materials typically possess defined slip systems that govern the movement of dislocations, which is critically important for plastic deformation. Research shows that such materials can exhibit complex interactions in plastic flows, like established turbulence, which suggests a correlated but chaotic microscopic response in their stress behavior (Acharya et al., 2012). Amorphous materials deform through a more heterogeneous flow process, in which shearing transformation is carried out in local volumes. The absence of plasticity under dislocation-mediated states in amorphous structures leads to non-smooth stress and strain profiles, dismantling traditional models of engineering that are based on smooth material response. The influence of external states such as temperature and pH further emphasizes the differential mechanical relaxations and behavior of crystalline versus amorphous materials. For instance, the conformational changes of polymers at varying pH values illustrate how environmental conditions might significantly influence mechanical properties. Differential scanning calorimetry studies reveal that pH significantly affects the viscoelastic characteristics of materials like LM-pectin, with transitions observed over the glass transition range (Al-Ruqaie et al., 2015). Therefore, during material design for a specific application, mechanics of these microstructures become significant, as these affect how these microstructures respond to environmental changes and mechanical loads and eventually the selection process of the material in engineering and product design (Figure 2.4).

2.1.2. Role of Grain Boundaries

The role of grain boundaries within microstructures lies at the heart of understanding the mechanical properties of materials. Grain boundaries are boundaries between different grains of a polycrystalline material and contain the secret to understanding mechanical strength, ductility, and performance. Whether or not the boundaries exist and what their nature is, it determines the way dislocations interact within the grains, which, in turn, affects the capacity of the materials to withstand applied stress. Also, grain boundaries can inhibit the motion of dislocations, providing a strengthening mechanism whereby the material is able to achieve higher performance under mechanical stress. As researchers dig deeper into the mechanisms of these boundaries, they discover their immense impact on other properties, which is why they are a unifying theme in materials science and engineering. In more complex materials such as high-entropy ceramics (HECs), things are even more complex and interesting at grain boundaries. Heterogeneity of structures in HECs allows for unique grain-boundary interactions that can lead to unique mechanical properties. For instance, the nature of such boundaries can facilitate not only hardening mechanisms but also enhance ionic conductivity and electromagnetic absorption phenomena, as discussed in recent reviews published on the subject (Xiang et al., 2021). Grain boundary behavior in such materials is crucial to be realized to make predictions regarding their mechanical performance and to engineer them towards specific applications. The challenges involved in establishing the nature of such grain boundaries and establishing the contribution to the overall microstructure are the subject of current research, which may open up material design and functionality breakthroughs. Furthermore, the interaction among grain boundaries and other microstructural components plays an important role in the mechanical performance of solid-state batteries, and particularly for lithium metal anodes. Current perspectives suggest that grain boundary dynamics can affect lithium-ion transport and interfacial layer growth, both of which are critical to battery performance. Because grain boundaries become pathways for lithium-ion diffusion, the understanding of how grain boundaries interact with solid electrolytes is vital to attaining optimal battery cycling and lifespan (Kelsey et al., 2020). This interaction depicts the inherent function of grain boundaries to enable proper ionic conductivity while overcoming problems such as dendrite growth. Thus, it is essential to study these interfaces since they hold essential information for constructing the technology of energy storage devices, resulting in improved performance parameters and reliability (Figure 2.5).

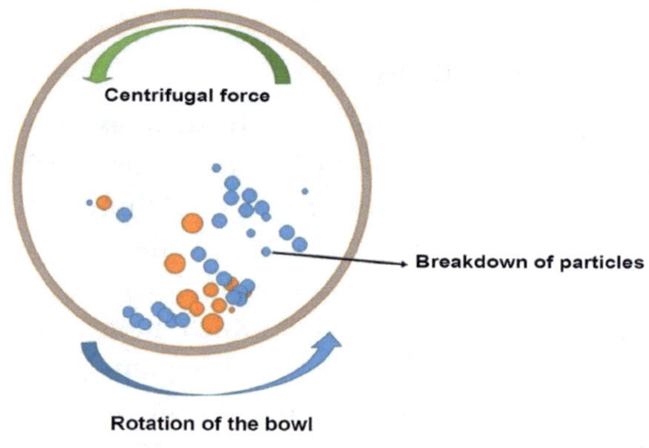

Figure 2.5. In advanced materials such as high-entropy ceramics (HECs), the role of grain boundaries becomes even more complex and intriguing.

Source: https://www.mdpi.com/ceramics/ceramics-07-00089/article_deploy/html/images/ceramics-07-00089-g003.png.

2.1.3. Influence of Phase Transformations

The function of phase transformations is an essential contributor to the mechanics of microstructures that has a significant impact on material properties and performance of various alloys and ceramics. Phase transformations have the tendency to produce different unique microstructural characteristics that can have either favorable or unfavorable effects on mechanical performance. For instance, phase transformations have the ability to modify grain size and orientation, which have significant roles in determining the strength and ductility of the material. The possibility of phase transformation control through processes such as severe plastic deformation (SPD) favors metastable phase stabilization, which enables materials to acquire a high strength and ductility combination. High-pressure torsion and equal-channel angular pressing are some of the SPD processes that facilitate the transformations, which in turn influence the mechanical properties of processed materials and condition them for high-level uses in various demanding environments (Edalati et al., 2022). In addition, the creation of high-entropy ceramics (HECs) is an example of how phase transformations can be harnessed for the creation of new directions in material property design. Unlike traditional ceramics, HECs are created with multiple principal elements, making it simpler to modify the band structure and phonon behavior. This new composition makes the phase transformation routes more complex, allowing for new functionalities such as superionic conductivity as well as exceptional thermal and electromagnetic performance. These complex interactions among various phases in HECs are the reasons behind the enhanced performance, especially in strength and thermal stability in varying conditions. Comprehending the phase transformation dynamics in HECs is therefore essential for making the best use of their potential in new applications, a paradigm shift in material design concepts (Xiang et al., 2021). Acknowledge of the significance of phase transformations within microstructures emphasizes the need for precise processing techniques that can effectively manage such transformations. Identification of new processing routes that can introduce particular phase transformations with no loss of material integrity remains essential as researchers continue to delve into the world of materials science. The ability to create nanoscale microstructures via SPD routes has already shown the potential for providing materials with improved mechanical properties. It is important that future studies focus not only on the development of such microstructures but also on their characterization as far as their phase composition and stability are concerned. This integrative methodology will give a better understanding of the coupling between mechanical performance and phase transformations, enhancing the development of materials for optimal performance against ever-evolving applications in the industry (Edalati et al., 2022; Xiang et al., 2021).

2.2. MECHANICAL PROPERTIES OF MATERIALS

The mechanical properties of materials are important parameters in the determination of their application and performance in various engineering fields. Some of these properties include tensile strength, ductility, hardness, and fatigue resistance, and they are closely linked to the microstructure of the materials. For example, the metal grain structure can have a great impact on its ability to deform when subjected to stress, hence altering its tensile strength. In the majority of composite materials, microstructural element heterogeneity, as in the case of the study of flax nonwoven bio-composites, may lead to complex interactions which affect mechanical performance.

This requires advanced numerical models able to forecast the mechanical behavior of the material under operating conditions correctly, and this underlines the necessity for a robust theoretical basis for the mechanics of microstructures. The complexity of microstructural interactions calls for the development of innovative methods of evaluating mechanical behavior. As an example, the synchronization of phenomenological and numerical multiscale approaches, demonstrated in composite material science, is a novel means to study the mechanical properties of nonwoven composites.

This methodology not only minimizes computational cost but also captures the various interactions of composite components efficiently, leading to more reliable predictions of failure in the structure. With these advancements, engineers and materials scientists are able to design materials to optimize performance in specific applications, thereby improving the use of advanced materials in modern engineering activities. The uses of these techniques extend beyond theoretical explanation, allowing engineering solutions that capitalize on the strength of complex microstructures. In addition, understanding the mechanical behavior of geological materials, such as the carbonate rocks of the Kawagarh Formation, illustrates the importance of microstructural analysis in diverse environments. Since these materials have significant diagenetic alteration and fossil content, their mechanical behavior can be extremely variable. The results of the uniaxial compressive strength and tensile strength tests confirm that the mechanical properties of these limestone aggregates fall within acceptable limits for construction applications. Through the integration of petrographic analysis with mechanical testing, the work offers a structured way of establishing how microstructural features influence the engineering characteristics of geologic materials. This approach not only further supplements knowledge about material behavior but also contributes towards decision-making concerning material selection for engineering and construction purposes, insofar as the interdependence of microstructure and mechanical property in materials science is encompassed.

2.2.1. Strength and Hardness

The concepts of strength and hardness are crucial in the case of microstructure mechanics in material science. Strength refers to a material's ability to withstand a load under it without breaking, whereas hardness is a material's ability to resist deformation or indentation. Both are very closely related to a material's microstructure, which comprises the structure of its individual phases, grain size, and defect concentrations. For example, a well-developed grain structure typically enhances strength due to the grain boundary strengthening mechanism. This relationship is illustrated in the situation of severe plastic deformation (SPD), which produces ultrafine-grained material with exceptional strength as well as enhanced ductility. The new approaches developed within the NanoSPD have led to materials that not only retain excellent strength but also enhance other functional characteristics, demonstrating the profound significance of microstructure modification in influencing mechanical performance (Edalati et al., 2022). The interrelation between strength and hardness is also important in the context of many applications of modern materials. Application of machine learning techniques to material design is now a revolutionary way of predicting and enhancing such properties.

Figure 2.6. Strength refers to a materials ability to withstand an applied load without failure, while hardness is the resistance of a material to deformation or indentation.

Source: https://fecision.com/wp-content/uploads/2025/06/Material-Properties-1-1024x828.jpg.

From the interaction of large databases of microstructural properties and the corresponding mechanical responses, predictive models of new material compositions can be generated. These advances highlight the importance of charting the microstructural constituents responsible for strength and hardness. Specifically, the present study focuses on optimizing processing procedures and alloy compositions to synthesize materials with selective performance properties in harsh environments. This integration of advanced computational methods not only accelerates the discovery of high-performance materials but also expands the understanding of intrinsic phenomena underpinning strength and hardness in microstructures (Guo et al., 2020). In short, the mechanics of microstructure basically dictate the properties of materials, most notably their strength and hardness. The continuing evolution of SPD technology and the advent of machine learning in materials science provide a powerful paradigm for further developing these mechanical properties. With the field in advance, it is increasingly important to narrow the gap between theoretical prediction and experimental proof to manufacture materials that are not only stronger but also richer. Studying the interplay between microstructural properties and mechanical behavior will undoubtedly offer innovations that meet the complex requirements of future technological uses, solidifying the central role of strength and hardness in materials science. Collectively, these advances will be of great help towards more effective and sustainable use of materials in industry and engineering processes (Figure 2.6).

2.2.2. Ductility and Toughness

The relationship between toughness and ductility is an important aspect of material mechanical behavior, especially for microstructures. Ductility is used to quantify the capacity of a material to deform significantly before failure, while toughness is used to quantify energy absorption with or without plastic deformation without fracture. These properties become critical in applications in which materials are subjected to dynamic loads and changing environments. Increased ductility typically accompanies increased toughness, since those materials with large deformation capacity can dissipate energy more effectively, and that will decrease the likelihood of abrupt failure. It is important for engineers developing components which must not only endure stress but additionally maintain their integrity when loaded cyclically to recognize this relationship. The bainitic steels provide an excellent example in considering ductility and toughness within microstructural mechanics. These investigations highlight that a specially composed alloy can exert a very significant influence on their mechanical properties. Specifically, it has been seen in research that it is possible to improve fracture toughness greatly by varying the level of boron in bainitic steels. With increased boron content, there is a substantial decrease in interlath martensite content, which promotes void coalescence. Such microstructural change dramatically enhances the toughness of the material so that it deforms prior to breaking. The findings reveal that minimal alterations on the microstructural scale can have major impacts on overall performance, as reflected by enhanced fatigue life that is observed with higher boron levels in the steels (Marrero et al., 2016). The universal applicability of these findings is in wide-ranging applications, particularly within structural and mechanical engineering, where material performance under adverse conditions matters. Because businesses are utilizing materials that provide both strength and resilience more so than before, the process of microstructures providing insight into how a balance can be achieved is priceless. Fatigue life, as well as methods such as Paris' law, illustrate the value of comprehending ductility and toughness thoroughly in the framework of microstructures. Through studying the microscopic interaction and behavior, researchers can design materials that meet the needs of the current engineering challenges, ultimately leading to secure and efficient designs in fields where material failure would have catastrophic consequences (Carpenter et al., 2025) (Figure 2.7).

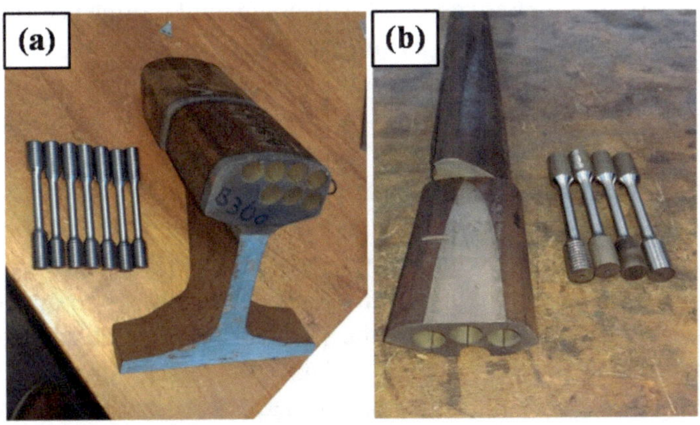

Figure 2.7. Bainitic steels serve as an exemplary case for analyzing ductility and toughness in microstructural mechanics.

Source: https://www.mdpi.com/metals/metals-09-00778/article_deploy/html/images/metals-09-00778-g001.png.

2.2.3. Fatigue Resistance and Failure Mechanisms

The study of fatigue resistance and associated failure mechanisms is a core materials science discipline, and more precisely, how microstructure influences the durability of materials under cyclic loading. Fatigue failure is a process of accumulation where microscopic damage occurs that can progress towards crack initiation and final fracture. The mechanisms involved typically entail mixed modes of crack growth as a reaction to microstructure factors such as grain size, phase composition, and inclusions. Most of the work, including that reviewed in a detailed review of structural material life prediction, emphasizes the essentiality of knowing how these interactions work in an integrated manner (Carpenter et al., 2025). By correlating microstructural features with fatigue strength, researchers and engineers can develop means to enhance material performance and translate principles to applications demanding durability in cycling stress. Processing operations and their ensuing microstructure interact to control fatigue strength. New developments in additive manufacturing processes, like laser power bed fusion (L-PBF), have become the focus because of their unique microstructural outcomes. L-PBF 316 L experiments demonstrate that the choice of build orientation and subsequent post-processing operations like annealing or hot isostatic pressing significantly alter fatigue and fracture behavior. Alteration is able to affect plastic deformation properties and lead to fatigue strength differences through mechanisms like fatigue-ratcheting interaction. The coupling between cyclic plasticity and fatigue is reflective of the manner in which microstructural evolution improves fatigue resistance through an optimally efficient process, adhering strictly to the axioms of microstructure mechanics in a convincing way (Hardacre et al., 2018). Furthermore, the mechanisms of fatigue failure become increasingly complex under conditions of long life, where the primary issue is elastic deformation. Under these conditions, it is increasingly important to understand nonlinear interactions between applied stress and microstructural response. Classical theories relying only on stress are not effective in making predictions, especially when cyclic plasticity and the governing variables are in the picture. New extensions to classical theories, such as a modified Goodman relation, have been proposed to account for the above intricacies and been shown to agree well with experiments under various loading conditions. This advancement highlights the requirement of a multidisciplinary approach to fatigue analysis that synergizes microstructural mechanics knowledge and empirical data to guarantee innovation in materials design for the optimization of fatigue resistance and durability of performance.

2.3. CHARACTERIZATION TECHNIQUES

In materials science, the characterization techniques are the foundation for the understanding of mechanical properties and structural integrity of microstructures. They are general methods that provide information regarding the composition of the materials, microstructural evolution, and response upon application of stresses. A major advancement has been the collaboration between experimental techniques and computational modeling such that scientists can develop predictive models from data derived from experiments. This combined strategy not only enhances structure-property knowledge but also facilitates innovation in the design and characterization of intricate materials (Dingreville et al., 2015). Quantitative analysis, for instance, can yield pivotal mechanical parameters that guide the selection of material for specific applications, maximizing performance and longevity. Furthermore, the utilization of different characterization techniques can reveal significant differences in microstructure such as grain size, porosity, and phase distribution, which have direct impacts on a material's mechanical properties. A case study example for the deposition of thin metal foils using self-induced ion plating presents the case here. The cleanliness of the zinc and copper foils obtained through this novel process defines that microstructures have a whole different character compared to what is produced conventionally with electroplating and rolling processes (Giraud et al., 2014). Such findings point toward the need for applying the correct characterization techniques unique to the character of material and processes under consideration so that the mechanical properties displayed match closely with the characteristic microstructural aspects. Ultimately, continuous advances in characterization procedures improve not only the scope of material analysis but also place a dynamic relationship between application and research. As greater, more rigorous demands are put upon contemporary materials, advanced characterization is increasingly critical. Through extensive research and fresh methodologies, like those highlighted in recent research, improved comprehension arises regarding microstructural mechanics (Dingreville et al., 2015). The differentiation between various production methods, as presented by way of research into ion plating methods, supports the extent to which microstructural properties can have a very real impact upon material characteristics like Youngs modulus and hardness (Giraud et al., 2014). Through continued examination of characterization measures, doors are opened to future advancement in material engineering and science, especially in situations where the need is for a specialized approach to the optimization of microstructure.

2.3.1. Scanning Electron Microscopy (SEM)

Scanning electron microscopy or SEM is now an essential technique of studying the intricate mechanics of microstructures in various materials. SEM is a high-resolution imaging method with the use of a focused electron beam to produce high-resolution surface morphology of the samples, allowing researchers to observe structural details at nanometer scale. The advantages of SEM over traditional optical microscopy are immense; it has an increased depth of field, better resolution, and can image surface structure in three dimensions. With SEM, scientists can not only study the physical properties of materials but even their compositional properties, which are crucial to understanding how microstructural variations influence macroscopic properties. With this technology comes greater insight into microstructural mechanics, thus paving the way for future developments in materials engineering and science. Adding to the importance of SEM, other recent studies have applied this technique to study the microstructural integrity of various composites, particularly following degradation using thermal stress. For instance, research conducted on epoxy glass-fiber reinforced composites demonstrated the double nature of fire-retardant additives, where increased levels of ammonium polyphosphate (APP) and inorganic silicate (InSi) facilitated char formation at the expense of diminishing flexural properties after exposure to fire.

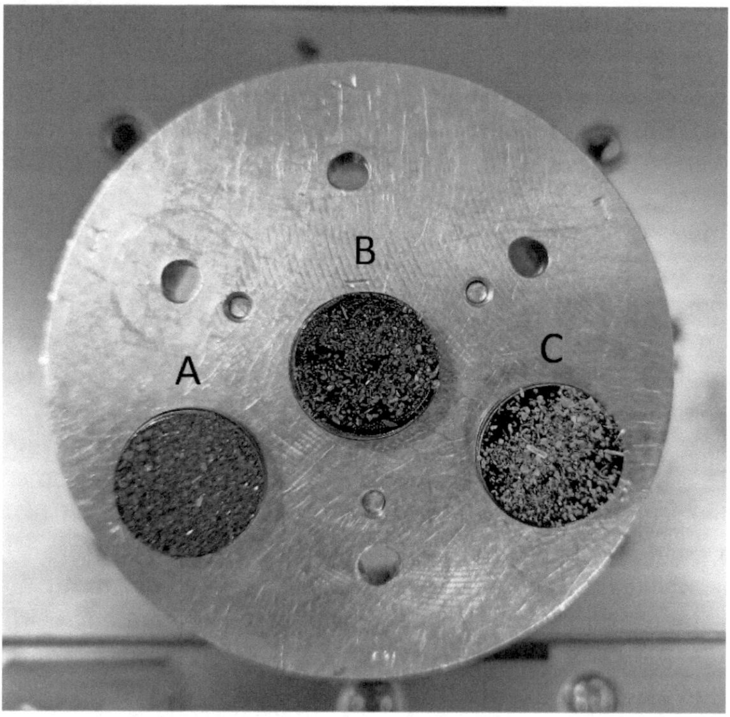

Figure 2.8. Scanning electron microscopy (SEM) has emerged as a fundamental technique in analyzing the intricate mechanics of microstructures across various materials.

Source: https://www.mdpi.com/applsci/applsci-13-12600/article_deploy/html/images/applsci-13-12600-g011.png.

The consequent SEM imaging of the changes afforded important microstructural changes such as fiber breakage and delamination, and these were important visual evidence for the mechanical deterioration witnessed in composite materials (Sunder et al., 2025). This ability of connection of microstructure with performance provides avenues for optimization of material properties by design and selection guided by knowledge according to the laws of mechanics of microstructures. In another application, SEM plays an important role in the agricultural sector with the clarification of soybean genotypes' surface characteristics, which are pivotal to cultivation and breeding. The technique provides unprecedented data on the micro-morphological characteristics of soybean seeds, such as surface texture and roughness, to facilitate a system of classification based on these factors. The systematic examination enabled by SEM facilitated researchers to classify 12 soybean genotypes into defined categories based on microscopic attributes, illustrating the effectiveness of such tests in agricultural studies (Sezer et al., 2024). Having insight into the microstructural elements that influence seed performance and nutrient provision, SEM not only facilitates agricultural production but also complements attempts at sustainability through plant maintenance. In general, the use of SEM in varied disciplines signifies its importance in dealing with microstructure mechanics comprehensively (Figure 2.8).

2.3.2. Transmission Electron Microscopy (TEM)

The contribution of transmission electron microscopy (TEM) to microstructural mechanics cannot be overstated. As a key characterization method, TEM makes possible observations of materials with unparalleled resolutions, allowing scientists to examine the internal morphology of materials to the atomic level. This usage is particularly critical in outlining the impact of microstructure on macroscopic mechanical properties. For instance, TEM can characterize the distribution of dislocations, grain boundaries, and phases in a material and thereby make way to understanding how these contribute to phenomena such as strength, ductility, and toughness. Integration of TEM findings with computational models, researchers can gain a wider insight into structure-property relations that allow for the creation of advanced materials with desired mechanical properties. In recent years, there has been a major shift towards the interface of advanced experimental techniques such as TEM and computational modeling in materials science. By applying TEM to inform multiscale models, scientists not only verify computational predictions but also enhance the discovery of new materials. The coupling is necessary in the analysis of complex behaviors observed by materials under various loading conditions. For instance, the interaction of mechanical properties and grain size can be better predicted using models that include experimental data derived from TEM. Such cooperative interactions provide deeper exploration of the processes governing material behavior, well in keeping with the current trends in materials research where experiment and modeling are no longer viewed as separate endeavors, but as interdependent procedures which contribute significantly to scientific progress (Dingreville et al., 2015). Also, TEM is a key tool for exploring micro-mechanisms controlling fracture and crack extension in materials. The advantages of gradient structures, where the grain size varies along the material, causing higher toughness compared to uniform grain structures, have been demonstrated by recent studies. For instance, considering the development of cracks in nickel with various grain sizes reveals that crack-initiation toughness is higher when cracks propagate from the coarse-grained side towards the nano-grained side. The procedure underlines the importance of understanding microstructural evolution at different scales, a task only made possible through TEM. With continuing studies on such microstructural effects, TEM remains essential in uncovering the intricate relationship between microstructure and mechanical performance, thereby influencing the creation of resilient materials suited for use in numerous diverse applications (Cao et al., 2020) (Figure 2.9).

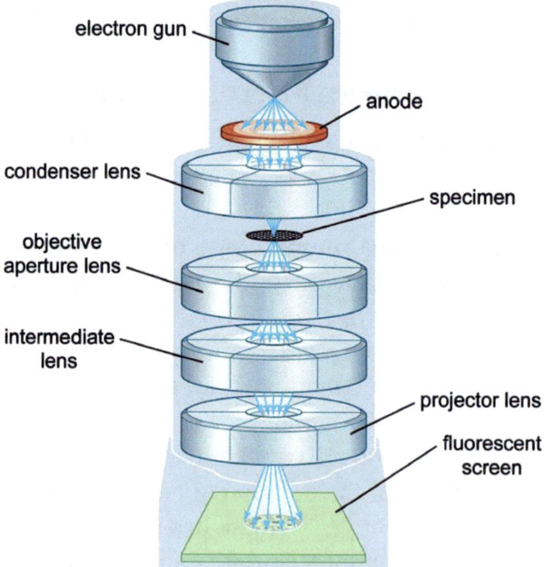

Figure 2.9. The significance of transmission electron microscopy (TEM) within the framework of microstructural mechanics cannot be overstated.

Source: https://cdn.britannica.com/90/113690-050-D614944D/Transmission-electron-microscope.jpg.

2.3.3. X-Ray Diffraction (XRD)

X-ray diffraction (XRD) has a vital role to play in microstructure analysis because the technique provides valuable information about crystallographic structures and material properties. XRD relies on X-ray scattering from crystalline materials and provides a non-destructive means of determining phase composition, crystallite size, and lattice strain. By shining X-rays on a sample and observing intensity and angle of diffracted beams, researchers can generate a diffraction pattern unique to the material. These patterns serve as a fingerprint to identify phases and provide information about the structural characteristics that control mechanical behavior. Quantitative relations regarding microstructure can be derived from diffraction peaks, thereby contributing to principles of material design and optimization. Thus, XRD is the center of such a study of structure-property relationship, particularly in novel composites where mechanical performance is linked directly with their microstructural features. In advanced materials, XRD plays a vital role in determining the efficacy of novel composite structures.

For example, bio-inspired architectures made from Mg-Ti composites have undergone phenomenal improvement in ductility and mechanical strength under XRD analysis. These composites were found to possess new spatial structures like nature, resulting in effective stress transfer and damage delocalization, as indicated in recent research (Zhang et al., 2022). XRD can be employed to identify how such microstructural structures influence the mechanical properties, highlighting the importance of knowing the crystallographic phases involved. By aligning diffraction data with mechanical performance standards, engineers design materials strategically for improved toughness, expanding their application across various industries. This intersection of XRD analysis and new composite design is a revolutionary technique in material science. In addition, XRD application is not only limited to conventional materials but also into the field of severe plastic deformation (SPD) processes, which have made it possible to produce ultrafine-grained and nanostructured materials. The SPD processes fundamentally alter the microstructure of the materials, in most instances resulting in a high concentration of lattice defects with substantial effects on their mechanical properties and functional behavior (Edalati et al., 2022). By using XRD, scientists are able to track these microstructural changes in real-time, determining the impact of changes during processing on phase stability and crystallinity.

The results of XRD research can be used to derive optimization strategies that will improve the desired attributes, such as strength, ductility, and environmental resistance. Accordingly, the interaction between XRD and SPD processes represents the seminal position of advanced characterization techniques in narrowing the gap between microstructural design and the achievement of high-performance materials.

2.4. THEORETICAL MODELS OF MICROSTRUCTURE MECHANICS

Theoretical models of microstructure mechanics offer a robust foundation for the properties and behavior of materials at the micro-level. The models can bridge the gap between the material microstructure and macroscopic mechanical properties, hence delivering valuable information on material design and optimization for application in various uses. Elevating to prominence in these models is the concept of scale, in which events at the molecular or atomic level influence the bulk properties of materials. The early classical theories, such as continuum mechanics, have laid the groundwork for the development of multiscale modeling methods. These methods allow researchers to examine interactions between different scales—ironically, acknowledging that events previously considered to be independent are frequently interrelated. Such a perspective reflects the essence of materials science as understood today, especially when the distinctive properties of materials, such as ductility and yield strength, can be understood in terms of their microstructure. Topology optimization in multi-scale structure design has been one of the greatest developments in theoretical models. Often inspiration is taken from structures found in nature. These forms of structures offer improved performance in their light yet strong characteristic, especially valuable in fields like bioengineering and material design. The initial seminal publications, most notably by Bendsøe and Kikuchi in 1988, have spurred research into further optimization of these structures through detailing how microstructural geometry influences mechanical performance. Continuous advancements in computational techniques now permit the simulation of complex loading conditions and environmental interactions. By understanding the principles of such models, researchers can not only predict failure mechanisms but also engineer materials with specific improvements in certain properties, thus ending up with improved overall material functionality and efficiency, as introduced by Wu et al. (2021). Despite major developments, there are several challenges that still exist in developing and applying theoretical models of microstructure mechanics. One of the basic difficulties is the lack of standardization in experimental protocols, leading to variability in outcomes and difficulties in using theoretical predictions directly in practical scenarios. As is the case of lithium metal anodes and solid-state batteries, an understanding of nonequilibrium dynamics at microstructural interfaces is necessary for the development of battery technologies. Breakthroughs in multimodal imaging and modeling techniques are necessary to unravel these complex interactions.

Moreover, by cross-leveraging ideas from other fields, researchers are able to gain a better understanding of microstructure behavior in various operating conditions. Ultimately, an interdisciplinary research culture that prizes robust reporting and discussion will advance the field, resolving the key questions yet to be resolved, as presented by Kelsey et al. (2020).

2.4.1. Continuum Mechanics Approach

Continuum mechanics is the underlying framework in explaining behavior of material at a macroscopic level, and it tends to offer valuable information regarding the mechanics of microstructures. Continuum mechanics offers the advantage of simplifying complicated interactions of microstructure into manageable mathematical models. By approximating materials as continuous rather than discrete, scientists can subsequently properly analyze stress, strain, and their effect on material performance under various loading conditions. This process becomes especially crucial in the case of dealing with phenomena such as crack propagation in inhomogeneous materials, where microstructural configurations become inextricably linked with macroscopic mechanical properties. As emphasized by recent studies, continuum mechanics' flexibility can reveal transitions in crack dynamics that are most critical for defining material failure and stability and furthermore establish a bridge between microscopic behaviors and macroscopic response (Barbier et al., 2013). In addition to revealing how cracks propagate, the continuum mechanics approach is also in the forefront of time-dependent material behaviors such as creep rupture analysis. The inclusion of continuum damage mechanics in predictive models applies most aptly to high-temperature use of advanced materials like ceramics. The need to accurately model time-dependent behavior in such materials has challenged some research to consolidate continuum mechanics based on principles and statistical methods, such as Weibull analysis. The dual method enables accuracy in understanding reliability and hazard rates of material degradation with respect to time. The analogies encountered between continuum and microstructural kinetic models reveal the versatile applicability of continuum mechanics to description of complex time-dependent behavior, a formidable tool in the hands of both engineers and materials scientists (Duffy et al., 2025). With each improvement in comprehension of the microstructures, mechanics continuity enhances to encompass further insight gleaned from experimental as well as computational methods. This evolution can be observed in the research of dynamic phenomena such as crackling dynamics, wherein loading conditions and specimen parameters determine the failure processes in materials based on their functionality. The fact that it is possible to make scaling laws based on dimensionless variables allows researchers to design accurate experimental guidelines for observing rapid crack dynamics. By illuminating these correlations, the continuum mechanics framework not only broadens our theoretical understanding but also empowers practical application, including optimization of material performance in engineering. Thus, the continuum mechanics framework remains an integral component in the study of microstructures, creating innovations in material design and utilization (Barbier et al., 2013) (Figure 2.10).

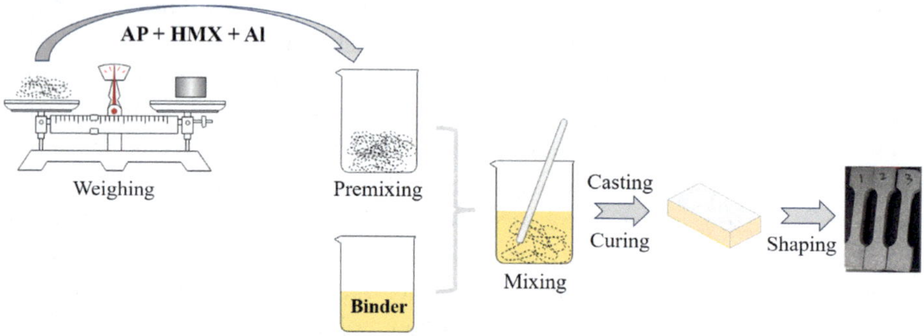

Figure 2.10. The continuum mechanics approach serves as a fundamental framework in understanding the behavior of materials on a macroscopic scale, often providing critical insights into the mechanics of microstructures.

Source: https://www.mdpi.com/materials/materials-16-06695/article_deploy/html/images/materials-16-06695-g001.png.

2.4.2. Discrete Models and Simulations

The study of discrete simulations and models in microstructure has broad implications for material behavior understanding at varying scales. Such convergence enables an advanced route to structure-property research, where theoretical comprehension converges with experimental information. Using computational approaches such as finite element analysis and molecular dynamics, researchers can simulate and study the microstructural subtleties of materials and reveal phenomena that are usually hidden in macroscopic observations. These discrete models allow the high fidelity and precision, considering the heterogeneity of the material properties and behaviors resulting from complex interactions at the atomic or molecular level. This approach enables a complete image of material systems that holds the promise of revolutionary applications in engineering as it emphasizes integrating computational predictions and empirical observations in materials science (Dingreville et al., 2015). Discrete models' flexibility is exposed when thinking about specific phenomena such as plasticity and failure in materials. Discrete simulations may uncover the correlated but chaotic motions in plastic flows, as in other complex systems like turbulence. A case in point is a simple mesoscopic model for describing scale-free behavior in plasticity that demonstrates how fluctuations arise from marginal stability of the elastic system. By focusing on such special models, researchers can uncover underlying mechanisms governing material response to stress that may not be captured by conventional smooth fitting functions. This detailed simulation model not only assists in preserving statistical properties of phenomena like avalanches but also contributes to the overall comprehension of dislocation behaviors in crystalline materials. This data is highly beneficial while creating and optimizing advanced materials since it gives a prediction framework for the performance of materials in real-life situations (Acharya et al., 2012). The integration of discrete models and simulations in the mechanics of microstructures also leads to new avenues of research, particularly multiscale modeling. The dynamic nature of material science requires the correlation of computational mechanics with experimental validation to drive advancements in the description of complex material systems. With predictive model capabilities continually refined, there emerges a symbiotic process where simulations are discovery tools that complement material engineering activities. This mutualism forges new avenues for hypothesis formation and experimentation, offering a platform for next-generation materials with properties tailored to individual applications. Through concerted efforts on the issues this field faces, such as understanding complex behavior at various length and time scales, researchers attempt to bridge the gap between the predictions of theory and reality, which represents a significant step in the mechanics of microstructures and the creation of higher-grade materials for various industrial applications (Dingreville et al., 2015).

2.4.3. Scale Effects in Microstructural Analysis

Microstructure examination plays a critical role in establishing the mechanical properties of materials, particularly because micro-scale interactions can significantly influence behavior at the macro-scale. This aligns with scale effects; whereby material response varies with structural feature size and arrangement. Grain size, phase distribution, and defect characteristics are microstructural factors that control material strength, ductility, and toughness. Therefore, successful microstructural analysis will require to be coupled with a broad-based understanding of how such attributes interplay across scales, finally revealing the underlying mechanics that dictate material performance. With scientists currently applying high-end characterization techniques, the complexity of microstructural influences is now becoming apparent, and hence, the need for an integrative approach that provides for these scale-dependent effects. Quantitative techniques significantly enhanced our understanding of scale effects in microstructural analysis, especially through methods correlating microstructural features and mechanical behaviors. A case in point is a lamellar-architectured, nacre-inspired composite that has been thoroughly examined to investigate the correlation between its microstructure and mechanical properties. The inclusion of phase-dependent hardness, modulus, and residual stress measurements at micro-scale sizes provides a better insight into the way localized differences affect bulk mechanical responses. The application of a shear-lag model here points to the potential for microstructural effect transfer onto macro-scale behaviors. These findings indicate that effective microstructural designs may lead to enhanced mechanical properties, which in turn suggests that bioinspired synthetic architectures can serve as rudimentary models in future material design research (Mo et al., 2022).

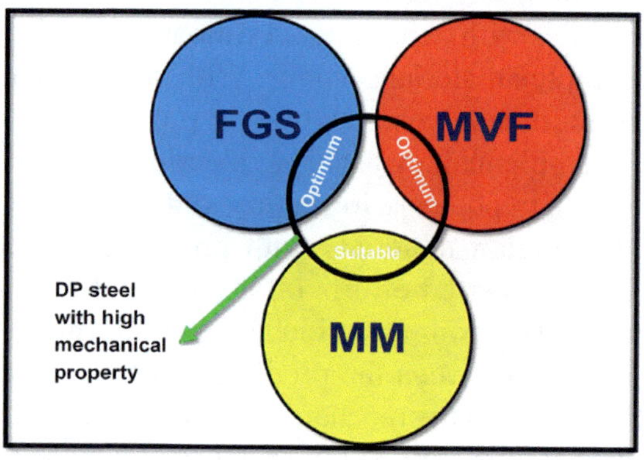

Figure 2.11. Microstructural elements such as grain size, phase distribution, and defect characteristics play pivotal roles in determining material strength, ductility, and toughness.

Source: https://www.mdpi.com/metals/metals-12-00101/article_deploy/html/images/metals-12-00101-g009.png.

Further, advancements in machine learning have opened new avenues to study scale effects in microstructural analysis. With data-centric approaches, researchers can understand complex interactions in microstructures that may be hard to see with traditional techniques. Machine learning routines can sift through large data sets, recognizing key microstructural features that correlate with desired mechanical properties. This revolutionary marriage is highlighting the potential for designing materials with optimized performance maps using improved knowledge of microstructural mechanics. Along this line, the integration of advanced computational approaches with experimental microstructural studies is yielding possibilities for more precise material innovation with prospects unfolding towards future materials science breakthroughs (Guo et al., 2020) (Figure 2.11).

2.5. INFLUENCE OF PROCESSING ON MICRO-STRUCTURES

The relationship between processing techniques and the ensuing microstructure in materials science is an important area of research. The intricate interdependencies that control the mechanical behavior of materials are based on their microstructural characteristics, which themselves are created as a direct consequence of the processing techniques employed. Various processing techniques, including casting, forging, and additive manufacturing, impart distinctive microstructural characteristics, influencing aspects such as hardness, ductility, and fracture toughness. For instance, high-speed solidification of materials can yield more fine grain structures that enhance strength based on grain boundary strengthening mechanisms. While scientists research the microstructure mechanics, they illustrate how these aspects influence structure-property relations, hence informing design-specific material design for their designated application. This supports the need for synergistic strategy since recently published works point toward cooperative experimental and computational approaches to explain these complex interactions (Dingreville et al., 2015). Processing conditions not only determine the phase distribution, but they also control defect formation and evolution in a microstructure of materials. Techniques such as heat treatment have the ability to modify the precipitate formation and dislocation density, which are essential for controlling a material's mechanical behavior. For example, in SiC whisker-reinforced ceramics, the fracture toughness and strength have the ability to be significantly modified by the processing parameters, essentially controlling the SiC whisker dispersion in a matrix (Duffy et al., 2025). This underlying appreciation of the processing-microstructure relationship allows materials scientists to manipulate microstructural features to maximize performance. Furthermore, advances in real-time characterization methods enable scientists to track such changes in real time, an understanding of how processing manipulates fundamental mechanical properties at the microscopic level. The marriage of computational modeling with experimental methodologies offers paradigmatic possibilities for prediction and optimizing microstructural properties across a variety of materials. With multiscale modeling approaches, it is possible to forecast microstructural evolution in the presence of specified processing conditions, thus bridging the gap between experimental observations and theory predictions. This is significant in deciphering the complex mechanisms involved in microstructural evolution generated by different processing techniques. Evidence shows how computational models can be used as tools of discovery in the form of a framework for investigating how processing strategies can be optimized to create improved material properties. In the future, as this field continues to develop, the collaboration between experimental investigation and computational simulation will inevitably play a significant role in furthering the progress within the mechanics of microstructures, ultimately expanding the frontiers of innovation in material application and design (Dingreville et al., 2015).

2.5.1. Effects of Heat Treatment

Heat treatment plays an important role in altering the mechanical properties of materials through making significant changes in their microstructures. Heat treatment involves well-controlled heating and cooling patterns that can potentially manage phase distributions, grain size, and material performance as a whole. For example, heat treatment processes like steel tempering and quenching can lead to the development of various microstructural phases like martensite, ferrite, and austenite. These phases not only determine the material's hardness and strength but also its toughness and ductility. Comprehending thus the intricacies of heat treatment is paramount in achieving maximum performance of structural materials exposed to high temperature conditions, as foregrounded in the summary of landmark papers on life prediction under high temperature conditions (Carpenter et al., 2025). The effect of heat treatment is quite large in advanced high-strength steels (AHSS), such as Transformation Induced Plasticity (TRIP) steel, where microstructure plays a vital role in the attainment of desired mechanical properties. TRIP steel is a combination of ferrite, bainite, retained austenite, and martensite, which are formed through heat treatment. The transformation of metastable retained austenite into stable martensite enhances ductility and strength due to the simultaneous transformation strain. This provides better formability and increases the performance of materials under stress and is why it is in high demand in automotive applications where strength and weight are top priorities (Geijselaers et al., 2011). Thus, not only does heat treatment change previously existing microstructures but also actively participates in the creation of new material properties. Moreover, the correspondence between heat treatment and microstructure evolution is nonlinear; it involves complex interactions between phases and their mechanical responses. Varying parameters such as thermal cycles, composition of alloy, and cooling rates allows engineers to produce an enormous range of desired microstructural morphologies. These changes lead to particular mechanical behaviors optimized for their individual applications, such as engine components requiring high strength and moderate ductility. An understanding of these relationships is significant in anticipating how materials will respond under real loads, and continued research pushes these processes further for improved performance. This research emphasizes the importance of heat treatment in materials engineering, proving to be crucial to the microstructure mechanics (Carpenter et al., 2025) and the future material design (Figure 2.12).

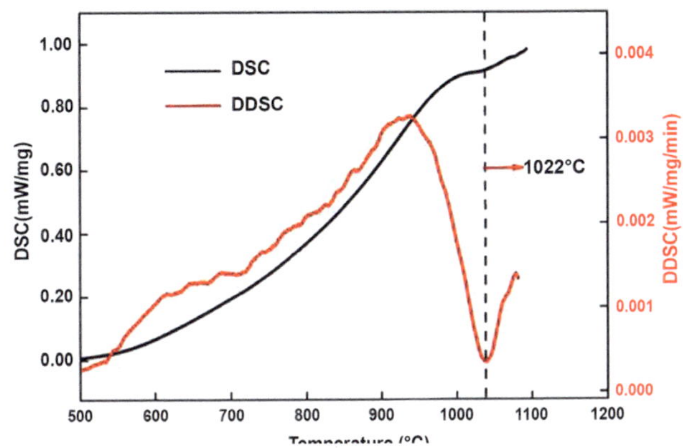

Figure 2.12. Heat treatment plays a pivotal role in altering the mechanical properties of materials by inducing significant changes in their microstructures.

Source: https://www.mdpi.com/materials/materials-16-00160/article_deploy/html/images/materials-16-00160-g001-550.jpg.

2.5.2. Impact of Mechanical Working

Mechanical working has a profound influence on microstructures, both influencing material properties and the performance of engineering applications. By forging, rolling, and extrusion, microstructural evolution of materials can be fundamentally altered. For instance, mechanical working increases dislocation activity, grain refinement, and the chances of texture formation, which collectively raises strength and ductility. Such improvement is due to reorganization of the internal microstructure, with processing-induced defects serving as sites for further deformation. Their interaction with the rest of the microstructure determines the ultimate mechanical properties and fatigue strength, thus having a profound effect on the overall material performance when experiencing service loads. In refractory materials, mechanical working can produce both positive and adverse effects on microstructure and performance. Current studies highlight that some compositions, e.g., those containing fused zirconia aggregates, behave differently to mechanical cycling. For example, optimal levels of stabilizers may enhance refractory stiffness and toughness, but elevated levels can result in the formation of secondary phases by interaction with impurities. Such secondary phases may result in microstructure degradation, thus strength and durability reductions. Engineers therefore must carefully heed the ratio between stabilizer composition and impurity concentration in the design to ensure consistency in refractories in high-temperature applications, emphasizing the need for precision in mechanical working processes (Quadling et al., 2015). Technological advances in manufacturing, particularly additive manufacturing, also bring to light the importance of mechanical working and its resultant microstructural implications. With the advancement of geometric algorithms for 3D printing, it is possible to utilize these to tune materials and processes for building desired microstructural characteristics. This is facilitated with the capability of conducting localized mechanical working aimed at enhancing specific properties during the build. The intricate interaction between process parameters, material deposition, and microstructural performance points towards the significance of mechanical working in achieving desired outcomes in processed parts. As the demands from industries keep increasing for higher performance, knowledge about the mechanics of microstructure and mechanical working effects remains significant in both theoretical studies and industrial applications (Attene et al., 2017).

2.5.3. Role of Alloying Elements

The role played by alloying elements to change the mechanical properties of steels is extremely significant in enhancing their performance in engineering applications. Alloying elements such as boron and carbon play a significant contribution towards microstructural evolution, which plays an extremely powerful role on the strength, toughness, as well as fatigue resistance of a material. As an example, the addition of boron in bainitic steels increases toughness and fatigue life considerably. Research indicates that minor variations in boron content can lead to significant variations in fracture toughness, basically by reducing interlath martensite components and promoting void coalescence, thereby enhancing the overall ductility of the steels. This versatility illustrates how single components of alloying can improve qualities of performance to make them valuable for use in engineering applications that need high reliability and mechanical strength, as the evidence documented about bainitic steels indicates the importance of compositions that can be tailored for specific demands of operation (Marrero et al., 2016). Aside from boron, the role of retained austenite also quite clearly demonstrates other alloying elements' effects on the mechanical properties of steels. More recent advancements, including the Quenching and Partitioning (Q&P) heat treatment technique, use alloying elements to stabilize retained austenite in the microstructure. Stability of retained austenite is crucial since retained austenite can significantly enhance toughness and ductility. The phase-field method has been employed to simulate these transformations in polycrystalline Fe-C systems and demonstrate how variations in carbon compositions can alter the volume fraction and distribution of retained austenite significantly. These findings highlight that in the Q&P process, the compression of carbon within the austenitic phase controls its stability, which is vital in optimizing the mechanical response of steel when applied practically (Amos et al., 2018). The intricate interaction between the alloying elements and the evolution of microstructure demonstrates the important role of alloying in material property design for a specific application. As exemplified by the research on bainitic steels and the Q&P process, the control of the microstructural phases and their transformations allow engineers to push the boundaries of steels' strength and toughness. Lastly, such knowledge is the foundation upon which next-generation materials are to be designed and optimized to be able to support complex operational regimes. As industries make demands for more performance from material technology, the role of alloying elements in enhancing mechanical properties will gain greater significance, giving rise to advancements in material science and engineering (Amos et al., 2018; Marrero et al., 2016).

2.6. CONCLUSION

In brief, the mechanical behavior of microstructures is at the core of the establishment of our understanding of material deformation at the micro scale. As we have been arguing throughout this essay, the interplay between microgeometries plays an important role on the composite's effective elastic properties. Specifically, the description of the set GU_f, which corresponds to the composition of materials with prescribed elasticity tensors, reveals a multifaceted relationship between structure and mechanical performance. The ability to minimize energy totals with the application of correct configurations brings into relief the importance of microstructural design. Such findings not only contribute to theoretical models but also have practical implications in material development for various engineering applications, pointing to a need for more research in this area (Briane et al., 2016). Improvements in fabrication techniques, such as dry etching, also suggest that novel techniques can be applied to control microstructures. The DEEMO process allows for the production of micromolds with aspect ratios of high values, facilitating intricate designs heretofore unobtainable. The process focuses on the synergy between microstructure engineering and process manufacturing by delivering the ability to manufacture materials with desirable properties for applications in electroplating and embossing. The meeting of theory and practice provides an opportunity for scientists and engineers to design materials that are not only highly efficient but also adapt to particular performance needs (Ehrfeld et al., 1995). Last but not least, microstructure study is the union of material science and engineering, and it offers critical insights into existing practices that may revolutionize them. The emphasis on accurate study of structural mechanics and manufacturing processes seals the notion that microstructural dynamics must be understood in a holistic manner to design sophisticated materials. With further research in this area, it is possible that we will see additional advancements that not only enhance the functionality of materials but also generate even greater possibilities for use in different fields. Understanding and using the physics of microstructures will definitely reshape the future picture of material science and engineering (Briane et al., 2016; Ehrfeld et al., 1995).

2.6.1. Summary of Key Points

While learning microstructure mechanics, pay attention to the twofold nature of rigidity and flexibility in various research studies. The interplay between the two specifically is significant in material behavior in the case of material such as shape-memory alloys. Research in this context talks about how maximal regularity in solutions to convex integration is analogous to lower bounds in variational models with surface energy involved. This is a sure indication that arrangements of microstructure can significantly impact material design and consequently expand the understanding of how microstructural mechanics can lead to advantageous properties in materials. It is a fact that variational models are a fundamental selection mechanism, giving insight into which microstructure formations are good or bad for the overall performance of material such as alloys (Rüland et al., 2018). Furthermore, the investigation of particle-reinforced composites, notably like Mg-PSZ refractories, sheds light on the influence of populations of aggregates on material durability over time. In service cycles, these refractories respond differently to thermal and mechanical loading, with their implications directly affecting their microstructural integrity. The research reveals that different aggregate configurations lead to very different outcomes in relation to stiffness and toughness following cycling. Interestingly, the aggregates with greater stabilizer content in refractories experience unfavorable secondary phase formations, and their toughness and strength are undermined through thermal expansion mismatches. The aggregates with less stabilizer content maintain their structural coherence, pointing out the importance of microstructural design to the behavior of the material (Quadling et al., 2015). Finally, the R-curve behavior exhibited in the refractories in question points out one characteristic of fracture mechanics. This performance demonstrates how materials may exhibit greater resistance to crack growth with increasing defects. The quasi-stable fracture curves in bend tests point to the fact that microstructural evolutions, e.g., the interaction between phases and their orientations, greatly increase the mechanical response of strained materials. The findings necessitate strict care during the design stage, particularly in terms of stabilizer concentration and impurity levels. Such parameters can effectively alter microstructural response and, as a result, the mechanical behavior of the final product. It is due to this reason that engineers who aim to optimize material performance by means of well-designed microstructural mechanics need to understand these dynamics (Quadling et al., 2015).

2.6.2. Future Directions in Microstructure Research

The destiny of microstructure research is firmly embedded in the advancement of characterization techniques, particularly through the utilization of equipment like atomic force microscopy (AFM). With an increased understanding by scientists of the microstructural properties of materials, AFM offers a special window for the observation of microscopic material behavior detail, particularly in testing applications like asphalt binder. It enables the study of micromechanical behavior and morphology relationships, both being critical in order to forecast macroscopic behavior. The ability to precisely determine material composition-structure-mechanics relationship using AFM might lead to breakthrough advancements in material science. Future research needs to focus on the further development of such AFM capabilities, generating new sample preparation methodologies, and exploring the applicability of these techniques to high-performance materials design, especially in infrastructure, where durability is paramount (Zhang et al., 2023). Besides AFM, the integration of electro-chemo-mechanical studies is necessary to further enhance microstructure research, especially in energy storage. As there is more demand for next-generation batteries, it becomes critical to describe the intricate interactions across material interfaces. The electrode mechanical stability and electrochemical interface behavior under various operating conditions should be studied by researchers. Understanding solid–liquid and solid–solid interfaces explain how heterogeneities at the microstructure level contribute to degradation mechanisms in electrode materials. Promoting such awareness is essential to address issues such as dendrite growth in lithium metal batteries and to develop better overall cycling stability. Hence, new approaches to simulating and characterizing these interactions are able to develop more durable batteries and other devices used for energy storage (Kaustubh et al., 2023). Lastly, the integrated approach that combines microscopy techniques, mechanical tests, and computational simulation will be paramount in future microstructure science. Combining experimental research with theoretical models enables improved predictions of intricate microstructure behavior under load. Such synergy is particularly valuable in failure mechanism analysis, e.g., the growth of cracks in inhomogeneous materials. Future studies need to strive to create coupled frameworks that include microstructural evolution due to environmental conditions and mechanical loads. Interdisciplinary teams that incorporate machine learning as well as data-based approaches can further accelerate the discovery in microstructure mechanics. Through the adoption of a multidisciplinary strategy, the field of microstructure research can transcend existing limitations and unlock potential in material design and application, ultimately paving the way for advancements in a number of industrial sectors, including the construction and energy sectors (Kaustubh et al., 2023; Zhang et al., 2023).

2.6.3. Importance of Understanding Microstructures in Engineering Applications

In contemporary applications of engineering, a good understanding of microstructures is crucial for enhanced material performance and optimized processing technologies. Microstructures, i.e., the structure of the material at the microscopic level, define inherently the strength, ductility, and fatigue properties. For instance, the intricate grain structures, pores, and phase arrangements can define the material behavior under varying stress conditions. Thereafter, expert engineers in micromechanics can design materials for a specific application, making them more efficient and durable. The enhancement of computational simulations and testing techniques further emphasizes the importance of microstructural analysis in predicting the behavior of materials under working stresses, thereby promoting innovations in any field of engineering. Furthermore, information regarding microstructures is of the highest importance in machining operations of sophisticated materials, particularly metal matrix composites (MMCs). The unique microstructural characteristics of MMCs create challenges in precision machining, leading to poor surface finish and tool wear. Systematic inspection of the microstructure interactions during machining will yield valuable information regarding cutting mechanics, material removal, and surface roughness. For example, studies indicate that the adjustment of machining parameters such as cutting speed and feed rate can significantly influence the machinability of surface quality in MMCs, indicating a notable relationship between microstructure and manufacturability (Niu et al., 2019). Therefore, engineers must integrate microstructural parameters into their machining to improve performance and sustainability in manufacturing operations. Last but not least, microstructural understanding has consequences for the development of novel engineering materials and quantum technologies. As technology develops, especially that of quantum simulation, there will be an increasing demand for quantum simulator performance-enhancing materials with engineered microstructures. It has been shown that carefully designed microstructures can improve the performance of quantum simulators and other complex systems (Georgescu et al., 2014). Moreover, the interaction between quantum mechanical properties and microstructural attributes can shatter areas like condensed-matter physics and quantum chemistry. Ultimately, mastery of the microstructure's mechanics gives engineers and scientists the right toolset for innovating and mitigating complicated issues, thereby driving development in theoretical and applied engineering areas.

REFERENCES

1. Acharya, A., Argon, A. S., Bak, P., Baret, J. C., Barnsley, M. F., Ben-Zion, Y., et al. (2012). On the critical nature of plastic flow: One- and two-dimensional models. http://arxiv.org/abs/1202.4753 (accessed on 14th October 2025).
2. Al-Ruqaie, A., Alba, A., Almrhag, H., Behrouzian, F., Bohmer, M., Espitia, M. J., et al. (2015). Influence of pH on mechanical relaxations in high solids LM-pectin preparations. https://core.ac.uk/download/30731867.pdf (accessed on 14th October 2025).
3. Amos, P. G., Kubendran, K., Nestler, B., Schneider, D., Schoof, E., et al. (2018). Phase-field analysis of quenching and partitioning in a polycrystalline Fe-C system under constrained-carbon equilibrium condition. http://arxiv.org/abs/1812.04306 (accessed on 14th October 2025).
4. Appolaire, B., Cottura, M., de Geus, T. W. J., Geers, M. G. D., et al. (2016). Fracture initiation in multi-phase materials: A systematic three-dimensional approach using a FFT-based solver. http://arxiv.org/abs/1604.03817 (accessed on 14th October 2025).
5. Attene, M., Ellero, S., Lefebvre, S., Livesu, M., et al. (2017). From 3D models to 3D prints: An overview of the processing pipeline. https://core.ac.uk/download/84977973.pdf (accessed on 14th October 2025).
6. Barbier, L., Barés, J., & Bonamy, D. (2013). Crackling vs. continuum-like dynamics in brittle failure. https://core.ac.uk/download/52684210.pdf (accessed on 14th October 2025).
7. Bergerhoff, G., Erhan, D., Gatys, L., Goodfellow, I., Kingma, D., Larsen, A. B. L., LeCun, Y., et al. (2019). Structural material property tailoring using deep neural networks. http://arxiv.org/abs/1901.10281 (accessed on 14th October 2025).
8. Briane, M., Harutyunyan, D., & Milton, G. W. (2016). On the possible effective elasticity tensors of 2-dimensional and 3-dimensional printed materials. http://arxiv.org/abs/1606.03305 (accessed on 14th October 2025).
9. Cabrera Marrero, J. M., Mestra Rodríguez, Á. M., Olaya Florez, J. J., Rodriguez Galeano, A., et al. (2016). Influence of boron content on the fracture toughness and fatigue crack propagation kinetics of bainitic steels. https://core.ac.uk/download/81580716.pdf (accessed on 14th October 2025).
10. Cao, R., Li, Y., Lin, Y., Pan, Z., et al. (2020). On the exceptional

damage-tolerance of gradient metallic materials. https://core.ac.uk/download/287622265.pdf (accessed on 14th October 2025).

11. Carpenter, J. L., & Stuhrke, W. F. (2025). Life prediction of materials exposed to monotonic and cyclic loading: A new technology survey. https://core.ac.uk/download/pdf/42889227.pdf (accessed on 14th October 2025).
12. Concheri, G., Meneghello, R., Rosso, S., Savio, E., et al. (2018). Geometric modeling of cellular materials for additive manufacturing in biomedical field: A review. https://core.ac.uk/download/154326170.pdf (accessed on 14th October 2025).
13. de Geus, T. W. J., Geers, M. G. D., & Peerlings, R. H. J. (2015). Microstructural topology effects on the onset of ductile failure in multi-phase materials–A systematic computational approach. http://arxiv.org/abs/1604.02858 (accessed on 14th October 2025).
14. Della Porta, F. (2018). Analysis of a moving mask approximation for martensitic transformations. http://arxiv.org/abs/1811.08247 (accessed on 14th October 2025).
15. Dingreville, R., Karnesky, R. A., Puel, G., Schmitt, R., et al. (2015). Review of the synergies between computational modeling and experimental characterization of materials across length scales. http://arxiv.org/abs/1509.03892 (accessed on 14th October 2025).
16. Duffy, S. F. (2025). Whisker-reinforced ceramic composites for heat engine components. https://core.ac.uk/download/pdf/42832667.pdf (accessed on 14th October 2025).
17. Duffy, S. F., & Gyekenyesi, J. P. (2025). Time dependent reliability model incorporating continuum damage mechanics for high-temperature ceramics. https://core.ac.uk/download/pdf/42827890.pdf (accessed on 14th October 2025).
18. Edalati, K., Bachmaier, A., Beloshenko, V. A., Beygelzimer, Y., Botta Filho, W. J., Bryła, K., et al. (2022). Nanomaterials by severe plastic deformation: Review of historical developments and recent advances. *Materials Research Letters, 10*, 163–256. https://doi.org/10.1080/21663831.2022.2029779 (accessed on 14th October 2025).
19. Ehrfeld, W., Elders, J., Elwenspoek, M., Jansen, H., et al. (1995). DEEMO: A new technology for the fabrication of microstructures. https://core.ac.uk/download/pdf/11454076.pdf (accessed on 14th October 2025).
20. Forest, S., McDowell, D., Müller, S., & Werner, E. (2024). *Mechanics of materials: Multiscale design of advanced materials and structures. Oberwolfach Reports.* https://www.semanticscholar.org/paper/69586e21

8e69da832b312c58e4ef11270ec08862 (accessed on 14th October 2025).

21. Gao, P., Wei, J., Ye, G., Yu, J., et al. (2015). Multi-scale simulation of capillary pores and gel pores in Portland cement paste. https://core.ac.uk/download/55735871.pdf (accessed on 14th October 2025).
22. Geijselaers, H. J. M., & Perdahcioglu, E. S. (2011). Prediction of the mechanical behaviour of TRIP steel. https://core.ac.uk/download/pdf/11481243.pdf (accessed on 14th October 2025).
23. Georgescu, I., Ashhab, S., & Nori, F. (2014). Quantum simulation. *Reviews of Modern Physics, 86*, 153–185. https://doi.org/10.1103/revmodphys.86.153 (accessed on 14th October 2025).
24. Giraud, E., Lecomte-Beckers, J., & Pace, S. (2014). Elaboration of thin foils in copper and zinc by self-induced ion plating. https://core.ac.uk/download/29137555.pdf (accessed on 14th October 2025).
25. Guo, K., Yang, Z., Yu, C. H., & Buehler, M. J. (2020). Artificial intelligence and machine learning in design of mechanical materials. *Materials Horizons, 8*, 1153–1172. https://doi.org/10.1039/d0mh01451f (accessed on 14th October 2025).
26. Hardacre, D., Li, H., Sun, C. N., Wei, L., et al. (2018). High cycle fatigue and ratcheting interaction of laser powder bed fusion stainless steel 316L: Fracture behaviour and stress-based modelling. https://core.ac.uk/download/228156756.pdf (accessed on 14th October 2025).
27. Hatzell, K. B., Chen, X. C., Cobb, C. L., Dasgupta, N. P., Dixit, M., Marbella, L. E., McDowell, M. T., et al. (2020). Challenges in lithium metal anodes for solid-state batteries. *ACS Energy Letters, 5*, 922–934. https://doi.org/10.1021/acsenergylett.9b02668 (accessed on 14th October 2025).
28. Hou, H., Yin, J., & Jiang, X. (2019). Smart patterned surface with dynamic wrinkles. *Accounts of Chemical Research, 52*, 1025–1035. https://doi.org/10.1021/acs.accounts.8b00623 (accessed on 14th October 2025).
29. Khanna, S. K., Roylance, D., & Jenkins, C. H. (2020). Innovations in teaching mechanics of materials in materials science and engineering departments. https://doi.org/10.18260/1-2--9388 (accessed on 14th October 2025).
30. Lebensohn, R. A., & Rollett, A. D. (2020). Spectral methods for full-field micromechanical modelling of polycrystalline materials. *Computational Materials Science, 173*, 109336. https://doi.org/10.1016/j.commatsci.2019.109336 (accessed on 14th October 2025).
31. Liu, J. (2025). A review of the applications of mechanical metamaterials

in extreme environments. *Advances in Engineering Innovation*. https://www.semanticscholar.org/paper/75682a9040d83d4dfb351837ac719917eae775f4 (accessed on 14th October 2025).

32. Martinez, X., Bachmann, J., Otero, F., Oller, S., & Bugeda, G. (2022). Novel approach combining two homogenization procedures for the analysis of nonwoven bio composites. *Mechanics of Advanced Materials and Structures, 31*, 1167–1181. https://www.semanticscholar.org/paper/11641e305281d98a707114e3d6512c726b5901d5 (accessed on 14th October 2025).

33. Merodio, J., Penta, R., Rodriguez-Ramos, R., Sevostianov, I., et al. (2017). Connection between electrical conductivity and diffusion coefficient of a conductive porous material filled with electrolyte. https://core.ac.uk/download/132211469.pdf (accessed on 14th October 2025).

34. Mo, J., Leung, N., Jargalsaikhan, U., Wan, H., Zhu, B., Su, B., & Sui, T. (2022). Advanced microscopic characterization of multi-scale high-resolution mechanical behaviour of a nacre-inspired composite. *Composites Communications, 35*, 101315–101315. https://doi.org/10.1016/j.coco.2022.101315 (accessed on 14th October 2025).

35. Naik, K. G., Vishnugopi, B. S., Datta, J., Datta, D., & Mukherjee, P. P. (2023). Electro-chemo-mechanical challenges and perspective in lithium metal batteries. *Applied Mechanics Reviews, 75*. https://doi.org/10.1115/1.4057039 (accessed on 14th October 2025).

36. Niu, Z., & Cheng, K. (2019). Investigation on the material removal and surface roughness in ultraprecision machining of $Al/B_4C/50p$ metal matrix composites. *The International Journal of Advanced Manufacturing Technology, 105*, 2815–2831. https://doi.org/10.1007/s00170-019-04553-w (accessed on 14th October 2025).

37. Quadling, A. G. (2015). The role of aggregates in the thermal stability of Mg-PSZ refractories for vacuum induction melting. https://core.ac.uk/download/76995524.pdf (accessed on 14th October 2025).

38. Rüland, A., Taylor, J. M., & Zillinger, C. (2018). Convex integration arising in the modelling of shape-memory alloys: Some remarks on rigidity, flexibility and some numerical implementations. http://arxiv.org/abs/1801.08503 (accessed on 14th October 2025).

39. Sajid, S. M. W., Ahmad, M. S., Khan, E., & Sabir, M. (2022). Petrographic and physio mechanical investigation of Late Cretaceous Kawagarh Formation Kahi Section, Nizampur Basin. *International Journal of Economic and Environmental Geology*. https://www.semanticscholar.org/

paper/87da57238dcae2c692460ae5e43f218a50f18111 (accessed on 14th October 2025).

40. Sezer, A. Y., & Uysal, T. (2024). Investigation of the structural characteristics of seed surfaces of some soybean genotypes by using scanning electron microscopy (SEM). *Selcuk Journal of Agricultural and Food Sciences*. https://www.semanticscholar.org/paper/9d7083eb59490f261d4d88ba48c771c61447e7ad (accessed on 14th October 2025).
41. Sunder, S., Jauregui Rozo, M., Inasu, S., Meinel, D., Schartel, B., & Ruckdaeschel, H. (2025). Effect of ammonium polyphosphate/silicate content on the postfire mechanics of epoxy glass-fiber composites using facile chocolate bar-inspired structures. *Fire and Materials, 49*. https://www.semanticscholar.org/paper/f7e7042f9547f0fe1729f4cfc205badfc547ab16 (accessed on 14th October 2025).
42. Torrecillas, R., Moya, J. S., Díaz, L. A., Gómez, B., et al. (2009). Nanotechnology in joint replacement. https://core.ac.uk/download/pdf/36093046.pdf (accessed on 14th October 2025).
43. Tran, A., Wildey, T., & Lim, H. (2022). Microstructure-sensitive uncertainty quantification for crystal plasticity finite element constitutive models using stochastic collocation methods. *Frontiers in Materials, 9*. https://doi.org/10.3389/fmats.2022.915254 (accessed on 14th October 2025).
44. Wu, J., Sigmund, O., & Groen, J. P. (2021). Topology optimization of multi-scale structures: A review. *Structural and Multidisciplinary Optimization, 63*, 1455–1480. https://doi.org/10.1007/s00158-021-02881-8 (accessed on 14th October 2025).
45. Xiang, H., Xing, Y., Dai, F., Wang, H., Su, L., Miao, L., Zhang, G. J., et al. (2021). High-entropy ceramics: Present status, challenges, and a look forward. *Journal of Advanced Ceramics, 10*, 385–441. https://doi.org/10.1007/s40145-021-0477-y (accessed on 14th October 2025).
46. Zhang, E., Shan, L., & Tan, Y. (2023). Utilization of atomic force microscopy (AFM) in characterizing microscopic properties of asphalt binders: A review. *Journal of Testing and Evaluation, 52*, 754–775. https://doi.org/10.1520/jte20220672 (accessed on 14th October 2025).
47. Zhang, M., Zhao, N., Yu, Q., Liu, Z., Qu, R., Zhang, J., Li, S., et al. (2022). On the damage tolerance of 3-D printed Mg-Ti interpenetrating-phase composites with bioinspired architectures. *Nature Communications, 13*. https://doi.org/10.1038/s41467-022-30873-9 (accessed on 14th October 2025).

CHAPTER 3

ADVANCED SENSING MECHANISMS

Contents

INTRODUCTION

High-end sensing mechanisms have experienced unprecedented growth, pushing the envelope of possibilities in communication, navigation, and environmental monitoring. Such mechanisms, distinguished by their ability to sense and respond to far-reaching stimuli, play important roles in driving smart technologies that are increasingly pervasive in modern society. With sophisticated hardware and powerful algorithms, high-end sensing mechanisms can break down complex data streams in real-time, enabling intelligent decision-making processes. It is thus imperative to learn about the underlying fundamentals and working models of such sensing devices, especially as industries continue to move towards automation and the Internet of Things. Prioritizing security within such systems, as has been addressed by Alhakami et al. (2014), reveals underpinning aspects of technology integration that are not only inventive but also necessarily open to a multitude of threats. A prime example of the implementation of high-end sensing mechanisms is cognitive radio networks, which utilize dynamic spectrum access to address the pressing shortage of communications channels. By leveraging artificial intelligence and real-time spectrum sharing, these networks can make dynamic adjustments to shifting demands, optimizing the use of available frequencies. But the success of cognitive radio depends vitally on the health of its common control channel, the foundation upon which such improvements in operation are built. The advent of new security threats—such as licensed user emulation attacks—highlights the complexity of the threats these networks pose, calling for robust protection strategies (Alhakami et al., 2014). These factors form the premise for more research and development in the future, ensuring that innovation comes coupled with adequate protections. Furthermore, the use of micro systems technology (MST) is a fine example of how advanced sensing mechanisms are evolving in space and terrestrial applications. Clusters like MISAT are setting the pace through new satellite platforms that seek to utilize miniaturization to improve dramatically technology capabilities for use in space missions. Such innovations include the development of wireless sensor networks with plug-and-play characteristics that support easy deployment and usage. As such technologies mature, they provide viable options to transcend difficult data acquisition challenges in scientific and industrial applications. The integration of high-end sensing with MST not only propels space exploration but also redefines earth-bound applications, with enhanced multi-domain interoperability and efficiency in data collection and processing. Taken together, these elements show the revolutionary promise of cutting-edge sensing mechanisms in today's increasingly interconnected world.

Definition of Sensing Mechanisms

At the sensing mechanisms level, the definition of sensing mechanisms itself is a building block. Sensing mechanisms are processes or systems that permit detection and measurement of physical phenomena, usually by converting environmental stimuli into useful signals. These phenomena can be everything from a wide range of variables such as light, temperature, pressure, and chemical concentrations. The significance of effective sensing mechanisms is evident in applications as diverse as industrial automation to environmental monitoring, where precise data acquisition is paramount. Employing a variety of transducers, which convert physical parameters into electrical signals, sensing mechanisms play a crucial role in providing access to the real-time analytics necessary for decision-making within complex systems.

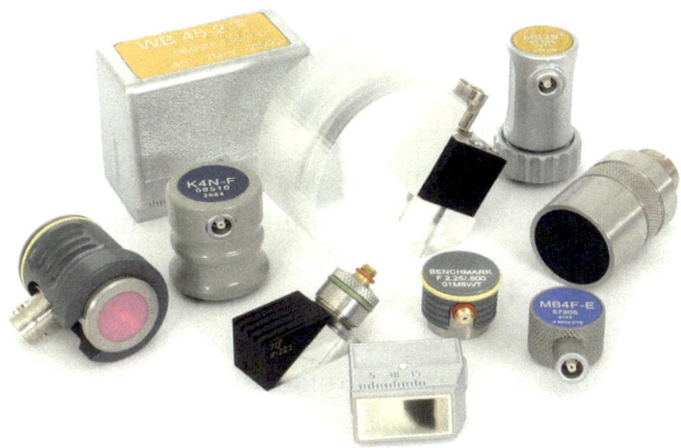

Figure 3.1. By employing a variety of transducers, which convert physical quantities into electrical signals, sensing mechanisms play a crucial role in facilitating the real-time analytics required for decision-making processes in complex systems.

Source: https://www.ndtproducts.ca/wp-content/uploads/krautkramer-ultrasonic-transducers.png.

The new advancements in sensing technologies have introduced more intricate and more adaptable mechanisms. The shift from rigid fixed-channel infrastructures, as the case with Wavelength Switched Optical Networks (WSON), indicates the trend of improvement towards dynamic and adaptable solutions. For example, the concept of an elastic optical connection relies on dividing optical spectrum into slices of variable size, hence effectively accommodating various bandwidth requirements (Cassellas et al., 2013). It is required for today's high-data-rate communication networks due to the fact that it is capable of catering to the dynamic needs of users and applications. Having these advanced sensing mechanisms on board facilitates a robust platform for achieving high performance and reliability in the transfer of data, particularly for environments requiring agility and scalability. The relationship between sensing mechanisms and automation technologies cannot be overstated. The application of leading-edge robotics and automation, as recommended by NASAs Advanced Technology Advisory Committee, emphasizes the critical need for advanced sensing in operating automated systems in complex environments. Such high-tech sensing capabilities enable robots and automated systems to gather critical information about their environment, enhancing operational performance and safety. Accordingly, the integration of sensing technologies and automation not only fosters innovation but also enables the realization of systems that possess the potential to adapt to dynamic conditions in real-time. This integration plays a critical role in the development of smart technologies that characterize contemporary industrial and scientific settings (Figure 3.1).

Importance of Advanced Sensing in Technology

The significance of advanced sensing technology cannot be overstressed since they are one enabling technology that promotes the enhancement of different aspects of technology through better data collection and decision-making. The technology is based in communication networks, particularly in developing cognitive radio systems. Advanced sensors assist in solving the universal problem of having small amounts of spectrum through dynamic spectrum access and the capacity of devices to share a frequency. This aspect optimizes resource use as well as introduces an intelligence level through cognitive radio capabilities offered by artificial intelligence functionalities. However, this innovation has accompanying security concerns and measures of protection have to be robust to counter potential threats, including licensed user emulation attacks and typical control channel vulnerabilities (Alhakami et al., 2014). In essence, advanced sensing mechanisms are a cornerstone of communication technology innovation, striving for highest efficiency and ensuring safe data transmission. Moreover, the integration of advanced sensing into technology supports the advancement of environmental monitoring systems to enable proactive responses to environmental change. Utilizing opportunities for information generated with pervasive sensors, academics and professionals can access real-time information about local and regional environmental conditions. Such usages are essential in addressing global sustainability problems as they allow for enhanced cooperation between expert researchers and civic scientists. The shift from a Sensor Web to an Observation Web signals a paradigm shift towards better management of environmental data, which renders semantic information richness in collected information enhanced (Berre et al., 2011). This novel integration of various sources of data contributes to an integrated perspective of environmental trends, thus allowing stakeholders to devise appropriate conservation and resource management plans. Consequently, advanced sensing technologies, in addition to their role in enabling data collection, drive social change toward greater environmental awareness and action. Finally, applications of advanced sensing technologies extend beyond environmental and communication contexts to enter fields such as medicine, industry, and urban planning. In medicine, sensors for patient vital sign monitoring or for detecting anomalies facilitate interventions in a timely manner and thereby revolutionize patient care. Similarly, in industry, advanced sensing technologies enhance manufacturing lines by making intelligent automation and predictive maintenance possible. This leads to reduced downtime and increased efficiency. Secondly, city landscapes are increasingly adopting sensor-integrated systems in order to improve traffic management, upgrade public security, and maximize utility resource allocation. Such multiple applications attest to the multifaceted nature of next generation sensing technologies, supporting their importance in the development of smart cities and sustainable living environments. As the technologies continue to improve, they are likely to revolutionize technology interaction with ordinary life by focusing on efficiency, sustainability, and connectivity (Figure 3.2).

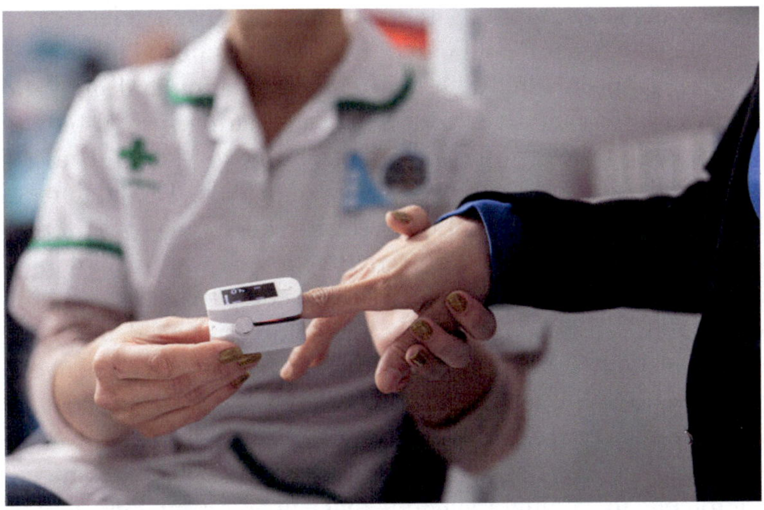

Figure 3.2. In healthcare, sensors that monitor patient vitals or detect anomalies facilitate timely interventions, thereby revolutionizing patient care.

Source: https://static.wixstatic.com/media/8c1b4c_62d98d4af0b149f095fe78186daf5fe5~mv2.jpg/v1/fill/w_800,h_533,al_c,q_85,enc_avif,quality_auto/8c1b4c_62d98d4af0b149f095fe78186daf5fe5~mv2.jpg.

3.1. HISTORICAL DEVELOPMENT OF SENSING TECHNOLOGIES

The history of sensing technologies has been the key force driving the contemporary society, beginning with simple observation devices, and progressing to advanced digital systems. Traditional sensors, such as thermometers and barometers, allowed human beings to quantify and understand environmental variables. As scientific knowledge grew, integration of mechanical and electronic components developed more sophisticated sensors with the capability to measure more variables. This historical trajectory shows a greater trend of moving from qualitative to quantitative data gathering, where it represents a change in the way knowledge is created and distributed. This elementary understanding of sensing devices lies at the core of informing sophisticated systems in contemporary contexts, which shows how historical developments are the precursor to today's intricate sensing mechanisms. The evolution of the Internet has also revolutionized sensing technologies and relocated traditional systems into networked systems. The Sensor Web illustrates this shift by enabling data exchange between various sensors on various platforms, significantly enriching the data space. Citizen science's evolution has also enabled the general public to provide environmental information using personal devices, speculating a shift in data generation and utilization. This transition toward collaborative work and data sharing in real-time indicates the growing consciousness of the importance of integration of data, as brought to fore through the quest towards an Observation Web. The attainment of this web is founded on environmental observation and simulations alongside data from users, thus promoting an integrative approach toward building sensing technology (Berre et al., 2011). In managing the complexities of meshing human and machine abilities, modern sensing technologies now identify themselves with their reliance on advanced computational methods. The convergence of different sources of data, technology, and human work gives rise to complex systems that are better able to manage society's problems. With advances in human computation, these systems employ cognitive processes together with digital equipment to facilitate synergy that enhances predictive models and analytics. This evolution charts the course of a holistic predictive organism with the ability to solve pressing problems on the world agenda, from climate change mitigation to conserving biodiversity (Clark et al., 2015). With the growing ability of sensing technologies, the importance of interdisciplinary collaboration and sustainability arises in increasing distinguishability, underscoring their traditional role in influencing future advancements.

3.1.1. Early Sensing Devices and their Applications

The start of sensing devices in the past was a milestone in technological advancement, providing basic functionalities that would later grow into the sophisticated sensing processes applied today. The devices were primarily mechanical and electrochemical to detect variation in physical or chemical factors. For instance, the thermocouple was a foundational device to detect temperature, and this allowed one to make great strides in industrial processes as well as scientific research. In addition, electrochemical sensors enabled the monitoring of environmental conditions to be possible, yielding crucial information to as wide-ranging industries as agriculture to health care. Through leveraging the chemistry and conductivity principles in chemical reactions, such early sensors created the foundation for advancements in the more evolved sensors, ultimately influencing fields of interest like chemical processing and atmospheric monitoring as the sensors were further developed and implemented in various uses.

Figure 3.3. The thermocouple emerged as an essential tool for measuring temperature, allowing for significant improvements in industrial processes and scientific research.

Source: https://www.processparameters.co.uk/wp-content/uploads/2019/07/explaining-how-does-a-thermocouple-work.jpg.

The application of early sensing technologies had a huge enlargement of human capability, most notably in sectors such as healthcare and the environment. In hospitals, sensors enabled patients' vital signs to be monitored in real time, with immediate intervention saving lives. Devices such as glucose meters revolutionized diabetes care, demonstrating the direct impact of sensing technology on quality of life. Furthermore, environmental sensors were also of interest in public health because they offered the ability to detect pollutants and pathogens, thus they informed public health interventions. These developments pinpointed how integrating early sensing devices into everyday practices not only enhanced safety and efficacy in health but also put attention on monitoring the environment in order to maintain public health levels. Ongoing advancement of sensing technology built upon the innovative designs and paradigms of early devices, leading to increased functionality and adaptability. As materials science and engineering improved, modern sensing devices now take advantage of state-of-the-art materials, such as nanostructures and conductive polymers.

Such materials possess properties that significantly increase sensor sensitivity and specificity, enabling a broader spectrum of applications, ranging from intelligent clothes to medical devices. For instance, the insertion of bio-responsive hydrogels into sensors enables dynamic responses to biological stimulants, which is a step towards personalized healthcare solutions. This advancement illustrates how critical techniques from trailblazing sensing devices propel contemporary design, focusing on making more responsive and efficient systems to address challenges facing us across various industries today, including healthcare and environmental sustainability (Oliver et al., 2009; Puigdomenech et al., 2018) (Figure 3.3).

3.1.2. Evolution of Sensors Through the 20th Century

The evolution of sensors during the 20th century was a revolution that was characterized by technology innovation and the desire for more accuracy and efficiency in numerous fields. Sensors at the start of the century were mostly mechanical in design and founded on concepts of direct measurement and simple physical interactions. As a case in point, the mercury thermometer and barometers were the first technologies utilized to gauge temperature and pressure of the atmosphere, respectively. As industrialization progressed, however, the need for more sophisticated and more accurate sensors arose. This prompted developments to introduce electrical sensors, such as resistive and capacitive sensors, in order to achieve greater accuracy and response time. The development of transistors during the mid-20th century continued to revolutionize sensor technology, paving the way for the integration of sensor systems into upper-level machines and devices, previewing the sensing devices we are familiar with today. The latter part of the century witnessed a sped-up acceleration in sensor technology, fueled by breakthroughs in material science and electronics. The development of semiconductor material enabled the development of more efficient and compact sensors. Of particular importance was the development of silicon-based sensors in the 1970s, which enhanced performance in numerous applications, from automotive systems to consumer products. Sensors began to incorporate digital technology at this time, which was more effective in data processing and communication. These developments led to the creation of intelligent sensors that could provide real-time feedback and analysis, effectively interacting with bigger systems to make more informed decisions. Moreover, devices such as fiber optics and microelectromechanical systems (MEMS) established a new model in the sensing capability, hence expanding the range of applications much more than before. At the closing of the century, sensor evolution shifted toward miniaturization and interfacing with digital networks, setting the stage for the Internet of Things (IoT). Beyond adding capability to conventional sensors, this process enabled new sensing modalities with the ability to interact in a wide range of environments. Among the most important developments were the addition of wireless sensors and adaptive smart materials that could react to stimuli, mirroring the complexity and demands of modern society. Furthermore, scientific pursuits came to emphasize the need for sustainability and circular economy thinking, leaving a permanent influence on sensor deployment and design. With further ventures into battery technologies, an essential step in powering such advanced sensors, the 20th century closed on a utopian vision of a highly interconnected world that still informs sensing strategies today, as detailed by Fichtner et al. (2021) (Figure 3.4).

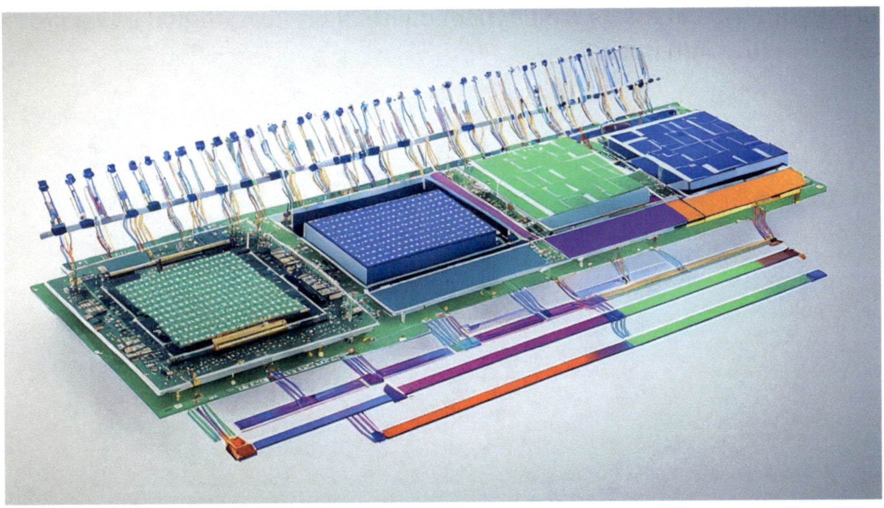

Figure 3.4. The development of silicon-based sensors in the 1970s enhanced performance in various applications, from automotive systems to consumer electronics.

Source: https://vlsiweb.com/wp-content/uploads/2024/02/The-History-of-VLSI.jpg.

3.1.3. Impact of Digital Technology on Sensing Mechanisms

Digital technology has revolutionized traditional sensing devices in a radical manner, leading to higher precision and efficacy in data collection and analysis. IoT, artificial intelligence, and similar new sensing technologies have reoriented the functionality of sensors, enabling real-time data collection and processing. For instance, IoT-enabled sensors are able to monitor environmental parameters and system conditions in real time continuously, enabling more proactive decision-making. All such improvements have shifted the terrain away from manual collection and interpretation of data towards more automated processes, which not only reduce the risks of human error but also increase operational capability. With organizations increasingly adopting these next-generation sensing technologies, they experience improved predictive capabilities as well as aggregate productivity, a game-changing phenomenon within industries like manufacturing, healthcare, and farming. The coming together of sophisticated analytics further enhances the capability to derive insightful meaning from vast amounts of data collected through these sensors. Further, digital technology has facilitated the innovation of new applications within sensing mechanisms, best represented by digital twin technology. Through real-time monitoring and modeling of physical assets with dynamic virtual replicas, digital twins provide greater visibility into operations.

Figure 3.5. Technologies such as satellite remote sensing, unmanned aerial vehicles, and IoT-based systems have enabled farmers to collect critical data regarding soil health, crop conditions, and climate variables.

Source: https://www.gim-international.com/cache/f/a/d/5/7/fad57201f8c2066308d-006cb5bfb98356f15393d.jpeg.

This innovative approach allows businesses to replicate various scenarios and thus improve strategic planning and risk assessment abilities. The integration of digital twin technology with advanced sensing systems not only supports real-time data synchronization but also promotes predictive and prescriptive analytics. This translates to businesses being able to move away from reactive tactics towards proactive simulation-driven decision-making. This shift is also sustained by a growing evidence base indicating that digital twins are extremely effective in delivering improvement in operational efficiency and response time, thereby conferring substantial competitive advantage across multiple industries, as indicated by successful installations and case studies in multiple environments (Puppala, 2025). In agriculture, the application of digital technology in sensing systems has revolutionized practices through enhanced accuracy and efficiency in farm procedures. Satellite remote sensing, UAVs, and IoT-based systems have been capable of enabling farmers to collect precious information on soil health, crop status, and climatic conditions. These technologies not just promote effective agricultural production but also contribute to sustainable activities by enabling climate-smart agriculture initiatives that maximize the utilization of resources. By implementing effective digital systems, as emphasized by initiatives supported by global institutions, producers can apply real-time information in making responsible decisions that impact directly on the level of production and usage of resources. However, there are still hindrances encountered, such as giving appropriate training to agricultural specialists and creating policies that facilitate the acceptance of the technologies. Surmounting these is important in a bid to harness the optimum potential benefit of digital sensing mechanisms in agriculture (Adamenko, 2025) (Figure 3.5).

3.2. TYPES OF ADVANCED SENSING MECHANISMS

Among next-generation sensing technologies, one of the most distinguishing groups is cognitive radio sensing technologies that make efficient reuse of spectrum possible. Cognitive radio is intended to dynamically access and manage radio frequency spectrum by breaking the limitation of available channels. It employs advanced algorithms to scan and detect unused spectrum—enabling devices to communicate in a non-interference way. Through the application of artificial intelligence, cognitive radio systems can wisely allocate channels based on real-time data analysis and therefore optimize communication. However, with this comes great security threats. For instance, cognitive radio networks can be susceptible to licensed user emulation attacks, which compromise the integrity of spectrum sensing. This puts a focus on ensuring the implementation of robust security in these networks to protect not only the spectrum but also communication channels relying on it, as seen in the extensive review of spectrum sharing security mechanisms in the literature (Alhakami et al., 2014). The second important type of advanced sensing mechanism is a product of space life sciences, and it consists of an array of sensing technologies that are specially designed for monitoring interactions between humans and the environment in space. As human exploration increases in space, the importance of novel approaches to studying the effects of microgravity on human beings cannot be emphasized enough. Advanced sensors used to monitor physiological changes in astronauts during spaceflight have therefore become imperative. These sensors do not only take readings of the health and safety of the crew but also data on other environmental factors that affect life in space and, by extension, life on Earth. Understanding how these sensors provide vital information facilitates future expeditions and provides room for them to be used on the planet. Work remains to refine these technologies, enhancing their capability to provide accurate and practical information in more advanced environments. In addition, advanced sensing technologies permeate other sectors, for instance, aviation and automotive, where sensor technologies become a necessity in environmental perception and security. In aviation, advanced radar and satellite-based sensing technology is employed to detect weather conditions in order to prevent hazardous occurrences. Similarly, in the automobile sector, sensors such as Lidar and radar become a necessity in developing autonomous vehicles that can perceive their surroundings in real-time. These systems function in concert, processing vast amounts of data to aid decision-making processes and enhance operational safety. With changing technologies, they must contend with sensor accuracy issues, sensor data fusion, and communications delays. Eliminating these challenges is essential for greater integration of sophisticated sensing mechanisms in enhancing safety and efficiency across domains. Thus, research and development investment remains at the core of advancements in future sensing technologies.

3.2.1. Optical Sensors and their Uses

Optical sensors bear critical roles in most technology industries due to their ability to quantify and monitor various phenomena with high precision. By employing light waves, the sensors can capture information that include temperature, pressure, and chemical compositions. Their applications are particularly evident in environmental monitoring, where they apply in sensing pollutants as well as measuring air quality. For instance, optical sensors are able to sense the concentration of greenhouse gases such as carbon dioxide or methane, which is vital in addressing the problems of climate change. The versatility of optical sensors allows them to be integrated into various systems, enhancing function and reliability of surveillance operations essential for secure and sustainable development. Optical sensors play an important role in aerospace and automotive technology in enhancing vehicle safety and performance. They are employed to track the structural integrity of vehicles in real time, which will prevent any emerging problems from escalating out of control. Their sensitivity is also extremely high, enabling them to sense minute fluctuations in environmental conditions, hence guide and steer vehicles. For example, the implementation of miniaturized optical sensors on non-dispersive infrared technology can improve CO_2 sensing in automobiles or air spaces a great deal.

The innovation demonstrates how optical sensors can be used towards user safety and comfort while simultaneously reducing energy consumption, which is very important towards improving efficiency in transport systems (Baets et al., 2019). Moreover, research into the development of advanced optical sensor technologies has grown their use in medicine and health monitoring. Optical sensors can be applied to non-invasive diagnosis, for instance, of parameters like blood glucose or oxygen saturation without the need to withdraw blood. Monitoring of these parameters is continuous, thus making patients equipped with real-time feedback, which is essential in managing chronic diseases. Current advancements, such as those presented in studies on the design of sensors for active flow control, emphasize meeting specific needs for practical use in the real world, where optical sensors are effective and efficient in various environments (Kopylov et al., 2015). Therefore, the continued development of optical sensors is a measure of their significant progress across various industries, and it has contributed to better quality of life and improved safety through new design and application (Figure 3.6).

Figure 3.6. Optical sensors have become pivotal in various technological fields due to their ability to monitor and measure diverse phenomena with high precision.

Source: https://www.azosensors.com/images/Article_Images/ImageForArticle_3121_17284967503378494.jpg.

3.2.2. Chemical Sensors and Detection Capabilities

Chemical sensor development has been a key to enhancing detection capabilities across many industries, from process industries to environmental surveillance.

Chemical sensors are essential tools for the identification of distinct chemical species, making real-time analysis and a better response to possible threats possible. New sensing mechanisms have endowed these devices with greater sensitivity and specificity. Despite tremendous advances in the underlying technology for such sensors, take-up in the market has been difficult. As suggested, large-scale deployment of greatly more adaptable and capable flexible sensors compared to conventional rigid ones is restricted by a variety of factors, including technical challenges and the lack of suitable interfaces with biological systems. Overcoming such bottlenecks is essential to optimizing chemical sensors, particularly in scenarios where real-time and accurate data sampling is required (Luo et al., 2023). The combination of nanotechnology with chemical sensing has ushered in a new era of multifunctional sensing opportunities over the last few years.

Nano-sensors, being extremely sensitive and robust, leverage the unique properties of different nanomaterials to identify compounds at extremely low concentrations. Nano-enabled sensors have emerged as an essential tool across all walks of life, from diagnosing diseases to ensuring food safety. The uniqueness of nano-armed platforms is that they possess numerous fabrication processes and chemical configurations, resulting in numerous sensing mechanisms that can be tailored to fulfill specific requirements. The literature notes not only the effectiveness of such nanomaterials, but also the role of standardized processes and regulatory frameworks in facilitating their commercialization and implementation in real-world environments (Malini et al., 2022). With science and technology continuously breaking new ground, opportunities for such cutting-edge sensors to improve safety and quality of life are significantly increased. Challenges associated with the commercialization of chemical sensors involve technical limitations but also regulatory and ethical challenges to be addressed so that sustainable development in the field is ensured. On environmental fronts, there have been concerns about sustainability and social responsibility associated with the production and disposal of sensors, particularly where there has been the use of nanomaterials. Experts have called for an integrated approach with contributions from multiple stakeholders to address the complexity of implementing such advanced sensing technologies. Collaborative efforts of academia, industry, and government departments can lead to more robust development plans that align technological innovation with social needs. By accepting such cross-disciplinary discussions and promoting sensor innovations in design, the scientific community will be able to take a crucial step forward in developing not just the functionalities of chemical sensors but also bringing them into acceptance and mainstreaming their applications into everyday life (Luo et al., 2023; Malini et al., 2022).

3.2.3. Mechanical Sensors and their Applications in Robotics

Mechanical sensors form a core component of robotic system development, the foundation of their performance characteristics, and enabling their interaction with the physical environment. Sensors employed to measure variables such as pressure, force, and acceleration are the key that enables robots to carry out tasks with precision and react to conditions. For example, tactile sensors trigger human-like feedback, which allows robots to sense different materials or respond differently to pressure variations. Integrating such advanced mechanical sensing capabilities contributes to further enhancement of more advanced robotic systems that can carry out intricate applications in manufacturing, health care, and exploration.

Figure 3.7. Mechanical sensors play a pivotal role in the advancement of robotic systems, underpinning their operational capabilities and enhancing their interactions with the physical environment.

Source: https://www.awerobotics.com/wp-content/uploads/2023/10/image-16.png.

This reliance on mechanical sensors signifies a paradigmatic shift from the traditional application of robotics to smarter and more sensitive systems capable of optimizing efficiency and safety across various work environments. Further advances in mechanical sensors have led to the evolution of sophisticated materials that closely mimic biological sensing ability. Iontronic sensors, for instance, exploit ion migration to create a skin-like interface, with enhanced tactility and multimodal feedback expanding the possibilities of human–robot interaction. Such developments open up the possibility of soft robotics, where mechanical sensors can be integrated into flexible and degradable forms to enable novel levels of movement and applications. These sensors not only enhance the capabilities of the robots but also enable artificial intelligence to progress since they offer high-value data to fuel machine learning algorithms. This integration is the basis for more advanced applications in a diverse number of industries, including robotic surgery and simulation engagement platforms, bridging the gap between humans' sensorial experience and robotics execution. With the new paradigm of robotics, the combination of mechanical sensors with extended realities such as virtual and augmented environments becomes more significant. The advancement of interactive systems with tactile feedback and sensations is crucial in applications such as teleoperation and robotic surgery where precision and responsiveness are critical. Vibro-haptic and thermo-haptic technologies are the future in this scenario to provide immersive experiences. These advancements facilitate deeper user engagement through authentic touch feedback that facilitates effective remote operation. Lastly, the integration of mechanical sensors in sophisticated robotic systems not only enhances operating functionalities but also promotes ethical aspects in design and use, such that these technologies enhance human well-being and safety in increasingly complicated worlds (Figure 3.7).

3.3. PRINCIPLES OF OPERATION

In sophisticated sensing systems, operating principles are manifestations of the integration of emerging technologies that enhance data intake and interpretation. Operating principles are often based on multisensory data fusion, which is the fusion of information from two or more sensor modalities to construct a coherent environment perception. For instance, in robots, LiDAR and cameras are combined to provide complete spatial awareness. This integration enables systems to act optimally in dynamic situations through the mitigation of uncertainties imposed by sensor noise and environmental harshness. Interconnectedness emphasizes the importance of robust algorithms that enable real-time processing, showing the way in which data-driven approach can lead to improved operational performance and effectiveness across a broad spectrum of applications. Familiarity with the operation principles also includes an understanding of the role of machine learning methodology in sensor optimization. Advanced algorithms allow systems to be taught by their interactions with the environment, allowing ongoing refinement of responsiveness and accuracy. For instance, machine learning algorithms can adjust sensor calibration in real-time through dynamic input, enhancing the ability of the systems to filter out irrelevant information and focus on essential measures. In this context, systems not only react to their surroundings but also have the ability to forecast change, further demonstrating the efficiency of these types of next-generation sensing devices. It has been clear that this level of adaptability makes operations more robust, particularly under complex conditions where other sensing processes tend to fail, as demonstrated by recent studies on cognitive computational models (Rahimi et al., 2025). Furthermore, use of novel ideas in very advanced sensing mechanisms offers room for addressing global challenges, such as resource management and environmental monitoring. The operating principles behind such mechanisms allow for designing more efficient systems that can provide informative insights across various industries. For instance, precision agriculture uses sophisticated sensing technologies to track soil health and crop condition, allowing farmers to make the best-informed decisions that maximize output while reducing wastage of resources. Such uses not only optimize productivity but also help in sustainability through environmentally friendly practices. With growing calls for smart solutions, the working mechanisms inherent in such technologies will continue to evolve, driving innovation that guarantees economic viability in conformity with environmental management, particularly in nations like Azerbaijan, where innovative entrepreneurship is becoming ever more significant (Figure 3.8).

Figure 3.8. Understanding the principles of operation also entails recognizing the significance of machine learning techniques in refining sensor performance.

Source: https://miro.medium.com/v2/resize:fit:1400/0*RMOgy1KTBoDrW6gD.

3.3.1. Basic Principles of Sensor Functionality

The operation of sensors is fundamentally dependent on their ability to perceive the changes in the environment and convert them into quantifiable signals. All of these operations are primarily characterized by three primary phases: sensing, signal transduction, and signal processing. Sensors initially perceive physical phenomena such as temperature, pressure, or light using materials and technologies that respond to them. For instance, piezoelectric materials would be applied in pressure sensors where deformation leads to an electrical signal output. Transduction of the physical change to an electrical signal is necessary to enable further analysis and interpretation. An understanding of these basics allows for the creation of responsive and adaptive sensors, something which is further reinforced by the use of intelligent materials in the latest sensing technology, including clothing and ecological monitoring systems (Oliver et al., 2009). Signal processing is next in the sequence of operations in a sensor where raw data is filtered and processed to extract useful information. This stage tends to utilize algorithms that improve data quality and reduce noise, providing accurate readings. Middleware can be used to implement this stage through the provision of higher-level abstractions, which can enhance interoperability across different sensor networks. Wireless sensor networks tend to program directly based on operating system functionality instead of higher-level middleware, with both challenges and opportunities. One needs to grasp the important requirements of their design so that one can more suitably match them to actual applications as well as predict possible pitfalls that might arise at the time of deployment. As sensors continue to advance, concerns regarding their functional design become paramount (Mottola et al., 2011). Finally, one needs to comprehend the implications of such sensing mechanisms in broader contexts, especially in applications that treat human-centering concerns. With the merging of high-sensing devices and smart materials, the potential for new applications arises across a wide range of fields such as energy management and medicine. One such example is the intelligent bandage that has the ability to sense the wound environment or energy-generating solar clothes, showing the vast impact of responsive sensor technology. These developments not only underscore the versatility of sensor capability but also suggest the significance of interdiscipline methods fusing material science with engineering. Consequently, a grasp of the basic principles governing sensor function forms the basis on which creativity in design and development of sophisticated sensing devices across disciplines can be fostered (Oliver et al., 2009).

3.3.2. Signal Processing Techniques in Sensing

Signal processing techniques in sensing evolve rapidly because of the need to enhance the accuracy and dependability of data acquisition in various applications. Such techniques involve the analysis and processing of signals to obtain useful information out of raw data, particularly where noise and interference prevail. By means of application of advanced algorithms, such as filtering, modulation, and feature extraction, sensor performance is significantly improved. Distributed acoustic sensing (DAS) technology is impressive in tracing acoustic signals along extended fiber-optic webs by detecting slight perturbations. The integration of machine learning (ML) with DAS systems has shown promising developments to facilitate intelligent processing of data that can be responsive to real-time conditions. This integration not only enhances the accuracy of measurement but also extends the use of sensing technology to critical infrastructure, transportation, and environmental monitoring (Shi et al., 2025). In exploring the convergence of advanced signal processing and sensing, a noteworthy trend is the application of deep learning algorithms, which have radically revolutionized how sensor data are processed. Traditional filtering methods have been ineffective in situations with an underlying complex noise pattern. For instance, wheel speed sensors in cars have to frequently handle high-intensity interferences that impair their signals. A new study on Enhanced Long Short-Term Memory (LSTM) models offers a new solution to handle this issue by incorporating an attention mechanism. The solution allows the model to focus on the key data points and eliminate the effect of noise effectively. Through comparative experimentation, the Enhanced LSTM performs better than existing algorithms and establishes a new benchmark for real-time automotive sensing activities. Such innovation not only verifies the resilience of deep learning within real-world applications but also illustrates attempts to continue developing within the signal processing field (Lin, 2025). The continued development within signal processing techniques in sensing also introduces substantial implications towards computational efficiency as well as data handling. With growing algorithmic sophistication comes the need for efficient computation power and management of large data sets. This is particularly evident in real-time analysis applications where processing latency may result in significant operational inefficiencies. With the advent of new technologies such as edge computing, researchers are looking at decentralized capability for processing to enable rapid decision-making close to locations where data is generated. This approach can mitigate the high bandwidth demands often associated with high quantities of data transfer to central systems. The future of sensing mechanisms thus lies in a balance between advanced signal processing techniques and novel data handling techniques in an effort to deliver high performance while maintaining operational viability. As they are further developed, they promise to revolutionize industries that rely upon timely and accurate sensory information, ushering in more intelligent integrated systems for numerous applications (Shi et al., 2025).

3.3.3. Calibration and Accuracy in Sensor Measurements

Calibration is a basic method of guaranteeing precision in sensor readings and is a basic application in a vast array of uses, ranging from environmental observation to industrial automation. Calibration involves the comparison of a sensors reading to standards that are known to determine whether there are any variations and to make modifications on the sensor's readings accordingly. Calibration is central because minor inaccuracy can lead to enormous errors in data expression. For instance, calibration of sensors like the Sea-viewing Wide Field-of-view Sensor (SeaWiFS) in remote sensing is essential to obtain credible ocean color data. In the absence of rigid calibration procedures, interpretations based on satellite images would be severely distorted, and marine science and policy-making efforts would be influenced. Therefore, an understanding of calibration nuances is not just a technical requirement but a necessity for the maintenance of sensor measurements' reliability and integrity in advanced sensing systems (Firestone et al., 2025). Sensor measurement precision does not depend solely on the calibration procedure but on other factors such as environmental conditions, the structure of the sensor, and context of use. For example, in precision micropositioning devices, material behavior and mechanical robustness are of critical concern.

The design and production of a closed-loop XY micro positioning stage based on parallel-kinematic mechanisms show sensor accuracy as being inextricably linked with engineering processes. In this instance, the 5.24 nm standard deviation in positional uncertainty shows accurate engineering to achieve reproducible sensor performance. Thus, it is critical to balance calibration with robust sensor design to enhance measurement accuracy. As sensors become more embedded in advanced systems, being capable of recognizing these factors can lead to higher data fidelity and operational performance across a variety of study disciplines (Koo et al., 2014). Moreover, the interaction among calibration and accuracy is interactively influenced by developments in sensor systems. Technologies such as adaptive calibration methods and machine learning algorithms are revolutionizing the domain of calibration with real-time compensations powered by data analytics. Not only are these methods enhancing accuracy but also enabling sensors to remain accurate even in dynamic operation environments. For instance, application of closed-loop control systems can significantly enhance the performance of micro-positioners, which enable adaptive feedback mechanisms for precise movement and position tracking. These advances throw the spotlight on a continued research and development of calibration techniques, thereby setting the stage for advances in sensor measurements. As these methods continue to advance, they do not just impact scientific investigation but also set the stage for the utilization of sophisticated sensing mechanisms in fields ranging from medicine to smart infrastructure (Firestone et al., 2025; Koo et al., 2014) (Figure 3.9).

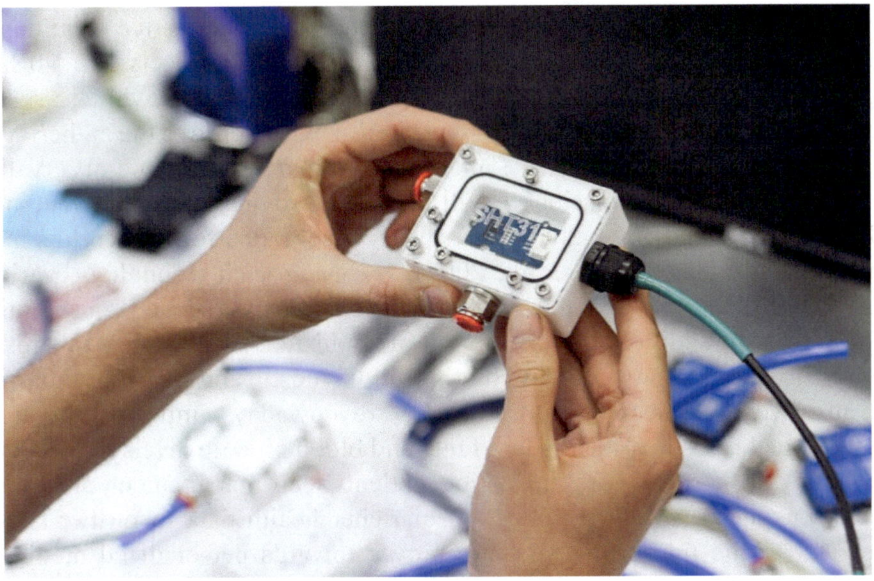

Figure 3.9. Calibration involves the comparison of a sensors output to known standards in order to identify any deviations and to adjust the sensors readings accordingly.

Source: https://static.wixstatic.com/media/8c1b4c_fa0569d19b394b1b92a3769b27c2c42d~mv2.jpg/v1/fill/w_954,h_639,al_c,q_85/8c1b4c_fa0569d19b394b1b92a3769b27c2c42d~mv2.jpg.

3.4. APPLICATIONS IN VARIOUS FIELDS

The applications of sophisticated sensing systems filtered into numerous fields, significantly influencing technological advancement and research methodology. In environmental monitoring, for instance, advanced sensors help gather and analyze data on weather patterns and pollutants. These advanced systems enable scientists to establish the impact of man's activity on the environment with improved accuracy. Various research programs, for example, those sponsored by NASA at the Jet Propulsion Laboratory (JPL), leverage advanced sensing technology to expand earth and space science through the processing and interpretation of real-time data. Through an integration of these mechanisms, scientists are able to not only monitor the well-being of the planet but also predict changes in the environment, thereby having an advanced level of climate patterns and their implications on global sustainability. In defense and aerospace, advanced sensing systems play a critical role in enabling operational effectiveness and safety. The development of satellite platforms using Micro Systems Technology (MST) is one of the innovative solutions in satellite architecture and payload design. For example, projects like MISAT, targeting micro-technologies, enable the integration of robust wireless sensor networks and autonomous platforms that significantly improve the gathering of data on earth and in space. These innovations allow for real-time observation and responsiveness in missions of all types, improving the effectiveness of utilization in space activities and offering new solutions for the world's most challenging problems (Gill et al., 2007). By using such technology, various aerospace projects can operate with higher precision and reduced consumption of resources. In addition, the integration of sophisticated sensing mechanisms into medical technology could revolutionize diagnostic and patient monitoring systems. Wearable sensors, for example, are nowadays imperative to track physiological measures such as heart rate and blood glucose levels in real-time. Not only are these devices enhancing patients' self-management of chronic illness, but they also allow collection of richer health data to further enable research. As the future of healthcare moves towards personalized medicine, the role of cutting-edge sensing technologies becomes more critical, enabling tailored treatment protocols based on a patient's unique data. The possibility for instant diagnosis and immediate health interventions underscores the revolutionary influence of such mechanisms in many areas of medicine, ultimately leading to better patient care and setting new paradigms in health management.

3.4.1. Medical Applications of Advanced Sensors

The incorporation of state-of-the-art sensors in medical devices has transformed patient monitoring and diagnosis, greatly enhancing the quality of healthcare. With the use of microelectronics and state-of-the-art biosensing technologies, these sophisticated systems can monitor a multitude of biological parameters continuously. Such an ability is particularly useful in chronic disease management, where data in real time can provide insights for timely interventions. As an illustration, diabetic patients are highly benefited from continuous sensors measuring blood glucose levels for more effective insulin dosing. This provision in real time not only provides patients with more independence but also provides doctors with basic information to modify treatment regimens more accurately. With advanced sensors in the forefront of the healthcare setting, their role in individualizing care and enabling proactive disease management is apparent, showcasing their priceless contribution to modern medical processes (N/A, 2016). The application of advanced sensors advances beyond routine monitoring, venturing into such areas as predictive analytics and remote diagnostics. With artificial intelligence algorithms working in tandem with such sensors, they can analyze streams of incoming data to detect patterns and predict potential health events before they occur. This is predictive capability that can usher in a paradigm shift for healthcare professionals to treat patient care as a proactive rather than reactive style of healthcare. Moreover, advances in miniaturization and biocompatibility of sensors have made it possible to develop wearable technology, which empowers patients through permanent health information. Apart from stimulating patient engagement, these devices also enable clinicians to tailor treatment based on objective, real-time information. The varied applications of sophisticated sensors, thus, constitute an essential medical technology breakthrough, which enhances clinical decision-making as well as patient satisfaction (Baj-Rossi et al., 2012).

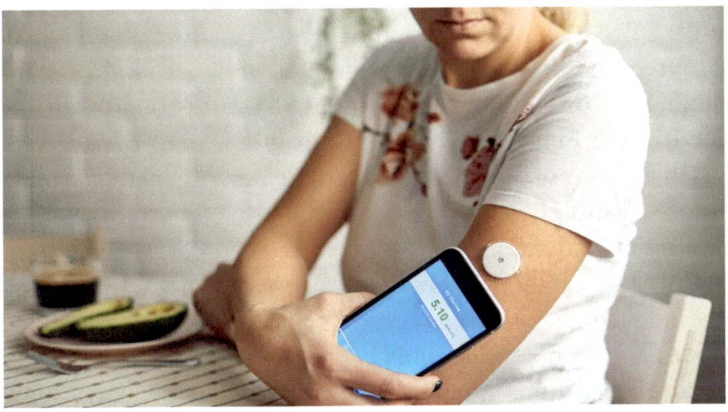

Figure 3.10. Diabetic patients greatly benefit from sensors that monitor glucose levels continuously, allowing for more precise insulin administration.

Source: https://images.everydayhealth.com/images/diabetes/type-2-diabetes/cs-innovations-in-type-2-diabetes-treatment-1440x810.jpg?sfvrsn=1deedc8d_5.

In the future, the potential for medicine with sophisticated sensors is immense, particularly as research continues to provide solution to existing challenges in the field. Requirements of high sensitivity and specificity in measurements, data privacy, and integration of devices have to be tackled in an orderly manner to ensure that widescale application is realized. For instance, the completely integrated and minimally invasive devices can make a significant impact on compliance and comfort of patients, allowing for continuous monitoring of health in a non-invasive and efficient manner. Continuous interaction between academia and industry, as brought to the fore in research efforts, will be pivotal in bringing about innovation and making such advanced sensing mechanisms both effective and accessible to the masses. With this in mind, the continuous development of medical sensors is set to not only improve healthcare provision but also reshape the therapeutic environment in general, exemplifying their vital importance to the progress of human health (N/A, 2016) (Figure 3.10).

3.4.2. Environmental Monitoring and Sensing

The integration of advanced sensing mechanisms in environmental monitoring is an unprecedented shift in our ability to perceive and respond to ecological variation. Through the use of enormous sensor networks, researchers can gather enormous amounts of near-real-time information regarding various environmental conditions such as air and water quality, biodiversity, and climate, in real-time. These systems capitalize on the increasing pervasiveness of sensors that allow for continuous observation across expansive scales. Furthermore, the possibility of synergistic interaction between citizen-generated data and traditional sources holds the key to a greater view of environmental health. Broader integration of the data is required if the vision for an Observation Web, outlined in the Future Internet PPP program, is to be realized (Berre et al., 2011). This program aims to harmonize heterogeneous data assets and deliver a more comprehensive view of dynamic environmental processes. Also, technological advancement in sensing is most critical to addressing impending global sustainability challenges. Environmental enablers in Future Internets include an extensive array of sensors such as various electronic sensing devices and web-service tools supporting data exchange. More reliable environmental observations are made possible by these technologies as they integrate knowledge from numerous fields — from ground ecosystems to atmospheric conditions. The issues highlighted in the novel discipline, such as the achievement of data fusion and uncertainty management, are central to enhancing the reliability and usability of aggregated data. Thus, as scientists proceed to solve such technical problems, the likelihood of constructing scalable and resilient environmental monitoring systems is increasingly possible, leading society toward improved environmental stewardship methods (Berre et al., 2011). In addition, dramatic advances in space-based technology further enhance environmental monitoring capabilities. Satellite remote sensing has totally revolutionized resource observation and environmental measurement, providing a macroscopic perspective that complements on-ground data. Space methods are applied in programs for benefiting from, which are striving to advance the capability to comprehend terrestrial conditions and encourage interdisciplinary applications. These encompass not only the monitoring of air pollution but also the management of ocean ecosystems and biodiversity resources. This interdisciplinary approach that involves direct earth observation and harmonization with ground-based initiatives has the potential to significantly enhance our ability to respond to environmental threats. As the S&T Applications Program grows, its emphasis on synthesizing plural disciplinary visions will be essential to designing balanced monitoring systems bridging the divide between space and Earth environmental science.

3.4.3. Industrial Automation and Smart Manufacturing

The transformation towards industrial automation and intelligent manufacturing is effectively transforming manufacturing ecosystems. Advanced sensing technologies take center stage in this revolution, with real-time data aggregation and process refinement being enabled. Such technologies employ sensors to monitor machinery condition, environmental conditions, and operation workflows, hence the ease of making effective decisions and being effective. For example, within injection molding, low-cost molding inserts produced through additive and subtractive processes have been responsible for lowering entry barriers for potential new customers. Advanced sensing technology aids in ensuring the prototypes will be up to specifications while eliminating flaws in surface finish and removal of parts that are essential in achieving quality standards for manufacturing processes (Ewart et al., 2019). Moreover, industrial applications of artificial intelligence (AI) emphasize the growing importance of smart manufacturing. Systems based on AI not only automate routine activities but also enhance security and efficiency by reducing the risks and burdens caused by human activities. For instance, in space manufacturing and assembly application, AI can monitor and diagnose in real time, with the processes being productive and enhancing human workers' capabilities (Firschein et al., 2025). This collaboration between advanced sensing and AI leads to tremendous improvements in operational autonomy and resource management. With such technologies being increasingly used by industries, the potential for greater agility and rapid response to market influences is apparent. Finally, the future of industrial automation and smart manufacturing is inextricably linked with continued technology research in sensors and analytics. Advanced sensors that can sense subtle variations in surroundings or equipment operation allow for predictive maintenance and real-time optimization of production operations. These technologies not only prolong equipment life but also significantly reduce downtime, which can amount to considerable cost savings. Further, the complementarity of big data analytics and smart manufacturing enables strategic decision-making based on data, enabling organizations to better predict customers' needs and adapt to shifting market forces. As these intelligent systems permeate manufacturing environments, they promise a more efficient, responsive, and sustainable industry worldwide, eventually reshaping the design and production of products worldwide (Figure 3.11).

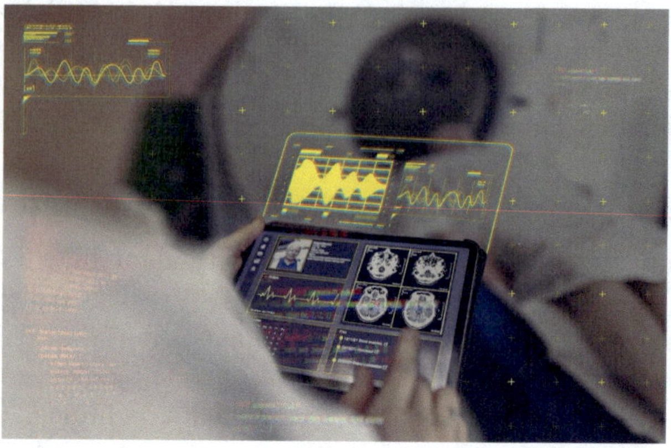

Figure 3.11. Advanced sensing mechanisms are pivotal in this transition, enabling real-time data collection and process optimization.

Source: https://www.ey.com/adobe/dynamicmedia/deliver/dm-aid--4c99107f-061c-4245-9fe5-c5fac83d69f5/ey-dotai-ai-helath-final.jpg?quality=85&preferwebp=true.

3.5. CHALLENGES IN SENSING TECHNOLOGY

In the high-speed environment of sensing technology, there are persisting challenges due to the nature of environmental complexity as well as the increasing demand for more sophisticated devices. The primary challenge is the constraint on the availability of spectrum, which inevitably impacts the efficiency of the sensing mechanisms. The traditional sensing methods are inefficient at adapting to changing environments, thus leading to underutilization of available channels. Cognitive radio technology promises resolution through dynamic spectrum access but introduces an enormous array of new security challenges. Because cognitive radio networks depend on cooperative device-to-device communication, they are susceptible to licensed user emulation attacks and misbehavior in the common control channel. These issues not only result in inefficient sharing of the spectrum but also in compromised network security, for which effective protection and detection mechanisms must be designed to counteract the vulnerabilities (Alhakami et al., 2014). Another pressing problem in sensor technology is the energy efficiency trend, particularly in cellular systems where the majority of the energy is utilized by base stations. The search for green technology calls for innovative measures to save energy without compromising operational performance. While recent efforts have sought to enhance power efficiency through techniques such as deployment of heterogeneous networks, including micro-, pico-, and femto-cells, there remain significant hurdles.

Integrating cognitive radio and cooperative relaying technologies, while promising, also requires further research to achieve sustainable gains in energy. Effective deployment of these technologies is a question of balance between energy consumption and service quality, which calls for holistic strategies with precedence for performance and environmental friendliness (Bhargava et al., 2011). Further, the incorporation of advanced sensing mechanisms is problematic in terms of data accuracy and adaptability to the environment. With the rising sensors, they must also contend with environmental parameters that could interfere with their operation, such as physical barriers and varying signal conditions. Their ongoing interaction impacts sensor reliability data, and thus there is a necessity for innovative solutions that will enhance the robustness and accuracy of sensing technology. In addition, the proliferation of intelligent devices and the Internet of Things (IoT) create an avalanche of data which must be processed in real time and analyzed. This type of flood of data will place unnecessary strain on existing infrastructure, necessitating the use of new algorithms and processing techniques to retain efficiency of handling data. Transcending these is necessary to harness the complete capability of advanced sensing mechanisms and facilitate further advancement of sensing technology.

3.5.1. Limitations of Current Sensing Mechanisms

There are significant limitations of existing sensing mechanisms, particularly wireless network mechanisms, which downgrade their effectiveness and efficiency. One such significant one is vulnerability to security threats, which has become more apparent as communication technology gets more complex. For instance, in cognitive radio networks, the sharing of spectrum creates unusual challenges. The common control channel, a building block to allow realization of spectrum sharing, is exposed to threats like licensed user emulation and control transaction misbehavior. These not only degrade the functioning of the networks but also compromise the data communication integrity between nodes, hence curtailing overall reliability of current sensing mechanisms. With more use of devices for cognitive radio communication, mitigation of such inherent security concerns is essential to enhancing the resilience of such sensing mechanisms and to ensuring secure data transmissions in such dynamic environments (Alhakami et al., 2014). The ever-growing application of the Internet of Things (IoT) also brings forth its own limitations in terms of current sensing mechanisms. While massive networks of interoperable devices generate gigantic amounts of data, matters of management, aggregation, and storage become central. Current mechanisms are usually inadequate to deal with the flood of data, causing inefficiency in the system. Furthermore, heterogeneity of devices in IoT systems complicates matters of communication even further due to the propensity of the devices to use different interfaces and operating systems. Middleware solutions have been proposed to bridge these variations and develop homogenous communication platforms. However, existing middleware technologies tend not to adapt to evolving IoT structures, widening further the limitations of existing sensing mechanisms in managing diverse and evolving data sources and communication channels efficiently (Farahbakhsh et al., 2017). Finally, the flexibility of existing sensing mechanisms is another area where limitations can be seen. The majority of modern technologies are designed with fixed parameters which fail to accommodate the rapidly changing contexts in which they operate. For instance, machine learning and artificial intelligence advancements are not being exploited properly to advance sensing activities. With no dynamic approach that can facilitate real-time analysis of information and predict behavior, current mechanisms will be unable to counter trends proactively or maximize performance across various circumstances. As adaptive and responsive sensing functions are becoming more inevitable, designing mechanisms that are fault-tolerant and adaptive is more important than ever. This evolution is not only required to optimize operational performance but also to maintain data integrity and security being conveyed through advanced sensing infrastructures (Alhakami et al., 2014; Farahbakhsh et al., 2017) (Figure 3.12).

Figure 3.12. As vast networks of interconnected devices generate massive amounts of data, challenges related to management, aggregation, and storage become critical.

Source: https://www.bics.com/wp-content/uploads/2025/04/AdobeStock_283063882.webp.

3.5.2. Issues related to Sensor Accuracy and Reliability

The development of sensor devices has been significantly retarded by ongoing issues with sensor accuracy and reliability. These types of issues are found in numerous applications, from robot systems to urban infrastructure. In robot-assisted minimally invasive surgery (MIS), for instance, the accuracy of continuum instruments relies on effective cable-driven actuation systems. Real-time tension feedback is essential for precise articulation; however, it is challenging to create compact units for actuation with the feedback. The addition of a torque cell to the actuation units, as proposed in recent research, aims to improve the reliability and precision of shape and external force sensing in an effort to advance the overall performance of robotic surgical systems (Zhang et al., 2025). This aligns with the overall need for innovative designs that emphasize sensor reliability under dynamic conditions, where varying conditions could affect measurement accuracy.

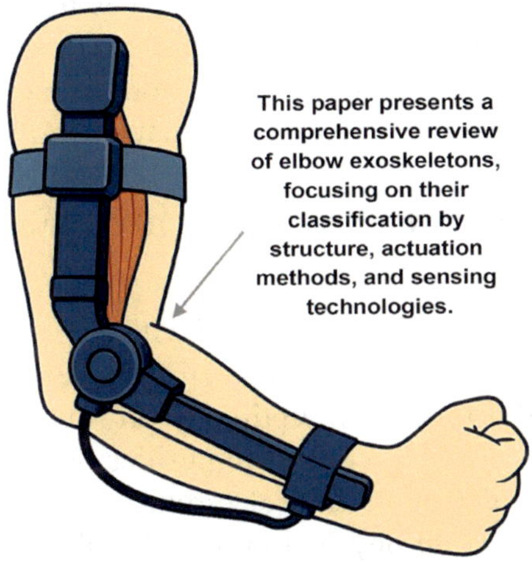

Figure 3.13. Real-time tension feedback is vital for precise articulation; however, creating compact actuation units that provide this feedback remains a challenge.

Source: https://www.preprints.org/frontend/manuscript/eb88561cdd25a09a638f-129ef6f55136/download_pub/figure/graphic_abstract.png.

In urban electrical infrastructure, increasingly sophisticated and interconnected, issues of sensor accuracy also have a very critical role. The availability of more sophisticated sensing technologies in combination with artificial intelligence (AI) not only improved fault detection but also emphasized the importance of accurate data acquisition. For example, studies aiming at the medium-scale city infrastructures have successfully applied smart meters in monitoring voltage measures, monitoring phase imbalances, and transient oscillations. Integration of AI, particularly through the application of models like the support vector machines (SVM), has had accuracy levels of over 99%

in the detection of system failures (Villarreal et al., 2025). This proves that sensor reliability must be enhanced to ensure stability in electric systems and thus necessitates creating fault detection mechanisms that are robust enough to adapt to the changing conditions of urban living. Sensor reliability is also highly linked to the operating environment and the quality of data interpretation algorithms. In most cases, sensor failures are due to environmental reasons, such as changes in temperature or electromagnetic interference that impact the accuracy of the measurements. Advanced algorithms that take these factors into account are then a necessity. As observed in intelligent sensing technologies integrated into city infrastructures, such interdisciplinary integration of sensor design and data processing capability remains pivotal for the attainment of resilience against outside interference (Villarreal et al., 2025). Further progress in these extremely advanced sensing mechanisms is leading us to the realization of more confident applications across all sectors as tackling the challenge of sensor accuracy and reliability emerges as a main priority (Figure 3.13).

3.5.3. The Impact of Environmental Factors on Sensor Performance

The operation of sensors is highly correlated with environmental factors, which can potentially affect their accuracy and reliability to a high degree. Sensory systems are required to perform optimally in fluctuating conditions like temperature, humidity, and presence of impurities. For instance, humidity can lead to sensor drift or degradation in electronic measurement, while extreme temperatures have effects on physical properties of materials used in sensors. Knowing these limitations is key to the development of dependable sensing mechanisms that are tolerant to harsh conditions. Furthermore, improvement of sensing technology also demands integration of multi-modal systems with the capability to overcome the limitation of single sensor failure susceptibility, hence enhancing data gathering overall dependability. Such integration is necessary, particularly in environments like estuarine and coastal ecosystems, where environmental variability condition can complicate effective resource management (Burghina et al., 2013). Among the environmental conditions that influence sensor performance, water levels and salinity are particularly significant in aquatic sensor applications. Fluctuations in these conditions can alter the conductivity and density of water, thereby affecting the readings from submersible sensors employed to measure ecosystem health. Using multi-parameter aquatic sensors, researchers can obtain a comprehensive insight into the way environmental conditions reshape sensor outputs. For example, the inclusion of visual sensing systems allows for simultaneous observation of salinity levels and the physical condition of the water, thereby developing a wider picture of aquatic processes. As determined by research on Dublin's River Liffey estuary in Ireland, it decreases the number of sensors with better quality data and more stability in networks, illustrating the efficiency of advanced sensing mechanisms in various environments (Burghina et al., 2013). The destiny of sensor technologies also relies significantly on understanding how environmental parameters affect sensor performance. Advances in modelling and data handling technologies can assist scientists to overcome inconsistencies related to environmental variability. The Space Life Sciences Program, for example, is geared towards making scientists aware of the effect of human factors and environmental parameters on health and sensor reliability in space missions. Consequently, effective control of environmental factors can lead to more sophisticated analytical models and improved sensor designs. These advances will not only complement scientific research but also improve resource management methods in different applications, further necessitating the integration of such environmental issues into practices of deploying sensors.

3.6. FUTURE TRENDS IN SENSING MECHANISMS

As technology progresses, particularly where sensing mechanisms are involved, the future portends a revolutionizing shift spearheaded by more integration and functionality. Stand-alone sensor networks are among the primary developments on this horizon, an advancement from the current state to a more integrated Observation Web in which several environmental observations get their data merged in an effortless manner. This paradigm is further facilitated by the potential to draw upon not only traditional sensors but also user-generated and citizen science contributions. This holistic method generates a rich stock of information that not only further enhances overall understanding of intricate environmental processes but also addresses central research issues with regard to data management and interoperability. The multidisciplinary integration is essential in developing robust and scalable sensing systems that would revolutionize environmental monitoring processes in different domains, thereby facilitating a deeper investigation of real-time ecological processes (Berre et al., 2011). Future advancements in sensing mechanisms also anticipate improved capabilities through innovative designs incorporating ideas from soft robotics. Perhaps most significantly, soft robotic grippers represent a leap forward in their ability to grasp unstructured objects with fewer sensing and control strategies. Despite their benefits, currently available speed and sensing limits restrict their application to dynamic environments. New work that explores tunable bistable behavior in these soft systems has the potential to address these weaknesses, enabling fast sensor-less interaction with higher efficiency. This mechanism not only offers greater speed of operation-capable of making grasping decisions in around 0.02 seconds—but also enables having variable design modifications that can optimize actuation strategies. With such robotic applications, sensing in the future will be marked by responsive and quick systems that can dynamically adapt to their surroundings (Abidi et al., 2019). The growing overlap between sensing mechanisms and digital technologies portends on developments centered on semantic enrichment and context-aware applications. As society is increasingly inclined towards data-driven decision-making, the ability to obtain worthwhile insights out of combined sensing networks is paramount. Future sensing mechanisms will employ machine learning and artificial intelligence capabilities to advance data fusion processes in order to enable systems to progressively refine their analysis based on variables that are context-specific. This advancement not only enhances the accuracy of sensor information but also unlocks possibilities ranging from environmental monitoring to city planning and disaster management. Understanding the potential of these technologies in conjunction with rigorous security protocols and standardization initiatives will be essential. These foundations will not only ensure the reliability and longevity of these sensor systems but also the ethical implications in various fields of use (Berre et al., 2011).

3.6.1. Integration of Artificial Intelligence in Sensing

The application of artificial intelligence (AI) in sensing technology is a significant development in the evolution of advanced sensing mechanisms. AI enhances the capabilities of ordinary sensors with the addition of adaptive learning, processing of data, and decision-making abilities that were previously out of reach. Applying machine learning algorithms, these advanced sensing mechanisms can process large amounts of data in real time to make quicker and more accurate responses to changing environmental conditions. For instance, in stress sensing, while the attention has been forged by recent advancements, the advanced functionalities provided by AI algorithms allow advanced interpretations of multimodal stimuli and stress distribution, basically reshaping the application of sensors, such as in healthcare and structural monitoring (Zhuang et al., 2021). Moreover, combining AI with sensing technology offers higher biocompatibility and user compatibility, especially in wearable devices. AI-based sensors can learn from the user's use, modulating sensitivity and data output based on specific requirements or conditions. It tailors the user's experience and ensures monitoring systems perform better in dynamic environments. Focusing on creating stretchable as well as remote sensing devices emphasizes the characteristics of such technology to function and be embedded conveniently across many applications. This technology can be especially useful in medical diagnosis, where real-time monitoring and tailored feedback are essential for patient health (Zhuang et al., 2021). In addition, AI capability to simulate intricate relationships in data enables the optimization of sensing materials and structures, which opens doors to the development of future sensing devices. The contribution of sophisticated programming frameworks like NumPy further accelerates the adoption of AI in sensing technologies. These interfaces facilitate easy data manipulation and calculation and allow researchers to automate tasks that involve tedious array-based computation. By providing a solid foundation for managing the huge data generated by integrated AI sensing platforms, NumPy significantly enhances visualization and analysis of this data. This compatibility is particularly useful in scientific communities where speed and precision are vital, allowing researchers to unearth insights more efficiently and innovatively. With continued development in AI, its combination with efficient programming tools will undoubtedly hasten the capabilities and applications of advanced sensing mechanisms, both enhancing performance as well as opportunities for revolutionizing discoveries across disciplines (Figure 3.14).

Figure 3.14. AI enhances the capabilities of traditional sensors by introducing adaptive learning, data analysis, and decision-making features that were previously unattainable.

Source: https://evelta.com/product_images/uploaded_images/sensor-motion-ab.jpg.

3.6.2. Development of Miniaturized and Wearable Sensors

The development of wearable and miniaturized sensors has been a primary driving force behind sophisticated sensing mechanisms, enabling one to transition towards highly integrated and personal health monitoring. These sensors leverage advancements in the field of nanotechnology and materials science to create devices that can be comfortably worn on the body for continuous monitoring of vital health information. The evolution of triboelectric nanogenerator (TENG)-based sensors is an example trend that offers self-powered capability in the form of contact electrification and electrostatic induction. This technology not only enhances energy efficiency but also gets used in a vast array of smart healthcare applications, where miniaturization potential can lead to a transformation in the role of wearables in monitoring physiological parameters continuously and preventively (Su et al., 2025). In addition to advancements in sensor technology, artificial intelligence (AI) has increasingly become more central to allowing scientists to enhance the processing and analysis capabilities of data collected with wearables.

By integrating smaller-sized sensors with deep learning algorithms, researchers are developing intelligent systems with the capability to decode complex datasets with increased precision. This synergy allows for more complex applications such as motion sensing and tailored health monitoring, facilitated through agile and ultra-lightweight designs without sacrificing user comfort. Hence, the interplay of AI and miniaturized but high-tech sensors is transforming the manner in which doctors and patients engage with health information, generating intelligent diagnostic equipment that adapts to individual needs and states (Su et al., 2025). In addition, wearables with built-in mechano-acoustic sensors have revolutionized body sounds measurement, thus opening new avenues for continuous monitoring of health. In addition to the non-invasive nature of these sensors in providing crucial health indicators, they also do away with the limitations that are a standard feature of traditional diagnostics like stethoscopes that are known to be bulky and subjective to diagnosis. Miniaturization and flexibility of the sensors allow for subtle wear and real-time monitoring of audio from systems such as cardiovascular and pulmonary functions. The technology is opening new possibilities for remote health monitoring and chronic disease management and ultimately improving patient outcomes through early detection and intervention by providing more dynamic and continuous monitoring.

3.6.3. Innovations in Sensor Materials and Technologies

The area of sensor materials and technologies is going through a revolution, primarily caused by nanotechnology developments and the emergence of novel fabrication techniques. These developments have given rise to a new generation of sensors that feature enhanced sensitivity, selectivity, and functionality in a myriad of applications. For instance, the integration of nanomaterials, such as carbon nanotubes and graphene, into sensor structures has significantly boosted the detection capability to facilitate real-time measurement of various environmental parameters. Not only does this innovation surpass the traditional performance of sensors, but it also introduces the possibility of applications in areas such as biomedical sensing and smart manufacturing. The result is a series of sensors able to detect infinitesimal changes in their environment, bringing promise to a new era of precision and dependability for sensing technologies with the potential to illuminate new understanding and more effectively control intricate systems across a wide range of disciplines. Furthermore, innovations in intelligent packaging systems provide a fascinating test case of the manner in which advanced sensing technologies can be used to advance food quality and safety management to unprecedented levels. Smart packaging, with advanced sensors, is employed especially to monitor food conditions and transfer vital information throughout the supply chain.

Figure 3.15. As the demand for smarter and more efficient products intensifies, the convergence of materials science and engineering has become essential in developing advanced sensor technologies.

Source: https://www.thedigitalspeaker.com/content/images/2023/04/materials-revolution-futurist-1.jpg.

These kinds of systems can detect spoilage signs such as temperature changes and gas emissions, alerting consumers as well as supply chain managers to potential safety hazards. By providing real-time feedback, these technologies not only increase the rate of pathogen detection but also greatly improve food quality assurance processes. With the growing demands of consumers for fresher and safer products, the use of sensor technologies to mitigate packaging issues will continue to rise, in the long run strengthening food safety infrastructure that bolsters the health of the public and bolsters the reduction of waste throughout the food supply chain (Carrasco et al., 2016). The motivating force for sensor material innovation is also supported by the collaboration of designers and scientists in the STEM field, who are increasingly aligning their objectives towards addressing some of the world's most pressing issues. As the world demands smarter and more efficient products, the crossing of material science and engineering has become essential in the development of new sensor technologies. These collaborative environments enable designers to incorporate cutting-edge materials with engineered attributes into product designs, thereby providing an environment where innovation can thrive. With understanding and implementation of the new properties these advanced materials offer, designers can develop sensing mechanisms that are not only cost-effective but also sustainable and resource-efficient. This design-led strategy for STEM innovation is a central method of dealing with the complexities of the 21st century, guiding sensing technologies' development for future purposes (Kapsali et al., 2014) (Figure 3.15).

3.7. CONCLUSION

In reflecting on the advances in sensing mechanisms, it is pertinent to note their applicability towards radical transformation in various industries. With cognitive radio technologies to soft robots, the merge of new ways of sensing has demonstrated enormous capability in enhancing operational effectiveness and flexibility. For instance, designing soft robotic grippers has been a groundbreaking development that enables unstructured objects to be gripped at unprecedented speed and precision. Not only does this technology embody the integration of new materials and design principles, but it also indicates the utility of passive mechanisms that are able to respond rapidly without the need for complex sensing systems (Abidi et al., 2019). As these advances are increasingly revealed, their influence on sectors from manufacturing to health will increasingly become more significant, solidifying more the need to invest in research and development of advanced sensing technologies. Furthermore, the research on cognitive radio networks emphasizes the need for advance sensing mechanism in solving problems of scarcity in the spectrum. By employing dynamic spectrum access and the application of artificial intelligence principles, cognitive radio technology has dominated communication in modern times. Effective spectrum sharing through advanced sensing eliminates harmful effects of traditional wireless networks, effectively utilizing available channels. However, these advances are accompanied by new security concerns unique to cognitive radio, such as vulnerability to licensed user emulation attacks and threats to shared control channel operations (Alhakami et al., 2014). Thus, whereas the promise of cognitive radio lies in its innovative application of sensing instruments, it also needs a strong framework to navigate the complexities of security in favor of protection of communication integrity. In short, sophisticated sensing mechanisms are not a development of existing technology but an evolution to more intelligent and responsive ones. As evidenced in soft robotics and cognitive radio, such mechanisms demonstrate the intersection of engineering, design, and computer science that enables applications that were previously bounded by technology. In the future, cross-disciplinary collaboration needs to see that these are addressed, particularly security and efficiency issues. Ultimately, sensing technology holds the potential not only for greater performance but also for a revolutionary capacity to reshape standards of operation across industries. As work in this field continues to advance, maintaining dual focus on innovation and security will be essential to realizing the full capabilities of next-generation sensing mechanisms.

3.7.1. Summary of Key Points Discussed

In the quest for advanced sensing mechanisms, much attention has been directed towards the advancement of technology in industrial usage. One example is how the establishment of the Cargo Transfer Vehicle (CTV) made Automated Rendezvous and Capture (AR&C) systems a necessary focus area, pointing to their effect in maximizing spacecraft operations. The research was a result of a comprehensive review in 1991, which brought together stakeholders from various government agencies, industries, and educational institutions. The large number of participations, as attested by over 100 abstract submissions, speaks to the growing interest in optimizing hardware systems, software programs, integrated operations, and supporting infrastructures. Systematic categorization of presentations emphasizes a cooperative thrust to create technology in critical areas, and one can see how earlier addresses have paved the way for current developments in automated systems as well as robot systems in aerospace and other industries. Also, while industries such as smart grid technology begin to emerge, there is yet another layer to sensing mechanisms. Smart grids utilize sophisticated communication technologies to enhance the security and reliability of energy, necessitating robust infrastructures for system management and information exchange. The advanced technologies highlight the core significance of advanced sensing abilities in measuring and regulating energy flows in order to facilitate efficient energy management. Digital systems also generate significant cyber security threats, whereby flaws can lead to operational disruptions and heightened risks. An overall evaluation of smart grid communications' security architectures thus becomes paramount in the provision of the efficiency and resilience of these systems. Offering protection to these systems against cyber threats is paramount for the consumer's trust and energy utility's honesty (Qian et al., 2012). Such observations and insights enable significant background information in the direction and future of advanced sensing mechanisms. The use of secure communication technologies is important not only for effectiveness but also for ensuring safety in various applications, including aerospace and energy. In the future, knowledge of existing research and findings assists in identifying gaps and potential areas to investigate further. The emphasis on collaboration among various stakeholders—from business executives to researchers—reflects the necessity for collaborative innovation in order to fight threats posed by infrastructure weaknesses and cyber-attacks. The convergence of these dialogues illustrates the necessity to keep advancing the boundaries of sensing technologies while fighting the complexities of security in order to build a robust technological platform across multiple domains (Qian et al., 2012).

3.7.2. The Significance of Continued Research in Sensing

The ongoing innovation in sensing technology supports sustaining research and development. Developments in manufacturing processes and materials create enhancements in sensor performance, enabling the development of sensors that can measure many physical, chemical, and biological stimuli with accuracy. As more industries rely on data-driven decision-making, optimization of such sensing mechanisms becomes a priority. Higher sensitivity, selectivity, and rapid response rates are important for application in environmental monitoring, diagnostics in healthcare, and industrial automation. Research activities on-going among sensor materials ensure that the devices can be able to meet the rising demands for precision and reliability, and this has resulted in better standards in industries across the board. The ongoing material and structural research for sensing purposes not only addresses present issues but also creates pathways towards ground-breaking innovations that are capable of producing far-reaching societal as well as economic impacts (Deepak et al., 2025). Beyond the material achievements, the combination with sophisticated computational approaches also plays an important role in boosting sensing system performance. The advent of Machine Learning (ML) and Artificial Intelligence (AI) technologies has revolutionized the processing and interpretation of sensor-sensed data. Utilizing such computing tools, scientists can optimize the performance of sensors along with the accuracy of interpretations obtained from sensed data. The use of frameworks such as the Mixture of Experts (MoE) within wireless networks is an example of how sophisticated algorithms can enable efficient resource management as well as enhanced scalability. This convergence of computing intelligence and sensing technologies upsurges the capabilities of sensors in dynamical environments, recommending that ongoing research is essential in guaranteeing unprecedented responsiveness and sensitivity in cutting-edge sensing mechanisms (Xu et al., 2025). Aside from technological advancement, the significance of ongoing research in sensing also extends to its ethical implications and societal impact. With more pervasive sensors, issues of privacy, data protection, and user consent come to the forefront. Researchers need to engage in discussion regarding the ethical use of sensing technology so that they contribute to the public good without encroaching on people's rights. Technical research should also align with these ethics-related issues to devise frameworks for responsibly using advanced sensing mechanisms. By prioritizing the ethical elements of sensor innovation, ongoing research can inform better policy and regulations that direct how these technologies are employed, making a society that can benefit from innovation while preserving fundamental human values.

3.7.3. Future Implications of Advanced Sensing Mechanisms

The evolution of advanced sensing mechanisms has huge potential to transform numerous industries, particularly aerospace and environmental monitoring. As these technologies advance, their applications transcend the traditional limits. Strong sensing enables real-time data collection and analysis, thereby an upgrade in decision-making in critical areas. For instance, the integration of sophisticated sensors in aerial activities allows for advanced monitoring of atmospheric conditions and performance of spacecraft, which can improve safety standards and mission effectiveness. This is facilitated by research that emphasizes the advantages of real-time data in flight crew behavior and performance, central elements for successful space missions. With the creation of these technologies, we anticipate that their integration into various systems will offer new avenues of exploration and discovery. For environmental science, new mechanisms for sensing will have a revolutionary effect on monitoring climate change and natural resource management. Such mechanisms offer unprecedented capability to monitor infinitesimal changes in environmental parameters, sharpening predictive models further. For example, with the help of sophisticated sensors, researchers can monitor health in an ecosystem and respond rapidly to changes predictive of climate stress. The potential for cross-disciplinary research with advanced sensing is great, as illustrated in projects sponsored by agencies like NASA. The consequences for observation and information systems can enable extreme insight into earth and extraterrestrial phenomena. This is evidenced by the initiatives that enhance integration of emerging technology into real-world applications, boosting inter-disciplinary research. Additionally, the future possibility for advanced sensing mechanisms is in direct correlation with the development in artificial intelligence and machine learning. These technologies can potentially process large amounts of data gathered using sensors to determine patterns and trends in order to inform decisions across disciplines. The ability of AI algorithms to read and process complex data sets enhances the performance and trustworthiness of sensing mechanisms, which will be incredibly valuable in autonomous vehicle applications up through smart cities. As researchers continue to enlighten us on how advanced sense technologies operate under dynamic conditions, we are likely to witness extreme transformations in the way these technologies disrupt societal challenges. The marriage of advanced sensors and AI will pave the way for the next generation of smart systems that behave in a reactive manner in response to real-world situations, totally redefining the technology environment and its applications.

REFERENCES

1. Abidi, S. H., Cianchetti, M., Falotico, E., Laschi, C., et al. (2019). A bistable soft gripper with mechanically embedded sensing and actuation for fast closed-loop grasping. https://core.ac.uk/download/543788444.pdf (accessed on 14th October 2025).
2. Adamenko, O. (2025). Foreign experience in digitalization of agricultural production and directions of its use in domestic practice. *Ekonomìka ta upravlìnnâ APK*. https://www.semanticscholar.org/paper/3ca85d0bbfcd45853c4c73b73a6833cb96fac2ba (accessed on 14th October 2025).
3. Alhakami, W., Mansour, A., & Safdar, G. A. (2014). Spectrum sharing security and attacks in CRNs: A review. https://core.ac.uk/download/29822044.pdf (accessed on 14th October 2025).
4. Baets, R., Jia, X., Roelkens, G., Roels, J., et al. (2019). On-chip non-dispersive infrared CO_2 sensor based on an integrating cylinder. https://core.ac.uk/download/237011556.pdf (accessed on 14th October 2025).
5. Baj-Rossi, C., Burleson, W., Carrara, S., Cavallini, A., et al. (2012). Fully integrated biochip platforms for advanced healthcare. https://core.ac.uk/download/286729684.pdf (accessed on 14th October 2025).
6. Bassegoda Puigdomenech, A., Ivanova, K. D., Portés, E. R., Tzanov, T., et al. (2018). Strategies to prevent the occurrence of resistance against antibiotics by using advanced materials. https://core.ac.uk/download/157810784.pdf (accessed on 14th October 2025).
7. Berre, A. J., Havlik, D., Mon, J. L., Mazzetti, P., et al. (2011). From sensor to observation web with environmental enablers in the future internet. https://core.ac.uk/download/8653076.pdf (accessed on 14th October 2025).
8. Bhargava, V. K., Boostanimehr, H., & Hasan, Z. (2011). Green cellular networks: A survey, some research issues and challenges. http://arxiv.org/abs/1108.5493 (accessed on 14th October 2025).
9. Briciu Burghina, C. C., Gualano, L., Heery, B., O'Connor, A., et al. (2013). Improving data driven decision making through integration of environmental sensing technologies. https://core.ac.uk/download/11311853.pdf (accessed on 14th October 2025).
10. Carrasco, R., Fuertes, G., Lagos, C., Sabattin, J., et al. (2016). Intelligent packaging systems: Sensors and nano sensors to monitor food quality and safety. https://core.ac.uk/download/288914520.pdf (accessed on 14th October 2025).
11. Cassellas, R., Cugini, F., D'Errico, A., Gerstel, O., et al. (2013). IDEALIST

control and service management solutions for dynamic and adaptive flexi-grid DWDM networks. https://core.ac.uk/download/17203288.pdf (accessed on 14th October 2025).

12. Clark, A., Dugas, A. F., Westphal, A. J., Woolley, A. W., Resch, B., Cardamone, C., Martin, C. C., et al. (2015). Human computation and convergence. http://arxiv.org/abs/1503.05959 (accessed on 14th October 2025).

13. Dang, T. B., Truong, T. A., Nguyen, C. C., Listyawan, M. A., Sapers, J. S., Zhao, S., Truong, D., et al. (2025). Flexible, wearable mechano-acoustic sensors for body sound monitoring applications. *Nanoscale*. https://www.semanticscholar.org/paper/58e21ea142ad4158ef8e893d5252248876b6c81a (accessed on 14th October 2025).

14. Estes, J. E., Smith, T., & Star, J. L. (2025). Remote sensing information sciences research group, Santa Barbara Information Sciences Research Group, year 3. https://core.ac.uk/download/pdf/42839628.pdf (accessed on 14th October 2025).

15. Ewart, P. (2019). A comparison of processing techniques for producing prototype injection moulding inserts. https://core.ac.uk/download/322841895.pdf (accessed on 14th October 2025).

16. Farahbakhsh, R., Farahzadia, A., Rezazadeh, J., Shams, Z., et al. (2017). Middleware technologies for cloud of things–A survey. http://arxiv.org/abs/1705.00387 (accessed on 14th October 2025).

17. Fichtner, M., Edström, K., Ayerbe, E., Berecibar, M., Bhowmik, A., Castelli, I. E., Clark, S., et al. (2021). Rechargeable batteries of the future—The state of the art from a BATTERY 2030+ perspective. *Advanced Energy Materials, 12*. https://doi.org/10.1002/aenm.202102904 (accessed on 14th October 2025).

18. Firestone, E. R., & Hooker, S. B. (2025). *SeaWiFS technical report series. Volume 6: SeaWiFS Technical Report Series Cumulative Index: Volumes 1–5*. https://core.ac.uk/download/pdf/42806235.pdf (accessed on 14th October 2025).

19. Firschein, O., Georgeff, M. P., Kautz, W. H., Levitt, T. S., et al. (2025). NASA space station automation: AI-based technology review. https://core.ac.uk/download/pdf/42844682.pdf (accessed on 14th October 2025).

20. Gill, E., Monna, G. L. E., Scherpen, J. M. A., et al. (2007). MISAT: Designing a series of powerful small satellites based upon micro systems technology. https://core.ac.uk/download/pdf/12941826.pdf (accessed on 14th October 2025).

21. Harris, C. R., Millman, K. J., van der Walt, S., Gommers, R., Virtanen, P., Cournapeau, D., Wieser, E., et al. (2020). Array programming with NumPy. *Nature, 585*, 357–362. https://doi.org/10.1038/s41586-020-2649-2 (accessed on 14th October 2025).

22. Jappe-Heinze, A. (2025). Explaining international co-authorship in global environmental change research. https://core.ac.uk/download/pdf/6710939.pdf (accessed on 14th October 2025).

23. Kamble, D. V., Patil, B., Patil, A. A., Metakari, A. D., Chonde, S. G., Makandar, G. A., Koli, V., et al. (2025). A comprehensive review on sensor materials, sensing mechanism and their applications. *Asian Journal of Chemistry*. https://www.semanticscholar.org/paper/e053f47adf6bd77d468b19eb208bd55aa57c99b7 (accessed on 14th October 2025).

24. Kapsali, V., & Toomey, A. (2014). D-STEM: A design-led approach to STEM innovation. https://core.ac.uk/download/41073312.pdf (accessed on 14th October 2025).

25. Ko, S. H., & Rogers, J. A. (2021). Functional materials and devices for XR (VR/AR/MR) applications. *Advanced Functional Materials, 31*. https://doi.org/10.1002/adfm.202106546 (accessed on 14th October 2025).

26. Koo, B. J. (2014). A 2 degree-of-freedom SOI-MEMS translation stage with closed loop positioning. https://core.ac.uk/download/29175019.pdf (accessed on 14th October 2025).

27. Kopylov, G. (2015). Overview of sensors suitable for active flow control methods. https://core.ac.uk/download/30294554.pdf (accessed on 14th October 2025).

28. Lin, S. L. (2025). Advancements in noise reduction for wheel speed sensing using enhanced LSTM models. *Scientific Reports, 15*. https://www.semanticscholar.org/paper/5e5508532d940c79d6e99b7e44a50e3da2806ab6 (accessed on 14th October 2025).

29. Luo, Y., Abidian, M. R., Ahn, J. H., Akinwande, D., Andrews, A. M., Antonietti, M., Bao, Z., et al. (2023). Technology roadmap for flexible sensors. *ACS Nano, 17*, 5211–5295. https://doi.org/10.1021/acsnano.2c12606 (accessed on 14th October 2025).

30. Malini, S., Roy, A., Singh Raj, K., Kumar, R., Ali, I. H., Mahesh, B., Yadav, K. K., et al. (2022). Sensing beyond senses: An overview of outstanding strides in architecting nanopolymer-enabled sensors for biomedical applications. *Polymers, 14*, 601–601. https://doi.org/10.3390/polym14030601 (accessed on 14th October 2025).

31. Mottola, L., & Picco, G. P. (2011). Middleware for wireless sensor

networks: An outlook. https://core.ac.uk/download/11435219.pdf (accessed on 14th October 2025).

32. N/A. (2016). *New Hampshire University Research and Industry Plan: A Roadmap for Collaboration and Innovation.* https://core.ac.uk/download/215517669.pdf (accessed on 14th October 2025).
33. Oliver, R., Tillotson, J., & Toomey, A. (2009). Bioengineered textiles and nonwovens – The convergence of bio-miniaturization and electroactive conductive polymers for assistive healthcare, portable power and design-led wearable technology. https://core.ac.uk/download/9587153.pdf (accessed on 14th October 2025).
34. Puppala, A. (2025). The role of digital twins in AI-driven enterprise BI: Transforming scenario simulation and strategic planning. *European Journal of Computer Science and Information Technology*. https://www.semanticscholar.org/paper/62d91e372372ff2485b664295494fa0bcaeaa163 (accessed on 14th October 2025).
35. Qian, Y., Sharif, H., Tipper, D., Yan, Y., et al. (2012). A survey on cyber security for smart grid communications. https://core.ac.uk/download/12211131.pdf (accessed on 14th October 2025).
36. Rahimi, M., Yaghoobzadeh, Y., Daliri, M. R., et al. (2025). Explanations of large language models explain language representations in the brain. https://www.semanticscholar.org/paper/847238f439aeb46284f72dc17ef2645181c924ee (accessed on 14th October 2025).
37. Shi, S., & Zong, Q. (2025). Benchmarking machine learning methods for distributed acoustic sensing. https://www.semanticscholar.org/paper/fdf2f3bd3c02b6ba356f280402c0409f3018b141 (accessed on 14th October 2025).
38. Su, Y., Yin, D., Zhao, X., Hu, T., & Liu, L. (2025). Exploration of advanced applications of triboelectric nanogenerator-based self-powered sensors in the era of artificial intelligence. *Sensors (Basel, Switzerland), 25.* https://www.semanticscholar.org/paper/f16abd231c6342d72c06e076c6cb44aef4659458 (accessed on 14th October 2025).
39. Villarreal, R., Chamorro-Solano, S., Vega-Sampayo, Y., Espejo, C. A., Cantillo, S., Gaviria, L., Paez, J., et al. (2025). A new approach to electrical fault detection in urban structures using dynamic programming and optimized support vector machines. *Sensors (Basel, Switzerland), 25.* https://www.semanticscholar.org/paper/e827480ac18072a8acd0d6b10d5b9d76f2cb02be (accessed on 14th October 2025).

40. Xiong, Y., Han, J., Wang, Y., Wang, Z. L., & Sun, Q. (2022). Emerging inotropic sensing: Materials, mechanisms, and applications. *Research, 2022*. https://doi.org/10.34133/2022/9867378 (accessed on 14th October 2025).

41. Xu, Y., Wang, J., Zhang, R., Zhao, C., Niyato, D., Kang, J., Xiong, Z., et al. (2025). Decentralization of generative AI via mixture of experts for wireless networks: A comprehensive survey. https://www.semanticscholar.org/paper/66af0b1428acecb817c577814032dd9406ea0c0f (accessed on 14th October 2025).

42. Zhang, G., Chen, Z., & Wang, L. (2025). Integrated shape and force sensing for continuum instruments via polynomial curvature modeling and torque-cell-based actuation. https://www.semanticscholar.org/paper/15debf9036c8463f49ceacf62fe49b81ef907eb4 (accessed on 14th October 2025).

43. Zhuang, Y., & Xie, R. J. (2021). Mechan luminescence rebrightening the prospects of stress sensing: A review. *Advanced Materials, 33*. https://doi.org/10.1002/adma.202005925 (accessed on 14th October 2025).

CHAPTER 4

ADVANCED ACTUATION MECHANISMS

Contents

INTRODUCTION

Actuation mechanisms' field has experienced a remarkable overhaul, professionally transforming numerous engineering disciplines, most importantly robotics and automation. Advances in technology have prompted researchers to develop more sophisticated control mechanisms that allow for more accuracy and functionality. These advances have seen high-level actuation mechanisms introduced, which not only maximize the efficiency of operations but also save energy. This chapter also discusses various types of actuation mechanisms, and the achievements are innovations such as MEMS (Micro-Electro-Mechanical Systems), where the mechanical parts are integrated on a microscopic scale. In general, these advancements reflect the crucial role of advanced actuation mechanisms in enhancing the capabilities of engineering and meeting the demands of modern applications. Of recent, integrating actuation mechanisms into photonic devices has gained a great deal of interest, synthesizing the mechanical and optical engineering in a highly intriguing merger. One good example is integrating MEMS-based components into silicon photonics systems, which offers several functionalities to devices along with low power consumption. This new approach leverages mechanical motion of wave-guiding structures and enables new paths for enhanced performance in photonic circuits (Bogaerts et al., 2020). These advances are especially significant in data communication technology, where efficiency and speed are critical. The integration of photonic systems and actuation mechanisms reflects the benefit that can be realized with interdisciplinary engineering and research strategies whose contributions to making newer and more sophisticated technologies are making worthwhile progress on a continual basis. Additionally, the creation and application of advanced actuation mechanisms also necessitate the need for redundancy and control protocols to ensure greater performance. By employing flexible mechanisms, i.e., those maximized with backup systems, researchers aim to eliminate issues like misalignments and production defects to deliver better positional precision. For instance, piezoelectric actuators' research in closed-loop systems offers a robust approach to position control that allows for precise path tracking regardless of disturbances (Acer et al., 2012). This is particularly significant in applications ranging from industrial automation to medical devices, where precision is crucial. Consequently, ongoing investigation into the actuation mechanism subtleties brings to the fore the significant role of the mechanism in enhancing both technology and application across different industries.

Definition of Actuation Mechanisms

The term actuation mechanisms is a term referring to the mechanisms employed to convert energy into motion, which enables mechanical systems to move. Actuation mechanisms are vital in numerous applications throughout various industries, from robotics and aerospace to biomedical implants. Actuation mechanisms are classified based on their energy source, i.e., electrical, pneumatic, or hydraulic systems, which differ in pros and cons. For instance, electric actuators are appreciated for their control and precision, while pneumatic and hydraulic systems are appreciated for the significantly larger force output, they deliver. An understanding of how these mechanisms work is crucial to enhancing the design and functioning of complex systems so that engineers and scientists can produce solutions that are meant for a specific job. As such, current research increasingly seeks to leverage these mechanisms for their power to optimize performance in dynamic and uncertain systems. Technology advances in the actuators have also inspired the development of soft actuators that resemble biological systems.

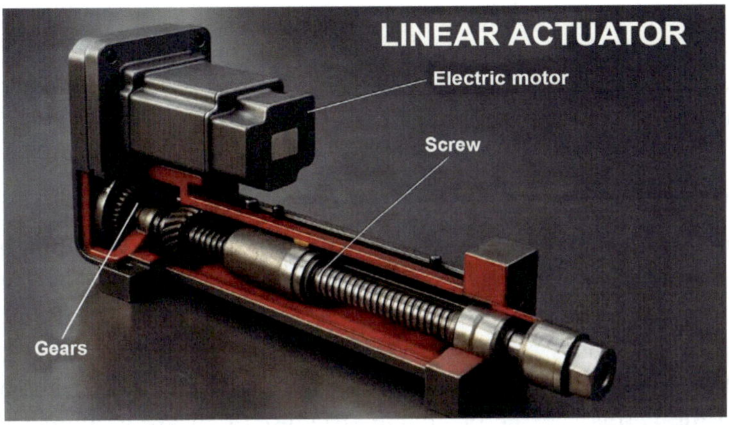

Figure 4.1. Actuation mechanisms can be classified based on their energy source, such as electrical, pneumatic, or hydraulic systems, each offering distinct advantages and limitations.

Source: https://www.mechdaily.com/wp-content/uploads/2025/05/what-is-Linear-Actuators.webp.

The new mechanisms utilize materials that capture the flexibility of soft-bodied organisms in an effort to enable complex motion and functionality in hostile environments. For example, soft robotic crawlers are designed to capture the locomotion behaviors of snakes and caterpillars through actuation mechanisms such as peristaltic and serpentine crawling. Such bio-inspired designs give clues to efficient actuation principles that hard conventional systems are unable to copy. By understanding the intricate interaction of motion, actuation mechanisms, and materials, researchers are unlocking the path to highly functional soft robots that can travel across a range of terrains. This research is a paradigm shift towards intermarriage between advanced biomechanics and material science, which underlies the imminent innovation in actuation mechanisms (Pan et al., 2025). In addition, the versatility of advanced actuation mechanisms is enhanced by using reconfigurable structures, for example, gold nanostructures (AuNSs). In comparison to fixed structures, reconfigurable systems have the potential to dynamically adjust to environmental conditions, representing the type of flexibility required in modern applications. By acquiring stimuli such as light, temperature, and chemical changes, these nanostructures can reshape their structure, providing critical feedback and control in systems ranging from biomedical use to materials science. Their functional efficacy relies greatly on the structure and design of the structures, making it possible to tailor responsiveness and enhance function. Such reconfigurable mechanisms not only provide the door to higher potential for innovation in sensing technology but also indicate the convergence of actuation design and functional application, favorably placing the field for further work on reactivity and adaptability in engineering solutions (Choi et al., 2025) (Figure 4.1).

Importance in Modern Technology

The role of advanced actuation mechanisms in contemporary technology cannot be overstated. The mechanisms provide the foundation for a wide variety of applications, particularly in aerospace, robotics, and biomedical engineering. The integration of light-weight structures within these systems has revolutionized design and fabrication procedures. For instance, light materials originally created in the aerospace sector enable structures that are as light as possible and as strong as possible, thereby improving performance and fuel efficiency. Developments based on sophisticated actuation mechanisms also enable the creation of intricate designs such as deployable and morphing structures crucial in enabling real-time applications in most industries. The integration of actuation technology with advanced materials enhances the operational functionality and is also primarily responsible for achieving sustainability and efficiency in modern engineering systems, as exhibited by the ongoing study of lightweight design (Hogr et al., 2024). The importance of using these actuation technologies is also evident in robotics. Advanced actuation systems incorporating smart sensors and embedded technology provide an increased degree of precision and responsiveness for robot motion. Such adaptability is necessary for applications ranging from manufacturing automation to medical robotics, where accuracy is of greatest concern. By integrating advanced mechanisms, such as those using shape memory alloys and innovative jointing techniques, robots can perform complex tasks with increased agility and lower energy consumption. These advancements reveal the importance of actuation mechanisms in driving the advancement of robotics into more intelligent and responsive applications, eventually advancing the technological landscape. With robotics advancing, the necessity for more advanced actuation through innovative mechanisms becomes increasingly evident, further ensuring their importance as a key component of modern technological advancement. Advanced actuation mechanisms achieve their importance within the biomedical field through sophisticated medical devices and treatments. For instance, the application of light structures maximizes prosthetic and biomechanical device performance, enhancing the comfort and mobility of their users. New actuator design enables smart systems to react in an optimal way to the human body, creating possibilities for customized medicine. While new treatments emerge, disease-free survival durations are extended—especially in complex diseases such as cancer—these mechanisms play a critical role in therapeutic delivery systems, high efficacy, and low invasiveness (Anand et al., 2022). As technology continues to evolve, the integration of emerging actuation mechanisms into modern medical technology will have the power to transform healthcare delivery, improve patient outcomes, and ultimately redefine medical technology boundaries. The ongoing discovery of these mechanisms will not only confirm their current applications but also drive future innovation in numerous fields (Figure 4.2).

Figure 4.2. Advanced actuation systems equipped with smart sensors and embedded technology facilitate a higher degree of precision and responsiveness in robotic movements.

Source: https://www.mikroe.com/img/images/How%20Robotics%20Is%20Evolving%20with%20Modular%20Embedded%20Hardware%20-%202.jpg.

Overview of Advanced Actuation Mechanisms

Sophisticated actuation mechanisms are instrumental in determining modern automated systems in various industries, including robotics, aerospace, and manufacturing. Sophisticated actuation mechanisms provide the means to convert various types of energy into mechanical motion and allow for precise movements essential for the operation of complex systems. Recent advancements in material science and engineering have provided actuators not just efficient but also energy-efficient. For example, integrating smart materials such as shape memory alloys and piezoelectric materials enables the creation of small and sensitive actuators that can operate in confined spaces. As the trend for automation progresses, understanding these advanced mechanisms is central to designing responsive systems that can function under unstructured environments, ultimately enhancing the performance and autonomy of machines. The advent of digital twinning technology is a break-through for the development and application of advanced actuation mechanisms. Digital twins provide a virtual representation of physical models, enabling real-time tracking, optimization, and prediction of actuator behavior. Employing advanced computational methods, designers can simulate varying operating conditions to ensure that actuation mechanisms are able to withstand real-world scenarios. Additionally, the technology facilitates stakeholders to engage in predictive maintenance, reducing downtime and extending the lifespan of mechanical systems. Digital twinning has broader implications than efficiency; it allows the development of deeper insights into inter-system behavior. This enhanced capability promotes the creation of actuators not just efficient but also synchronized into larger automated systems, responding to complex operating requirements and system resiliency as a whole (Rasheed et al., 2020). Furthermore, the development of sixth-generation (6G) technology is going to influence the actuation mechanism future more significantly. The broad scope capability of 6G, including ultra-reliable low-latency communication and pervasive artificial intelligence, will most likely drive the next generation of intelligent actuation systems. Such a high-level communication network will facilitate seamless interaction among smart devices, resulting in more advanced and coordinated actions. Moreover, the coexistence of ambient backscatter communications with other new methods will allow actuators to work optimally even in power-starved conditions. With enhanced smart system inclusion in our lives, there is a predicted profound transformation in the operation of actuation mechanisms. The marriage of advanced actuation with new communication technologies, therefore, paves the way for the creation of more sophisticated and networked automation solutions, which advance the effectiveness of modern systems significantly (Ian et al., 2020).

4.1. TYPES OF ACTUATION MECHANISMS

Actuation mechanisms are investigated to discover a diversified list of technologies, every one of them suited to specific applications and performance requirements. Piezoelectric actuation is perhaps one of the most important groups, where materials induce mechanical motion through a reaction to an electric field. Piezoelectric actuation is particularly suited for precision applications since it can carry out high-resolution displacements at a high frequency. In systems of application, redundancy in piezoelectric actuator systems enhances the performance of control by minimizing systematic errors such as parasitic motions and actuator misalignments. The performance has been proven through research by such systems with the enhanced performance facilitated by the redundancies and their incorporation into advanced mechanisms for the purpose of accurate positional control (Acer et al., 2012). This renders piezoelectric actuation an advanced mechanism with the ability to tackle the stringent demands of modern engineering applications. Soft actuators, with ionomeric polymer-metal composites (IPMCs) at their forefront, represent another extremely critical class of actuation mechanisms.

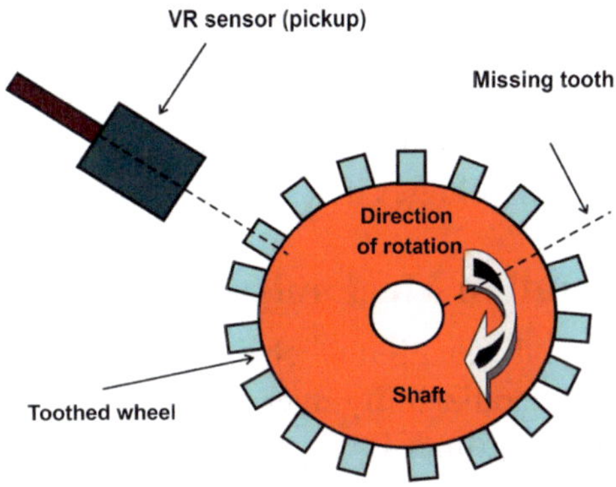

Figure 4.3. In practical implementations, the redundancy in piezoelectric actuator systems enhances control capabilities by minimizing systematic errors, such as parasitic motions and actuator misalignments.

Source: https://www.mdpi.com/sensors/sensors-10-06901/article_deploy/html/images/sensors-10-06901f3-1024.png.

These new actuators are defined by their remarkable ability to produce large displacements using low voltage, often requiring as little as approximately 1 V. Their flexural ability instead of translating through rigid motions holds a special strength in situations where precision of handling and flexural ability are important, i.e., in biomedical instruments and robotics. New advances in freeform fabrication have enabled one to create IPMCs with novel geometries, which permit more complex motion profiles than mere bending. These developments not only increase the functional capabilities of IPMCs but also bring their integration with other actuating systems, promoting more advanced and flexible technologies to be created (Lipson et al., 2005). This technological advancement in design and fabrication emphasizes the necessity to research advanced actuation methods in the areas of engineering and applied sciences. Other than piezoelectric and soft actuators, electromagnetic actuation mechanisms are often employed as well, particularly in environments where high performance and rapid response speeds are vital. Electromagnetic actuators involve electric energy conversion to mechanical energy through the aid of magnetic fields, widely utilized in devices such as electric motors and solenoids. These devices are praised for their simplicity in being constructed and relatively high reliability, and they represent a prevalent component in industrial automation and robotics alike. New technologies have been oriented to achieve maximum efficiency and controllability of electromagnetic systems, like the incorporation of digital control methods. Such incorporation makes motion control more accurate with the possibility of exercising adaptive behaviors suitable for various working conditions. As the technology of actuation continues to advance, interaction between various mechanisms, including electromagnetic systems, will definitely lead to highly advanced, efficient, and effective technologies in a highly automated world (Figure 4.3).

4.1.1. Electric Actuation Mechanisms

Amongst the sophisticated actuation mechanisms, electric actuation stands out for its efficiency and generalizability over a very broad range of applications. Electric actuation utilizes primarily electromagnetic forces to induce motion and is therefore an inherent field in applications ranging from robotics to medical devices. Employing electric signals to control the actuation process, the systems can yield controlled motion tailored for a specific task. The primary advantages of electric actuators include high response speed and low mechanical wear, leading to longer life during operation. Moreover, the use of intelligent materials, such as piezoelectric materials, enhances the ability of electric actuators in generating rich tactile feedback over simple motion. These developments in actuation technologies enable actuator engineers to design more sophisticated and responsive systems to respond to the evolving demands of modern applications, as described in ongoing work on tactile actuation mechanisms (Zhang et al., 2017). Among the new electric actuation systems, ionomeric polymer-metal composite (IPMC) actuators have been of particular interest due to their ability to achieve large displacements at low voltage. They are the current state-of-the-art of soft actuation systems, operating successfully with only ~1 V stimulus. This characteristic makes IPMC actuators particularly suitable for application in the biomedical field, where they can function in delicate environments without introducing additional risk. Moreover, past limitations, e.g., their actuation by bending, are now being addressed by advancements in freeform fabrication techniques. These methods allow the design of IPMC actuators with different geometries, facilitating linear and complex motion types. As a result, IPMC system functionalities and applications are expanding, providing new frontiers in robotics and medical devices (Lipson et al., 2005). The finding of electric actuation mechanisms not only enhances the convenience of devices but also assists the growing trend towards miniaturization and energy efficiency in technology. As the engineers continue to develop new concepts, integrating actuators with newer materials and fabrication techniques will be the center of developing electric actuation systems. Energy efficiency is highly sought after since there is an increasing demand for sustainable engineering solutions. As equipment continues to become more connected and complex, the integration of electric actuation mechanisms promises to revolutionize performance at greatly diminished power consumption. In the end, ongoing investigation and improvement of electric actuation mechanisms will lead to the further development of countless applications, setting the stage for responsive and sustainable technologies to come (Figure 4.4).

Figure 4.4. By employing electric signals to control the actuation process, these systems can achieve precise movements tailored for specific tasks.

Source: https://www.adatech.com.tr/wp-content/uploads/2025/01/actuator-scaled.webp.

4.1.2. Hydraulic Actuation Mechanisms

The role of hydraulic actuation systems in advanced systems is pivotal, providing unmatched force and precision to a wide range of applications. Hydraulic actuators convert hydraulic energy into mechanical energy, an aspect that has a unique benefit in environments where high strength-to-weight ratios are required. A hydraulic system's feature is the simplicity with which it can increase force, enabling it to transfer heavy loads using minimal input effort. This characteristic is most valuable in aerospace and industrial machinery applications, where reliability and efficiency are the highest priorities. In addition, hydraulic technologies continue to advance their performance, enabling them to be more flexible and responsive in actuation operations. As more industries bet on advanced automation, demand for hydraulic actuation mechanisms continues to be strong, proclaiming their sustained use in advanced actuation technologies. Another significant aspect of hydraulic actuation is its intricate relationship with control systems, which govern the performance and output of hydraulic actuators. The implementation of Prognostic and Health Management (PHM) strategies within hydraulic actuator systems also contributes to their increased efficiency and reliability. Using PHM techniques, the stakeholders are able to achieve advanced features such as real-time monitoring, anomaly detection, and predictive maintenance.

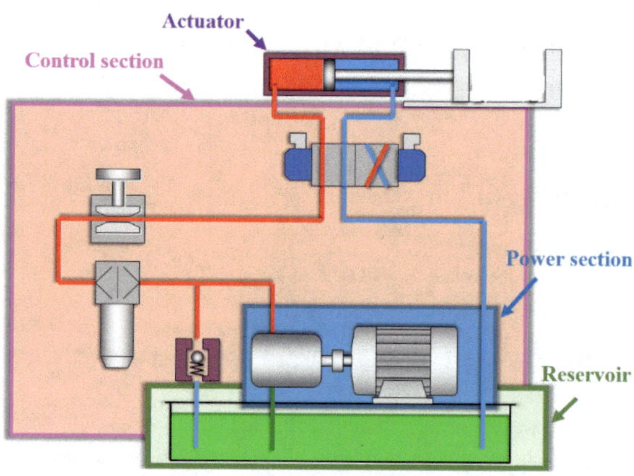

Figure 4.5. Hydraulic actuators operate by converting hydraulic energy into mechanical energy, which is particularly advantageous in contexts requiring high strength-to-weight ratios.

Source: https://rotontek.com/wp-content/uploads/2023/12/image-17.png.

This synergy is especially critical in the case of Electro-Hydraulic Actuators (EHAs), which are critical for safety-critical applications such as aircraft flight control systems. The stark absence of PHM technologies' application even in EHAs, despite the monumental power potential of these systems, reflects that there is still research and development needed to unlock their full potential. Addressing these gaps will further the technology of hydraulic actuation for more performance and reliability across a wide range of industries (Baldo et al., 2025). Furthermore, ongoing developments in hydraulic actuation systems are driving future applications, particularly in conditions demanding flexibility and robustness. Advanced modeling techniques, such as three-dimensional (3D) pore-scale microscopic modeling and two-dimensional (2D) macroscopic modeling, are on the cusp of revolutionizing the way hydraulic systems and porous media and complicated environments interact. Uptake of these techniques is vital not just for fluid dynamics understanding but also for improving system performance and design in hydraulic and coastal engineering applications. Further, mesh-free modeling platforms that integrate smooth particle hydrodynamics (SPH) with discrete element methods (DEM) provide complex simulation of fluid-solid interactions. It is crucial because as the research continues, it should incorporate these emerging technologies into practical applications, hence guaranteeing hydraulic actuators evolve in a way that meets modern-day engineering needs (Luo et al., 2025) (Figure 4.5).

4.1.3. Pneumatic Actuation Mechanisms

Pneumatic actuation mechanisms have profoundly influenced robotics and automation design, particularly in soft robotics. The compressed air powers actuators in pneumatic systems, which has characteristics such as lightweights and inherent safety benefits. Pneumatics contrastingly can easily mimic a variety of shapes and forms, like in biological organisms, compared to traditional rigid actuators. This flexibility is particularly important in applications involving delicate or precise handling, like in prosthetics or rehab equipment. Engineers have come to count on more and more pneumatic systems because they have the ability to offer smooth and natural motion, an ability which can be instrumental in improving user interaction and usability. Moreover, such mechanisms may be combined with smart materials to create responsive actuation according to the requirement of the user, hence optimizing the environment to a more interactive and optimized robotics system. Developments in pneumatic actuation are still following parallel developments in materials science, consequently broadening their fields of application. The combination of soft materials like elastomers with pneumatic systems has resulted in the creation of robots that can smoothly navigate complex environments. They offer greater flexibility and sturdiness compared to traditional rigid robotic systems. As an example, the functionality of biological systems like cephalopods has inspired engineers towards formulating compliant robotic systems with outstanding manipulation with soft bodies. As per the study, bioinspiration from nature has led to a new generation of actuation mechanisms having the capability to store elastic energy, improving their performances in dynamic applications. Consequently, the marriage of intelligent materials and pneumatic actuation positions these mechanisms at the leading edge of advanced actuation research with solutions that were once unimaginable. Besides, the use of sensory feedback in pneumatic actuation systems enhances their performance in action and usability. With the provision of tactile feedback mechanisms, such systems can provide immediate feedback about their environment and the materials they are handling. This is a key aspect required for use in applications that require high precision, like in surgical robots and precision assembly processes. As discussed in recent works, the development of tactile actuators capable of replicating various levels of stimuli, such as pressure and vibration, is an indication of how complex the systems are becoming. Tactile innovations provide more interaction between users and robot devices, enabling more intuitive control mechanisms. Lastly, pneumatic actuation systems, in combination with advanced sensory feedback, are a significant robotics milestone, paving the way for future applications that quietly reduce the disconnect between humans and automated systems, as attested by the ongoing advancements in the technology (Albu-Schaffer et al., 2014; Zhang et al., 2017) (Figure 4.6).

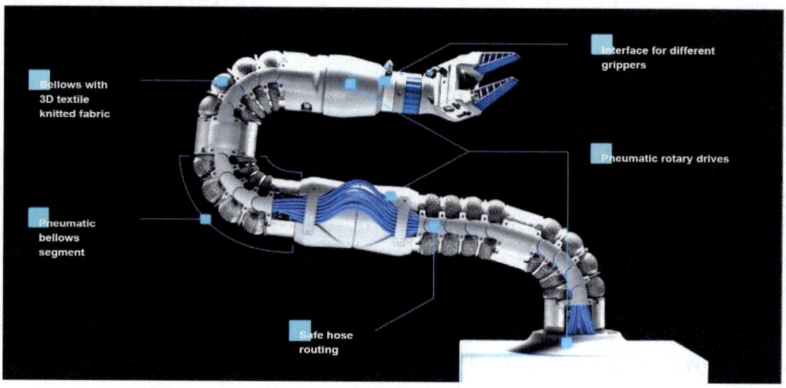

Figure 4.6. The evolution of pneumatic actuation mechanisms has significantly influenced the field of robotics and automation, particularly in the realm of soft robotics.

Source: https://www.mdpi.com/actuators/actuators-11-00092/article_deploy/html/images/actuators-11-00092-g007.png.

4.2. PRINCIPLES OF OPERATION

An in-depth understanding of the basic principles of operation of advanced actuation mechanisms is key to effective application of them in modern engineering. The principles are typically founded on the interaction among physical forces, energy transfer, and control systems that enable the intended motions or behaviors. For instance, in aviation, actuation mechanisms play a fundamental role in aerodynamic control of structures such as wind turbine rotors. UK Department of Energy-funded control system developments have demonstrated that new active control methods outperform traditional pitching systems, with enhanced efficiency and cost advantages (Bowles et al., 1992). This is representative of the overall trend in engineering, where operation principles being optimized can lead to significant improvements in performance, reliability, and sustainability in many applications, including renewable energy technologies. Furthermore, the actuation mechanism laws govern not only rigid bodies but also sophisticated soft materials, such as ionomeric polymer-metal composites (IPMCs). IPMCs' unique electromechanical properties enable them to achieve large displacements and rapid motion through extremely small electrical input, rendering them a driving force within industries ranging from consumer products to bio-medical equipment. The malleability of such materials demands a thorough study of their operational qualities, in particular their bending actuation, which limits their multiple applications (Lipson et al., 2005). Nevertheless, improvements in freeform fabrication techniques can enhance their use by enabling one to create individual geometries and embedded systems. Such innovation opens the vistas for IPMCs, encouraging them to be applied to complex actuation scenarios that involve multifunctional design solutions. Moreover, new actuation principles are being researched that are transforming the area of robotic systems and biomedical instrumentation. Such advanced mechanisms are designed to capture biological motions and capabilities, an important constituent of biologically inspired robot design. By integrating soft actuation and responsive materials, researchers are crafting devices capable of executing subtle and interactive functions, vital for rehabilitation and autonomous surgery. Recent developments in material science, particularly in freeform fabrication techniques, allow for the creation of complex actuators with programmable behavior, which can expand their operational capacities (Lipson et al., 2005). As these new modes of operation continue to be developed, not only do they propel existing technologies, but they also open new directions for investigation and use in a broad spectrum of fields ranging from soft robotics and medicine to many more.

4.2.1. Electromechanical Principles

The use of electromechanical concepts to advanced actuation mechanisms is an important consideration in the development of modern-day engineering applications. The clue to these concepts is the interaction between electrical energy and mechanical motion and can be used to produce accurate control in systems ranging from robotics to biomedical devices. Electromechanical actuation devices employ electrically conductive and mechanically flexible materials to generate a wide variety of actuation response. They include the ionomeric polymer-metal composite (IPMC) actuators, a recently discovered family of soft actuators that are capable of bending large amounts in response to low-voltage electrical stimulation. This property is particularly useful since it will enable the development of small and lightweight actuators to be utilized in complex mechanisms, where standard actuation methods may prove inadequate (Lipson et al., 2005). The comprehension of these concepts is thus significant in order to make innovations and enhance next-generation actuator design. The development of electromechanical systems highly relies on versatile building blocks that can be tuned to achieve high performance. One such method is the FlexIn approach, where evolutionary algorithms are employed to generate micromechanisms from piezoelectric materials in an effective way.

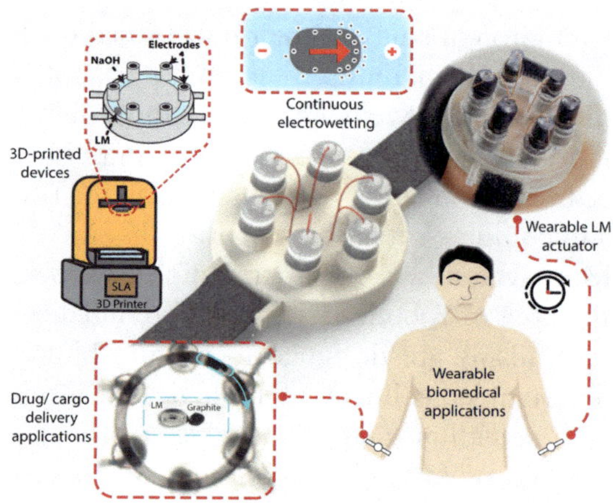

Figure 4.7. Electromechanical systems leverage materials that exhibit both electrical conductivity and mechanical flexibility, allowing for a myriad of actuation responses.

Source: https://media.springernature.com/lw1200/springer-static/image/art%3A10.1007%2Fs44258-025-00052-8/MediaObjects/44258_2025_52_Figa_HTML.png.

This transition from rigid traditional structures to compliant systems makes new configurations discoverable and analyzable that could respond more efficiently to control input. Combining electromechanical models with mechanical parameters in the optimization step enables engineers to obtain new structures that meet actuation and environmental requirements. The resulting planar monolithic compliant micro-actuator serves as the epitome of such principles' synergy, with increased modal controllability and observability that is crucial for applications in future robotics and other adaptronic applications (Boukallel et al., 2009). The adoption of such novel principles hence constitutes a paradigm shift in actuator technology. The implications of electromechanical principles extend to other areas, enhancing and fostering innovation and efficiency in design. For instance, biomedical applications leverage the unique characteristics of electromechanical materials to create devices that mimic biological systems. This bio-inspired design process is not merely a trend but a necessity when it comes to developing adaptive solutions that address complex interactions with living systems. IPMC actuators stand a high chance of succeeding in this direction because they can produce large displacements with minimal energy input, inducing more sustainable engineering. Furthermore, continued development of materials and techniques, such as freeform fabrication, continues to extend the abilities of these devices so that they may be part of working systems that function flawlessly in real-world conditions (Lipson et al., 2005). Coupled with the study of effective design methods, the potential for electromechanical actuation mechanisms is bright, paving the way for smart, effective technologies (Figure 4.7).

4.2.2. Fluid Dynamics in Hydraulic Systems

Fluid dynamics is an extremely crucial field that assists in the comprehension of hydraulic systems, particularly contemporary actuation systems. Fluid systems take advantage of the properties of fluids to transmit power and transport amounts of matter using principles such as Pascals law, which is founded on pressure exerted upon a held fluid and resulting in force conveyed throughout the fluid in all directions. This law allows hydraulic systems to achieve high mechanical advantage with relatively small input forces and is thereby ideal for a variety of applications, ranging from industrial machinery to space engineering. Additionally, the behavior of fluid under varying conditions—viscosity, turbulence, and flow rate—is crucial information that can be utilized to optimize system design and functionality in hydraulic actuation and thereby provide better reliability and efficiency for such complex mechanisms. In hydraulic circuits, fluid dynamics is involved beyond basic principles; it has a direct influence on actuator design and function. Engineers apply computational fluid dynamics (CFD) to simulate fluid flow in hydraulic components, allowing them to forecast the performance characteristics and potential issues, like cavitation or flow separation. The simulations enable actuator designs to be optimized through improved fluid flow and mechanical interaction. Engineers can design more sensitive and durable actuators by varying parameters such as pipe diameter, fluid viscosity, and pressure differentials. These developments are of special significance in cases where precision and reliability are most crucial, such as in aerospace technology (Gaugh et al., 2025) and biological-simulation robotic technologies (Albu-Schaffer et al., 2014). Furthermore, with the incorporation of advanced materials and design concepts in hydraulic systems, fluid dynamics can also be enhanced along with the performance of actuation mechanisms in general. Material science innovations have made possible the development of compliant material that allows for more flexible and adaptable hydraulic systems, duplicating strategies in nature. For instance, by the use of soft robotics technology, engineers can design systems which not only equal stiff robotic skeletons' functionality but also become more agile and sensitive, similar to living organisms like cephalopods. These sophisticated technologies underscore the revolution in hydraulic technology, demonstrating that fluid dynamics not only controls traditional hydraulic action but also produces new forms of actuation that push the boundaries of mechanical design and engineering possibilities (Figure 4.8).

Figure 4.8. The study of fluid dynamics plays a crucial role in understanding hydraulic systems, particularly in advanced actuation mechanisms.

Source: https://www.shuntecpress.com/wp-content/uploads/application-of-hydraulic-technology.webp.

4.2.3. Gas Laws in Pneumatic Systems

The functioning of pneumatic systems is largely based on the principles of gas laws, namely Boyle's law and Charles's law. Boyles law, which asserts that the pressure of a gas varies inversely with its volume when the temperature is held constant, is important in understanding how pneumatic actuators function. As air is compressed, its pressure increases so it can move or exert force. This fundamental concept allows engineers to design systems that leverage air pressure efficiently for many applications, including robotics and automation. Understanding these relationships is fundamental to gaining optimal performance in high-speed locomotion and adaptive mechanisms and allowing air to be not only a medium for actuation but also a strategic element in engineering design (Chi et al., 2022).

Figure 4.9. Boyle's law, which states that the pressure of a gas inversely correlates with its volume when temperature remains constant, is essential for understanding how pneumatic actuators operate.

Source: https://www.thoughtco.com/thmb/isMB36wVW0kzYYAJjoesuaCrY_Y=/1500x0/filters:no_upscale():max_bytes(150000):strip_icc()/GettyImages-476845985-15288f31328f40a1a4e65b131f497ab3.jpg.

Boyles law is supplemented by Charles's law, which specifies the contribution of temperature to pneumatic systems. According to Charles's law, volume increases proportionally with temperature if pressure is constant. Temperature variations in pneumatic systems can generate substantial differences in air volume and therefore impact actuator performance. By the capacity to foresee temperature variation and design systems that can accommodate such variation, engineers have the means to enhance the reliability and effectiveness of pneumatic actuators. Additionally, embedding materials that adapt dynamically to temperature opens a window for designing soft actuators that can change shape or function without needing complex external control, an occurrence growing in the area of soft robotics (Xiong et al., 2020). The combination of these gas laws has also resulted in novel approaches to developing pneumatic actuators that are multifunctional and adaptable. Bistable structures inspired by nature show the capability of pneumatic systems to produce sophisticated behaviors like rapid actuation and adaptive reconfiguration. Through the application of mechanical principles borrowed from gas laws, engineers can come up with actuators that mimic natural motion typical in organisms, which goes toward making innovation safer in the development of soft robots. Apart from offering operational flexibility, the advances also offer safer new paths of human-robot interaction, hence making the systems more intuitive and sensitive to the environment (Chi et al., 2022; Xiong et al., 2020). Thus, understanding gas laws is essential not only for current applications but also for the cutting-edge development of next-generation soft actuator and robot technologies (Figure 4.9).

4.3. APPLICATIONS IN ROBOTICS

In robotics, advanced actuation mechanisms hold a pivotal role in maximizing performance and capability across an extensive range of applications. One key innovation is the soft robotic gripper, which is used for manipulation of unstructured objects. Traditional robotic grippers struggle to manipulate objects with extremely high variability of shape, size, and material. However, soft robotic grippers leverage tunable bistable properties by which dynamic grasping is enabled in the presence of no complex sensing or control mechanisms. The technology allows for rapid responses, as evidenced in experimental observation where sensor-less dynamic grasping occurs in a mere 0.02 seconds. The flexibility of the actuation mechanism renders it highly efficient in various applications, including automated assembly lines and sorting systems, where rapid and reliable object handling is necessary (Abidi et al., 2019). Another fascinating use of advanced actuation mechanisms can be found in the application of ionomeric polymer-metal composite (IPMC) actuators in robotics. These actuators are particularly appealing due to their ability to produce large displacement and rapid motion with the application of low voltage. Developed first for use in commercial environments such as toys and biomedical devices, IPMC actuators showed potential for use in robotics applications that require soft and adaptive movement. Their inherent bending capability, although limited in certain respects, can be further enhanced with freeform fabrication techniques. With the creation of IPMC actuators with novel geometries, more elaborate motions are achievable, leading to more robotic uses, from complicated biomimetic devices to bendable surgery assistants (Lipson et al., 2005). In addition to robotic dexterity being enhanced through the use of state-of-the-art actuation mechanisms, new ways of solving problems in spaces that need to be flexible are introduced. For instance, the confluence of soft robotics and IPMC technology can lead to innovations in soft robotics that can operate in dynamic and unstructured environments. With robots increasingly taking active roles in dynamic and unstructured environments, demand for adaptive actuation will only grow stronger. Emerging areas of interest for future research include improved performance of such actuation mechanisms via multi-material techniques and sophisticated control algorithms. Ultimately, the evolution of actuation technology will transform the potential of robotics to power complex applications across various industries, including manufacturing, healthcare, and exploration (Abidi et al., 2019; Lipson et al., 2005).

4.3.1. Robotic Arms and Manipulators

The development of robotic arms and manipulators has transformed industries across the board by enhancing precision and efficiency in several operations. Described in terms of imitating the movement of the human arm, these robots consist of several joints and actuators with the capacity for complex movements. With the demand for automation growing, robot arms have gone beyond being merely ordinary manufacturing processes, with application now extending to industries such as healthcare, telecommunication, and aerospace. Coupled sensing technologies with machine learning algorithms help the devices adapt to changing surroundings, thereby improving their functionality. The developments in the control system, such as in haptic technology, ensure that the operators' level of dexterity and responsiveness is high for efficiently performing complex tasks that involve fine motor. In pharmaceutical application, the robotic arms have revolutionized the processing and handling of delicate components. Current automated systems are designed to handle hazardous materials without putting the operator's safety at risk, a consideration in this area. An application of a manually operated two-degree-of-freedom (DOF) maneuverable robot arm is a good illustration of such technology development. The new device employs DPDT switches and servo motors, operated by an Arduino microcontroller, to enable precise movements. These configurations present for real-time control, meaning operators are able to carry out complicated tasks, such as assembling medication and conducting complex quality checks without having to deal with potentially dangerous materials. Since the pharmaceutical sector prioritizes efficiency and precision, such robotic solutions play a crucial role in ensuring high levels of safety while reducing production processes (Yashawanth et al., 2023). Haptic glove technology extends the functionality of robotic arms further by enabling a sense of touch and feedback while performing manipulation tasks. Through the integration of the tactile feedback of the haptic glove with the dexterity of robotic arms, operators are able to perform operations in remote and hazardous environments with unmatched control and precision. This technological complementarity makes use of advanced sensors built into the glove to capture the motions of the user's fingers, which are in turn translated into robotic motions. Also, the application of feedback mechanisms increases not just people's capacity for recognizing shape and texture but also overall interaction with robot systems. As industries increasingly deploy such innovations, the application of robotic arms continues to evolve to provide innovative applications that rely on enhanced human-robot collaboration across diverse environments (Cyriac et al., 2025) (Figure 4.10).

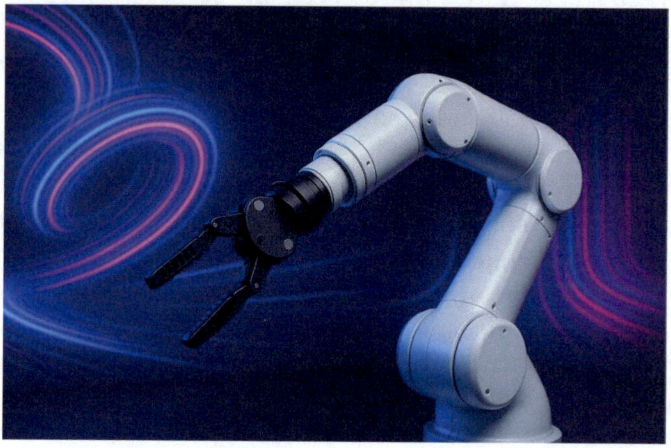

Figure 4.10. As the demand for automation grows, robotic arms have found applications beyond traditional manufacturing settings, extending into fields such as healthcare, telecommunications, and aerospace.

Source: https://www.amtec.us.com/wp-content/uploads/2025/04/robotic-arm-1.jpg.

4.3.2. Autonomous Vehicles

The advent of self-driving cars has transformed the face of modern transportation with advanced actuation mechanisms for maximizing operational effectiveness and security. The backbone of such a technological revolution is actuation systems that support high-precision control of various functionalities such as steering, braking, and throttle control. These devices are central to making self-driving cars run safely in complex environments based on dynamic conditions. For instance, a well-designed actuation system provides the ability for hardware and software to cooperate seamlessly together, enabling vehicles to process information from their surroundings and execute commands with minimal latency. As autonomous vehicle design progresses further, the performance and reliability of such actuation systems will increasingly contribute to their large-scale deployment. Of significance, farm business is a great example of how autonomous vehicles can be integrated into daily operations, particularly in low-cost applications in small-scale farming. Automated farm solutions have been expensive, leading most farmers to rely on conventional practices. However, research has demonstrated the feasibility of repurposing existing farm vehicles into multi-purpose platforms that can operate autonomously (Corpe et al., 2016). By adding mechanical actuators and controlled systems to such platforms, agronomists have been able to automate simple but repetitive jobs, making it more productive and efficient. This invention shows how advanced actuation mechanisms can make operation costs significantly lower while bringing technology within reach for small-scale agricultural companies, thus democratizing automation as an aspect. Furthermore, autonomous system design and production, like robots designed to walk efficiently, also illustrate the versatility of actuation mechanisms in order to achieve complex motion and functionality. The development of Dribbel, a passivity-based walking robot, for example, illustrates a sophisticated synergy between mechanical design and precise electronic control (Dertien et al., 2006). With the use of advanced actuation techniques, researchers can develop walking gaits that take inspiration from nature, enhancing the capability of the robots to move in diverse terrains. Not only do such initiatives contribute to an advancement in robotics technology but even in the broader field of autonomous vehicles for the purpose of maximizing their control systems. With the development of the market, a consistent focus on the innovation of actuation mechanisms will be crucial in producing effective vehicles with the capability to perform diversified tasks autonomously.

4.3.3. Drones and Aerial Robotics

Agricultural use of drones represents a significant advancement in mechanisms for actuation, optimizing both pest control and crop monitoring. Existing research emphasizes the application of small unmanned aerial vehicles (UAVs) in order to enhance yield prediction and reduce pesticide use. Through the application of advanced imaging technologies, drones can detect plant physiological alterations indicative of pest infestation, hence facilitating early action. These drones can be equipped with sensors to measure canopy reflectance, which are processed to create digital maps of hotspots of pest infestation. Furthermore, the actuation aspect is highlighted as drones are capable of autonomously applying precision applications and treatments, such as selective spraying of pesticides or autonomous release of natural enemies. This drone convergence of actuation and sensing is significant for the promotion of sustainable agriculture and calls for interdisciplinarity partnerships between technicians and agronomists (de Lange et al., 2020). Beyond the agricultural context, the drone has potential to be utilized elsewhere, for instance, in emergency response and environmental monitoring. Advanced actuation mechanisms facilitate the former switching of the aerial robots from one mode of operation to another, making their use in dynamic settings more valuable. A case in point is the SQUID (Streamlined Quick Unfolding Investigation Drone), showing how ballistically launched configuration can stabilize itself when flying on its own and remain operational integrity even in hostile conditions such as GPS-denied environments. This capability particularly proves useful for gathering critical situational data in crisis scenarios, where deployment at high speed and stable stabilization are of top priority. The presence of an onboard sensor suite also defines the actuation complexity of such drones to not only be responsive in an adaptive way to environment feedback in real time but also perform responsibilities with great precision and agility demands (Anderson et al., 2020).

Figure 4.11. The integration of drones in agriculture represents a significant advancement in actuation mechanisms, optimizing both pest management and crop monitoring.

Source: https://cdn.wikifarmer.com/images/thumbnail/2024/02/Drones-in-Crop-Monitoring-and-Management-Uses-Types-Advantages-and-Limitations-500x500.jpg.

The future of agricultural and broader applications of drone technology relies on the continued innovation of high-performance actuation technologies that are efficient and effective. As drones become more entrenched in various industries, the focus on maximizing their actuators will ensure that drones will be capable of more complex autonomous tasks. This evolution is not merely a matter of improving the physical form and functionality of drones but also converges with advancements in machine learning and artificial intelligence techniques. The collaborative effort of engineers, software developers, and domain experts will breed innovations that will make it possible for drones to navigate and interact with their environment in a smart manner. Ultimately, ongoing advancements of drones and aerial robots are a turning point towards automating processes in countless applications and opening a new era of operational capability and strategic resource management (Figure 4.11).

4.4. APPLICATIONS IN AEROSPACE

Application of advanced actuation mechanisms in the aerospace sector has, in turn, transformed satellite technology to enable miniaturized and more efficient spacecraft. By research projects like MISAT, the space application potential of micro-systems technology (MST) is being understood; the project aims to utilize miniaturization for novel applications in satellites. The distributed system technology, such as wireless sensor and actuator networks, facilitates plug-and-play technologies that enhance operational flexibility in space missions. Furthermore, energy-aware technologies, such as the technologies shown in autonomous digital Sun sensors and re-configurable radio front ends, represent an essential step towards green spacecraft that lower power consumption while enhancing functionality. Not only do these technologies address current industry requirements, but they also allow for new ideas of future missions that require compact solutions to complex engineering issues (Gill et al., 2007). In addition to satellites, actuation mechanisms are important in numerous other aerospace devices, including aircraft control systems and propulsion systems. The precision offered by existing actuators is required for ensuring stability and performance in aircraft design. These systems are designed to respond rapidly to pilot input or automatic control, enhancing maneuverability and safety. Advances in electrical and mechanical actuation technologies are making it possible to make airplanes more responsive yet lighter, an increasingly important issue in increasing fuel efficiency. For example, fly-by-wire systems, which employ electronic controls rather than traditional mechanical linkages, rely heavily on advanced actuation mechanisms that not only provide greater precision but also facilitate maintenance and reliability across the lifespan of an aircraft. As these types of technologies continue to evolve, their integration will revolutionize aerospace design and operations standards. The advanced actuation mechanism research also has shining prospects in the next-generation aerospace technology, especially that of autonomous systems and unmanned aerial vehicles (UAVs). With UAVs gaining wider use in commercial and military operations, there is a greater demand for efficient and versatile actuation systems. Technologies such as integrated microfluidic control systems, demonstrated in recent research, have the potential to provide enormous performance potential such as real-time monitoring and fluid transport. These systems utilize hybrid wave modes to achieve multi-aspect functionality, with direct integration of sensing and actuation on a single platform. These advancements not only expand operational capacities but also follow the trend towards increased automation and reducing human interaction in aerospace missions. The principal advancements in actuation mechanisms are therefore the cornerstones for transforming the future aerospace technology and operations environment (Dodd et al., 2020).

4.4.1. Flight Control Systems

Flight control system development is inextricably linked with actuation system innovation, which forms the basis of an aircraft's maneuverability and stability. Today, flight control systems employ a variety of actuation systems, including electrical, hydraulic, and mechanical systems, to effect precise control of multiple aircraft surfaces, such as ailerons, elevators, and rudders. With a growing demand for better system performance and dependability, the incorporation of advanced actuation systems is essential. For instance, investigations into the weight and cost benefit of incorporating active controls in aircraft structures indicate that the integration can be done effectively with significant increases in efficiency without compromising structural integrity. The processes discussed define the balance that must be achieved between technological advancement and operational constraints, which will allow aircraft to achieve regulatory needs and safety standards in more competitive environments (Anderson et al., 2025). Additionally, high-technology actuation devices are a key component of achieving novel flight control strategies, which have the potential to improve the responsiveness of aircraft within dynamic flight conditions. Technologies such as fly-by-wire systems replace traditional mechanical linkages with electronic controls, which move towards a more streamlined design ethic. Not only is weight saved, but control input fidelity is also enhanced so that pilots can react faster and more accurately to dynamically changing flight scenarios. Advances in such systems have also spurred research into new materials and geometries, further improving performance and safety. With the implementation of embedded technologies that involve active control methods, scientists are already investigating the implications of such developments on passenger safety as well as overall aircraft resilience, opening up new paths to safer flight (Bement et al., 2025). Additionally, integration of explosively actuated openings in composite panels proves to be an example of the new solution being sought for enhanced passenger survivability in aviation. These devices enable rapid and efficient emergency egress as well as fuselage venting and fuel dump scenarios. Incorporation of a stable explosive cord into composite structures enables structurally positive designs that may create defined exits in times of emergencies. These developments are a significant leap in thinking regarding traditional flight control systems by not only considering the flight dynamics but also taking into account the most vital consideration of passenger security. With this technology advancing, consideration of the safety provisions and acceptance within the industry is crucial, thereby making the new solutions compatible with current aviation practice and enhancing overall system performance and reliability (Bement et al., 2025) (Figure 4.12).

Figure 4.12. Modern flight control systems employ a variety of actuation technologies, including electrical, hydraulic, and mechanical systems, to achieve precise control of various aircraft surfaces, such as ailerons, elevators, and rudders.

Source: https://cdn1.genspark.ai/user-upload-image/gpt_image_generated/851e798a-1d94-4437-a6af-48a21ab9a3a2.

4.4.2. Landing Gear Mechanisms

Development of landing gear mechanisms is a crucial field within the wide topic area of advanced actuation mechanisms, as such mechanisms are at the center of providing smooth transition of the aircraft from flying to ground and vice versa. The landing gear serves numerous functions like supporting the aircraft during takeoff and landing, offering stability during the grounded state, and providing resistance to shock in case of touchdown. Improvements in landing gear design have focused on various actuation systems, such as hydraulic, electric, and pneumatic actuators with varying advantages and compromises. Hydraulic, for example, offers high force with reliability, while electric offers better control with minimal maintenance needs. With the aviation industry trending towards energy efficiency, electric actuation for landing gear systems is also increasingly becoming popular as a measure to further enhance overall operating efficiency and reduce aircraft weight, as pursued by current research work. The use of advanced actuation systems in landing gear is also aided by recent research that focuses on integrated digital and electric systems. This type of research, for example, from the Integrated Digital/Electric Aircraft (IDEA) project, explores the promise of innovative configurations of landing gear for improved fuel efficiency and operational performance. The IDEA aircraft, through the demonstration of integration of digital systems and distribution of electric power, showed considerable improvement in fuel use, hinting at the possibility of re-examining traditional landing gear configurations to integrate modern actuation technology. Through the incorporation of these mechanisms in conjunction with cutting-edge technological advancements, the landing gear not only assists in improving performance but also finds its way as a key element in the design and functionality of forthcoming aircraft.

Figure 4.13. Landing gear serves multiple functions, including supporting the aircraft during takeoff and landing, ensuring stability when on the ground, and providing the necessary shock absorption during touchdown.

Source: https://upload.wikimedia.org/wikipedia/commons/thumb/0/0a/Main_landing_gear_of_747-8F.jpg/1100px-Main_landing_gear_of_747-8F.jpg.

The shift to advanced actuation mechanisms is more scope for research and development with the aerospace sector anticipating solutions to higher demands for sustainability and efficiency. Lastly, design and verification of advanced landing gear systems require a systemic approach to system functional requirements. Research into formal design methodology frameworks, e.g., frameworks developed based on Yourdon's methods, aims to introduce innovative control system requirements necessary for landing gear actuation optimization. Development of these requirements begins with thorough mission analysis, wherein safety, reliability, and performance are prioritized over varied flight regimes. The use of systematic design paradigms ensures that the outcome landing gear mechanisms meet high standards, thereby enhancing system verification processes. With the technology progressing towards higher and more complex aircraft configurations, a focus on the optimization of the landing gear actuation mechanism will become vital to the exploitation of next-generation aeronautical design capability, and in direct accord with ongoing efforts to create and incorporate advanced flight control systems that enhance the performance and safety of aircraft (Bailey et al., 2025; Cohen et al., 2025) (Figure 4.13).

4.4.3. Spacecraft Deployment Systems

Spacecraft deployment mechanisms are the epitome of cutting-edge aeronautics today, combining mechanical engineering and the intricacies of space exploration. Sophisticated actuation mechanisms are at the heart of the system, facilitating accurate action when it counts, such as satellite release or solar array extension. Actuation has to be extremely reliable when deploying because any malfunction could disrupt mission objectives or compromise the spacecraft's integrity. Technologies such as modular designs for space vehicles allow for flexible configurations, and multimission modular spacecraft (MMS) can be incorporated that can be refurbished and upgraded in space. This adaptability not only extends the lifetime of the satellite but also allows for easier assembly in space, effectively reducing the costs and risks of deploying satellites (Bergonz et al., 2025). In recent years, telecommunications and remote operations have witnessed growing focus on autonomous deployment systems. Telepresence technologies have emerged as a significant development that offers the possibility of leveraging human control into extraterrestrial environments with minimal risks of direct human involvement. As per NASA's research, effective telepresence systems must be strong in hardware and superior in software platforms to allow intricate operation modes, enhancing ground-based operators' capability to respond to problems occurring in space in real time (Akin et al., 2025).

Improved actuation mechanisms operating in conjunction with telepresence can provide precise feedback to the operators, allowing them to make amendments and corrections during deployments that would otherwise become uncontrolled. As missions to space become more complex, the application of advanced actuation mechanisms in spacecraft deployment systems is of highest priority. Moving towards on-orbit servicing, wherein elements can be exchanged or repaired as modular units, is one way such systems can be utilized to support sustainability in space activities. With the use of orbital maneuvering vehicles (OMVs) and high-precision actuators, spacecraft can be controlled more precisely such that resupply and modular enhancements become achievable without Earth return (Bergonz et al., 2025). Not only are launch costs reduced but also the operational life of various space assets is enhanced, resulting in significant variations in planning and conducting a mission. The revolutionary potential of these technologies provides a glimpse into the future where flexibility and adaptability in space exploration are the norm, and it is an important step toward the sustainable operations of space.

4.5. INNOVATIONS IN ACTUATION TECHNOLOGY

The technology of actuation has transformed radically, particularly due to the emergence of intelligent materials that dynamically alter their characteristics based on the environment. Shape memory alloys (SMAs) and polymers are some of the materials that have provided new possibilities for advanced actuation systems across a broad spectrum of applications including aerospace engineering and robotics. The ability to process these materials dynamically not only enhances the performance of actuators but also supports more compact and space-efficient designs that are lighter, which is important for applications like aircraft and space exploration. Also, the integration of micro-electronic technology plays a vital role in that it enables the development of systems with precise control over actuation processes, which has a lot of potential in solving problems such as the reduction of drag in turbulent boundary layers and hence boosts aviation performance and fuel consumption (Abbas et al., 2017). Today's computational simulation breakthroughs have revolutionized the understanding of flow patterns under turbulent conditions to open up the door to new control schemes. A typical example of how these computational schemes can lead to tremendous skin-friction drag reductions is in the application of turbulent boundary layer control (TBLC). Active control strategies allow us to address fundamental aerodynamic problems that once constrained fuel efficiency in airplanes. The convergence of simulation, material sciences, and micro-actuation technologies foresees a strong basis for overcoming such challenges. Even though at a low technology readiness level (TRL), the focus of the industry on active control methods foresees high future potential in the area. As more abilities are brought forward for such technologies, the capacity for more efficient aerial systems becomes ever more within reach (Abbas et al., 2017). Furthermore, studies on shape memory polymers (SMPs) highlight the potential for new applications in specific industries, particularly with regard to space exploration. According to recent academic studies, an MCP suit integrated with SMPs could greatly increase the dexterity and usability of astronauts working on long-distance journeys, such as those slated for Mars. With the application of SMPs' inherent characteristics, researchers have developed initial designs that meet requirements for pressure distribution and environmental durability, which are highly critical in extravehicular applications. Simulations based on the coupling of the thermo-mechanical and textile mechanical behaviors are highly critical in the prediction of garment performance during working conditions. While durability and long-term performance remain concerns, these studies usher in a new generation of space suit technology as they emphasize the necessity of continued innovation in actuation systems.

4.5.1. Smart Materials and Actuators

Smart actuators and materials have emerged to be dominant forces behind the revolution of advanced actuation mechanisms primarily due to their unique properties to react to external stimuli. Of these, shape memory polymers (SMPs), particularly two-way shape memory polymers (2W-SMPs), are particularly noted to be able to experience reversible shape changes in the absence of external energy supplies. This inherent property causes them to return to their fundamental shape when exposed to a specific temperature, hence making them ideal for various uses in various industries. The lightweight feature, coupled with good deformation properties and reversible modulus change, renders 2W-SMPs key materials in the fields of medicine, robotics, and space technology. Moreover, the study of quasi-2W-SMPs extends even more the functional scope of the materials to allow them to be integrated into high-end actuation systems that respond in real-time to different operational conditions (Zou et al., 2025). Liquid crystal elastomers (LCEs) are yet another significant player in the field of smart actuators, which combines the anisotropy of liquid crystals with the elasticity of polymers. This unique composition allows LCEs to experience reversible, macroscopic shape transformations in accordance with various stimuli, i.e., temperature, light, and electrical fields. The magic behind LCE's capability lies in the highly ordered mesogenic molecules, whose structures alter in response to the stimulus, leading to various morphing responses. Surface anchoring, field effects, and mechanical alignment methods have been explored to enable the performance of LCEs in actuation. These methods enhance not only the efficiency of LCEs in generating motion but also facilitate their integration into new devices such as artificial muscles and reconfigurable optical elements, demonstrating their potential for advanced use in soft robotics and more (Cheng et al., 2025). Intelligent material integration, including 2W-SMPs and LCEs, in actuators contributes significantly to developing intelligent systems with advanced functionalities. With increasing numbers of industries embracing adaptive and multifunctional solutions, the engineering of these materials is ushering in new horizons for systems that can adapt based on real-time feedback from the environment. The future of actuators includes refining the design and use of the smart materials to maximize efficiency, responsiveness, and operating ability. Ongoing exploration of how to best optimize alignment methods for LCEs and how the multifunctional properties of 2W-SMPs can be further developed will propel development in fields from biomedical engineering to aerospace engineering. The interaction between smart materials and sophisticated actuation mechanisms will therefore define the future of engineering, leading to smarter, more efficient technologies (Cheng et al., 2025; Zou et al., 2025) (Figure 4.14).

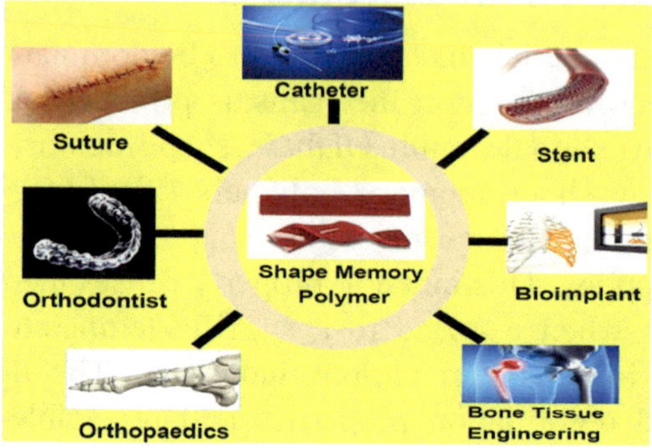

Figure 4.14. Shape memory polymers (SMPs), particularly two-way shape memory polymers (2W-SMPs), are noteworthy for their ability to undergo reversible transformations without the need for external energy sources.

Source: https://onlinelibrary.wiley.com/cms/asset/5950bc4a-6786-473b-acce-3024b5d4af8e/pat6138-fig-0003-m.jpg.

4.5.2. Miniaturization of Actuation Systems

Miniaturization of actuation systems has become an especially high priority focus on the development of advanced actuation mechanisms, providing enhanced performance in numerous applications, most significantly in medicine and robotics. Miniaturization of actuation systems to smaller scales and higher efficiencies allows for greater integration into small devices, facilitating functionalities that previously were not possible. For instance, in medicine, miniaturized actuators are of key importance in implantable drug delivery systems, where precise and controlled drug delivery is the determinant of clinical outcome. Systems based on advanced material and concept design enable these systems to overcome physiological barriers while delivering therapeutic effect. As witnessed with recent advancements, miniaturization not only increases bioavailability but also minimizes systemic side effects, making the devices patient friendlier. The continuous miniaturization of such systems brings to bear the importance of miniaturization as a method of responding to growing demands for effective and less intrusive medical treatment (Bono et al., 2025). The evolution of Micro-Electro-Mechanical Systems (MEMS) electromagnetic actuators also bears witness to how miniaturization has contributed significantly to the performance and versatility of actuation systems. These tiny devices, founded on such phenomena as the Lorentz force, are now an essential part in applications ranging from precision micromanipulation to microfluidic systems. Developments in techniques for their fabrication, such as microfabrication and additive manufacturing, have enabled the fabrication of highly efficient MEMS actuators that can function with very high accuracy and responsiveness at diminished scale. As discussed in recent literature, not only do such innovations enhance the functionality of MEMS devices but also pose the issues of power efficiency and reliability, which must be considered to unlock their true potential across all industries. By advancing the miniaturization methods further, researchers are unlocking the potential for a new generation of actuators, which will address the level of sophistication present in modern applications (Cheng et al., 2025). Finally, miniaturization has consequences extending beyond technical efficiencies; it also establishes theoretical models and operation paradigms in actuation system control. With the shrinking size of actuation components, traditional performance and control models need to be adapted to be able to tackle the unique dynamics of these miniature systems. Addition of smart control logic and adaptivity into compact actuation systems provides a potential to propel enhanced reliability and performance substantially. In particular, the fusion of MEMS technologies with new materials science assists in infusing this advancement with actuators that not only minimize size but also achieve greater functional features. The revolutionary nature of these innovations predicts a shift in perception and application of actuation systems towards various fields, and so it can be predicted that the future of engineering will increasingly depend upon the ability of miniaturization techniques to be applied (Bono et al., 2025; Cheng et al., 2025).

4.5.3. Integration with Artificial Intelligence (AI)

The integration of artificial intelligence (AI) within advanced actuation mechanisms is a force for innovation across various domains, particularly in autonomous systems. Using the capabilities of AI features, these mechanisms can maximize their operational efficiency, flexibility, and precision. A good example is the implementation of the Internet of Robotic Things (IoRT), which underscores the need for AI to manage the complex networking and coordination required by concurrent autonomous devices. The integration of reasoning capabilities allows for unbroken interaction and cooperation between diverse robotic platforms, enabling them to make time-critical decisions based on situational information. This paradigm shift not only increases the effectiveness of cooperative processes but also raises security, safety, and system reliability concerns, as robust AI algorithms can anticipate failure and facilitate self-healing behaviors among the networked topology to create a reliable operating environment for modern actuation systems (Bacciu et al., 2017). Moreover, the developments in AI also promote sophisticated actuation techniques that propel robotic performance in tough environments. The architecture of novel control paradigms, such as those utilized in teleoperated spherical tensegrity robots, demonstrates the capability of AI-based methods to optimize multi-cable actuation mechanisms to the fullest. The new method enables such robots to move on extremely inclined planes at speeds and stability that are not possible with conventional geometries. With the employment of AI, it becomes feasible to develop optimization algorithms that leverage the increased dynamic performance that the mechanisms provide.

Figure 4.15. The advent of the internet of robotic things (IoRT) underscores the importance of AI in managing the intricate networking and coordination requisite for multiple autonomous devices.

Source: https://media.assettype.com/analyticsinsight/import/wp-content/uploads/2018/03/IoRT-1.jpg?w=1200&h=675&auto=format%2Ccompress&fit=max&enlarge=true.

With the enhancement of task-specific control policies, robots are able to adapt dynamically to surroundings and exhibit greater agility, greatly enhancing their applicability across a wide variety of uses from search and rescue operations to autonomous construction in hostile environments (Agogino et al., 2017). Furthermore, the intersection of AI with advanced actuation systems opens up new avenues for research and exploration that further push the evolution of intelligent systems. The consistent growth of new paradigms such as the Autonomous Internet of Things (A-IoT) and the Autonomous System of Things (ASoT) identifies the critical role of AI not just in augmenting existing technologies but also in defining future trends in this field. Incorporating AI into actuation mechanisms enables devices to be able to operate efficiently with sophisticated tasks and dynamic resource configurations and context-aware service implementations. With researchers attempting to address integration problems such as discoverability and programmability of dynamic resources, sound AI frameworks become a necessity. With this ongoing enhancement, they not only offer more effective operational capability in smart environments but also re-emphasize the role of AI in triggering the overall revolution of advanced actuation mechanisms (Bacciu et al., 2017) (Figure 4.15).

4.6. CHALLENGES AND LIMITATIONS

With increasing actuation technology, numerous limitations and complications restrict growth. Integration of complicated materials and technology that can achieve desired performance in a way that is cost-effective is one of the primary concerns. The development of advanced actuators using compliant tendon actuation and micro-servo motors, for instance, requires functionality at the expense of affordability. Initiatives such as the partnership with Bhagwan Mahaveer Viklang Sahayata Samiti illustrate this conflict, as they have set a mission to manufacture low-cost, electrically powered prosthetic limbs that successfully meet the requirements of urban Indian amputees. However, the issue is how to make such devices not only function, but also be reliable and manufacturable in high volumes, thus reaching at the overall concern of scalability in actuator design and deployment (Ferris et al., 2019). Furthermore, although novel materials such as ionomeric polymer-metal composite (IPMC) actuators present new opportunities for actuation, their current limitations keep them from being used practically. While IPMCs achieve excellent displacements and high-speed movements through minimal electrical inputs, they respond primarily by bending and are therefore not very flexible and useful across various applications. The shortcomings of the actuators are reflected in their performance profiles, where, though they demonstrate improvements in freeform fabrication for new shapes, they remain incapable of equaling traditionally produced devices in output stress and efficiency (Lipson et al., 2005). These technological barriers necessitate ongoing research aimed at improving the effectiveness and multidisciplinary applicability of such materials, thereby overcoming the present limitations around their use in biomedical and robotics. The broader implications of the challenges extend far beyond the technical landscape, entering into the social and economic realms of actuator technology. Engaging with progress without resisting built-in limitations can lead to disparities in access, particularly in lower-income regions where cost is a significant barrier. The present effort to develop a functional prosthetic hand designed specifically for urban Indian amputees is an example of the social influence of actuator design, as it seeks to provide sophisticated technology to people in need. Still, the project reminds us that unless the social responsibility of making the manufacturing process reliable and scalable is addressed, even revolutionary designs will not lead to actual quality-of-life improvement for many individuals. The interaction between social responsibility and technology developments remains an area of major concern for research and innovation in advanced actuation mechanisms (Ferris et al., 2019).

4.6.1. Energy Efficiency Concerns

Since the demand for advanced actuation mechanisms is growing, the demand for energy efficiency has become a significant concern while developing and applying these systems. Traditional actuators usually require high energy input to work efficiently, thus raising operating costs as well as their environmental impact. In contrast, emerging technologies are being developed to consume less energy without compromising performance. For instance, soft robotics innovations have indicated the benefit of exploiting passive mechanisms in terms of energy efficiency. Such systems, exploiting adaptive bistability in dynamic grasping, are capable of storing and releasing energy from strain as necessary, resulting in an energy-efficient mode of actuation. By minimizing energy loss when operational, such innovations indicate a promising future direction in finding sustainable actuation solutions (Abidi et al., 2019). Another avenue through which energy efficiency can be greatly enhanced is the development of Ionomeric polymer-metal composite (IPMC) actuators. Deservingly, IPMC actuators are characterized by a high ability to create large displacements at extremely low voltage, typically 1 V, thereby becoming extremely energy-efficient in terms of power transformation. This feature allows for rapid movement and enjoys low input power, which is tremendously useful in fields such as biomedical applications and robotics. While bending actuation in IPMCs is not without its disadvantages, freeform fabrication research is pushing the creation of new geometries and actuation modes that will further facilitate energy use. As these technologies advance, the integration of IPMCs with other devices has the potential to make them more efficient, which can address important energy issues in actuator design (Lipson et al., 2005). Furthermore, the energy efficiency aspect is not just reserved for actuation mechanisms per se but also for the net impact of such technology on system performance and application environment. A case in point is the integration of advanced actuation mechanisms in robotics, which leads to simpler and lighter designs that can run on less energy. This is particularly crucial where responsiveness and flexibility are essential in dynamic environments. Hence, smart control systems that optimize the efficiency in energy utilization by adaptive actuation strategies are crucial. Through the optimization of not just the mechanism but also the control means, energy efficiency for the next-generation actuators can be maximized without affecting performance and functionality in the service of world sustainability goals. Thus, the future of actuation technology lies in unifying efficiency and effectiveness, responding to the rising energy consumption issues in innovative ways.

4.6.2. Reliability and Maintenance Issues

In the area of advanced actuation mechanisms, reliability and maintenance issues remain paramount to realizing peak performance and safety. Actuators must be capable of performing with high accuracy and resisting hostile environments of operation without inducing failure. Application of advanced materials and configurations, such as explosively actuated openings, is reliability demanding due to the inherent complexities of their operation. While integration of such mechanisms is capable of increasing functionalities, as in their proposed use in aircraft for emergency opening and fuel dumping systems, safety issues regarding explosive systems are elicited. Such requests for rigorously testing and proving processes to make such actuators reliable before widespread application (Bement et al., 2025). Producers thus must establish a harmonization between innovative design and sound performance to meet safety standards suitably. Furthermore, maintenance problems tend to complicate the deployment of advanced actuation systems, especially in industrial and aerospace applications where downtime can cause extensive operational delays and monetary consequences. The dependence on advanced components might raise the regularity and complexity of maintenance operations, which can place maintenance personnel in a dilemma of having specialized knowledge and resources. For instance, in the case of the operation of electric prosthetic hands that integrate state-of-the-art robotics with biointerfaces, regular maintenance is essential in an effort to sustain the performance level because these devices must be able to interact efficiently with users (Ferris et al., 2019). The process of maintenance must incorporate both hardware inspection and software updates, thus ensuring perpetual dependability while conforming to users' needs. Thus, proper maintenance strategy is essential to sustain the performance of state-of-the-art actuators for their entire life cycle. Additionally, reliability and maintenance management entails coordination among engineers, developers, and users to develop the culture of continuous improvement. Integration of users in feedback loops can enhance design plans to enable manufacturers to align actuators to real application conditions. In more advanced actuation mechanisms, such involvement would assist in identifying regions of failure or usability issues that are perhaps not found during initial testing phases. Additionally, the establishment of standard maintenance plans and training modules would assist in ensuring that technicians are properly equipped to deal with the complexity introduced by new technologies. With the advancement of the technology, these parameters will not only make the actuators more reliable but also extend their operating lifespan to ultimately meet diverse applications ranging from aviation safety to assistive devices (Bement et al., 2025; Ferris et al., 2019) (Figure 4.16).

Figure 4.16. Actuators must function with high precision and be capable of enduring harsh operational environments without failure.

Source: https://wevolver-project-images.s3.amazonaws.com/0.y9i6qhde7miwhat-is-an-actuator-cover-image.jpg.

4.6.3. Cost Implications of Advanced Systems

The incorporation of advanced actuation systems, such as ionomeric polymer-metal composite (IPMC) actuators, has significant cost implications that require detailed analysis. Though the potential of IPMC actuators lies in their light and flexible structure along with excellent displacement performance under low voltage, the manufacturing process continues to be a primary cost driver for the entire system. Conventionally fabricated IPMC actuators have performance characteristics that can differ enormously, with certain methods producing suboptimal stress outputs constrained by their manufacturing processes. Yet the more recent developments in freeform fabrication methods promise material and labor savings and enable more economically feasible production with innovative geometries engineered to a particular application. This innovation in production not only foresees a reduction of the cost per actuator but also increases flexibility and adoption of such technology in commercial vehicles, further increasing financial and operational viability in various industries (Lipson et al., 2005). In medical applications, the cost parameters of advanced actuation systems extend to costs outside initial production costs. A design case study of an electrically powered prosthetic hand shows how delicate design choices can balance functionality with cost. The prosthetic uses compliant tendon actuation and small linear servos, which results in a system that is affordable without compromising on the requirements of users. With consideration of biointerfaces that make interaction with users effortless, the prosthetic hand has the potential to significantly enhance the lives of amputees without being overly expensive. Thus, the development served to reiterate the importance of not only technology, but also strategic partnerships aimed at improving accessibility and distribution in resource-constrained environments, towards greater importance of a creative yet cost-efficient approach to advanced actuation mechanisms (Ferris et al., 2019). An analysis of the long-term cost implication of high-technology actuation systems also calls for equal regard for their operational performance and maintenance requirements. With technologies like IPMC actuators and prosthetics entering varied industries, stakeholders must understand their lifecycle costs, with factors for energy consumption, lifespan, and repair or replacement cost over time. For IPMC actuators, the prospect of lower energy consumption as opposed to conventional mechanisms can offset higher initial capital costs, a reversal that can reverse their economic viability. Similarly, cost-friendly prosthetics can also serve as blueprints for advancement to ensure wide accessibility and durability. Therefore, through a prudent analysis of initial expense and periodic cost, industries and organizations can better leverage next-generation actuation systems to deliver performance augmentation with financial restraint and societal benefit (Ferris et al., 2019; Lipson et al., 2005).

4.7. CONCLUSION

In conclusion, the presentation of advanced actuation mechanisms brings into perspective their central role in the functionality of modern robotic systems. As has been demonstrated through this chapter, actuation mechanisms such as soft robotic grippers and micro-positioning stages possess considerable advancements over traditional systems. Soft robotic grippers represent the pinnacle of innovation with the ability to perform dynamic grasping actions without the requirement of advanced sensing technology, illustrating the capability of bistable properties to effectively store and release strain energy. The successful implementation of these mechanisms not only widens the range of applications but also addresses the complexity of grasping unstructured objects in an effective manner. In addition, micro-positioning system design and analysis further point to the inclusion of advanced actuation techniques in the micro-scale. The exceptional XY micro-positioning stage through the application of electrostatic comb drives and parallel-kinematic mechanisms proves the precision possible in modern robotics. With increased motion control characterized by a closed-loop mechanism, these developments demonstrate improved precision and speed, needed for precise manipulation in applications. As illustrated by the introduction of such systems, the two-way fertilizing fields of control technology and microfabrication also have much to offer future robot device capabilities. Finally, the broad implication of such sophisticated actuation mechanisms is that they also have the potential to revolutionize several different industries, from healthcare services to manufacturing and service businesses. Technologies such as those found in soft robotics enable safer interaction with object manipulation that is crucial in applications where human-robot collaboration is vital. Additionally, micro-positioning stage accuracy, as with the case of modern fabrication techniques, emphasizes the necessity to further progress work in this field. Moving forward, sustained advancements in actuation mechanisms are bound to transcend traditional boundaries, paving the way for progressively adaptive, intelligent, and versatile robotic systems with the capability to conform to the dynamic requirements of their environment, as exemplified by successful sensor-less dynamic grasping techniques and the operating advantages of precision positioning systems (Abidi et al., 2019; Koo et al., 2014).

4.7.1. Summary of Key Points

The research into new actuation mechanisms has discovered considerable progress in various applications, from prosthetic limbs to fluid flow. A visible example is the development of a low-cost electrically powered prosthetic hand for Indian amputees in cities. The prosthetic hand utilizes compliant tendon actuation and a small linear servo with a biointerface to facilitate intuitive control. The design is not just functional and resilient but also presentable to give the device anthropomorphic looks and feel. Such innovations, as given through collaboration with Indian non-profit organizations, show how actuation mechanisms have contributed to enhancing the quality of life among individuals with disabilities while giving ease of production and distribution in different geographical settings (Ferris et al., 2019). In aerodynamics, advanced actuation systems have also facilitated new control approaches to achieve optimal aerodynamic performance. For instance, a model-free self-learning cluster-based control approach has been proposed to address high-fidelity simulation of post-stall separated flows on airfoils. Through clustering flow trajectories and performing an iterative evaluation process, this technique saves power consumption during flight. Employing a small number of sensor measurements for clustering and optimization allows for effective design of successful feedback control laws. These laws not only reorient flow trajectories but also shift the populations of clusters to aerodynamically favorable states, which consequently result in flight performance enhancement (Brunton et al., 2018). This synergy between emerging actuation mechanisms and fluid control substantiates the versatility and potential of these technologies to be implemented across different fields. The use of such advanced actuation mechanisms is indicative of an enormous advancement in medical technology and engineering, reflecting the potential of interdisciplinary research to deliver innovative solutions. For customized development of prosthetic devices, the users are provided with enhanced functionality that is tailored to specific cultural and bodily needs, and fluid dynamic applications illustrate the importance of study to deliver maximum performance and efficiency. The continued evolution in these disciplines signals a broader applicability of actuation mechanisms for the potential integration into innovative ideas to bridge human-focused design and technology. As researchers continue to push further into both disciplines, the impact of actuation mechanisms will only continue to push further development toward not just improving performance but also improving inclusivity and sustainability in engineering uses.

4.7.2. Future Trends in Actuation Mechanisms

With technological advancements, the future of actuation mechanisms will be predominantly determined by developments in material science and robotics. The direction of developing compliant actuators will be one where compliant actuator can mimic biological motion more than a rigid system can. These are actuators that employ materials with the capacity to bend and adapt to the environment, enhancing robotic systems' sensitivity, efficiency, and performance. One of the most important applications of this technology is in the development of prosthetic devices that communicate with users naturally. An illustration of a joint initiative to create an electrically powered prosthetic hand discusses how compliant tendon actuation can reengineer the technology to enable amputees to feel more anthropomorphic and have more functionality. These advancements bring a shift towards more user-centric designs that are customized to meet the singular requirements of individuals, thus opening wider doors for actuators in medicine and other areas (Ferris et al., 2019). Another significant direction in actuation mechanism innovation is integration into the Internet of Things (IoT), which is rapidly transforming the terrain of automatic systems. This convergence provides a new class of applications where actuators respond adaptively to environmental changes. For instance, the concept of Internet of Robotic Things (IoRT) would introduce a vision of a series of inter-connected devices that are able to interact and respond to context-based stimuli, providing a more interactive environment. With increasing sophistication of such systems, integration and coordination of heterogeneous devices emerge to be the most critical factors. The implementation of smart actuators in IoRT would have to be coupled with more programmability and communication capabilities so that they can be made to work collaboratively for optimal performance in smart homes, industries, and other applications (Bacciu et al., 2017). Finally, future actuation mechanisms will continue to witness more focus on the application of artificial intelligence (AI) to enhance the adaptability and decision-making capacity of such systems. With the integration of AI, actuators are not only able to execute pre-coded tasks but also learn from their surroundings and adapt behavior. This functionality is critical for applications like autonomous cars, where actuators need to react in real-time to changing circumstances. Moreover, AI-based systems can assist in predictive maintenance, keeping actuators running efficiently and effectively over a period of time. The coordination of intelligent devices in the IoT system will spur innovations, and actuators will be made more responsive, self-healing, and able to perform complex tasks independently. With the technology advancing, the integration of AI with actuation mechanisms will usher in unprecedented developments in many fields, eventually redefining our relationship with machines.

4.7.3. Final Thoughts on the Impact of Advancements in the Field

The evolution of advanced actuation mechanisms has stamped an irreversible imprint on various industries, particularly in the areas of robotics and automatic systems. These innovations enable greater precision and responsiveness in devices in various applications, hence enhancing their working ability. High-performing sensors and actuators enable dexterous manipulation, which is required in systems that demand human-like interaction and sensitivity. Research at facilities such as the Wisconsin Center for Space Automation and Robotics has contributed immensely to developing this area of study, with even more learned about how electrical, mechanical, and control aspects interact to influence performance. With the application of advanced test methods, the field continues to streamline its methods and hence innovative forces that not only set higher standards for robotic systems but also broaden their applications from space exploration to medicine. Ongoing research in actuation technology has demonstrated the necessity of multidisciplinary collaboration that transcends engineering, computer science, and cognitive sciences. The design of sophisticated actuation systems must take into account human factors, therefore honing user interfaces and control mechanisms. Such innovations have far-reaching implications, in particular enhancing the reliability and performance of telerobotic systems used in hazardous environments. By focusing on building upon human perception and tolerance of robotic interface, the field is making tremendous strides towards ubiquitous implementation of these technologies in everyday life. The emphasis on creating robust systems that can work in other operating environments is a proof of the transformation of the attitude of industries towards the potential of automation. Since mechanisms are becoming intuitive and simplistic to operate, their adoption in commercial as well as domestic applications will continue to gain momentum. In addition, ongoing technology reviews and development programs suggest the necessity of a formal process for further innovation in actuation mechanisms. As emphasized in recent research, some areas, such as cryogenic cooling and dynamic structural control, are the target of further research and funding. Eliminating these technological shortcomings not only optimizes current systems but also prepares the way for innovations that could possibly redefine space exploration potential and beyond. The quest for augmentation projects suggests an organized effort to develop a master plan for next-generation actuation technology development. As technology evolves, it attaches greater significance to technological foresight when it comes to the development of systems that not only function efficiently but also adapt to an evolving environment, ultimately enhancing the overall performance and reliability of advanced actuation systems across various disciplines.

REFERENCES

1. Abbas, A., Bugeda Castelltort, G., Ferrer Vaccarezza, E., Fu, C., et al. (2017). Drag reduction via turbulent boundary layer flow control. https://core.ac.uk/download/159237966.pdf (accessed on 14th October 2025).
2. Abidi, S. H., Cianchetti, M., Falotico, E., Laschi, C., et al. (2019). A bistable soft gripper with mechanically embedded sensing and actuation for fast closed-loop grasping. https://core.ac.uk/download/543788444.pdf (accessed on 14th October 2025).
3. Acer, M., Sabanovic, A., & Şabanoviç, A. (2012). Sliding-mode control of a flexure-based mechanism using piezoelectric actuators. https://core.ac.uk/download/pdf/11743451.pdf (accessed on 14th October 2025).
4. Agnew, D. L., & Jones, P. A. (2025). *Large deployable reflector (LDR) system concept and technology definition study. Volume 2: Technology Assessment and Technology Development Plan*. https://core.ac.uk/download/pdf/42810764.pdf (accessed on 14th October 2025).
5. Agogino, A. K., Agogino, A. M., Bronars, A., Cera, D., et al. (2017). Inclined surface locomotion strategies for spherical tensegrity robots. http://arxiv.org/abs/1708.08150 (accessed on 14th October 2025).
6. Akin, D. L., Kurtzman, C. R., Minsky, M. L., Thiel, D. A., et al. (2025). *Space applications of automation, robotics and machine intelligence systems (ARAMIS), phase 2. Volume 2: Telepresence Project Applications*. https://core.ac.uk/download/pdf/42850922.pdf (accessed on 14th October 2025).
7. Akyildiz, I. F., Kak, A. C., & Nie, S. (2020). 6G and beyond: The future of wireless communications systems. *IEEE Access, 8*, 133995–134030. https://doi.org/10.1109/access.2020.3010896 (accessed on 14th October 2025).
8. Albu-Schaffer, A., Stokes, A. A., Marchese, A. D., Marchese, A. D., Gaikwad, A. M., Mazzolai, B., Mosadegh, B., et al. (2014). Design, fabrication and control of soft robots. https://core.ac.uk/download/78063761.pdf (accessed on 14th October 2025).
9. Anand, U., Dey, A., Singh Chandel, A. K., Sanyal, R., Mishra, A., Pandey, D. K., De Falco, V., et al. (2022). Cancer chemotherapy and beyond: Current status, drug candidates, associated risks and progress in targeted therapeutics. *Genes & Diseases, 10*, 1367–1401. https://doi.org/10.1016/j.gendis.2022.02.007 (accessed on 14th October 2025).
10. Anderson, M., Bouman, A., Burdick, J., Izraelevitz, J., et al. (2020). Design and autonomous stabilization of a ballistically launched multirotor. https://

core.ac.uk/download/334986371.pdf (accessed on 14th October 2025).

11. Anderson, R. D., Flora, C. C., Nelson, R. M., Raymond, D., et al. (2025). Development of weight and cost estimates for lifting surfaces with active controls. https://core.ac.uk/download/pdf/42884138.pdf (accessed on 14th October 2025).
12. Bacciu, D., Bahr, R., Bröring, A., Cavallo, F., et al. (2017). Internet of robotic things: Converging sensing/actuating, hypoconnectivity, artificial intelligence and IoT platforms. https://core.ac.uk/download/84045908.pdf (accessed on 14th October 2025).
13. Bailey, A. R., Irish, L. A., & Tagge, G. E. (2025). *Systems Study for an Integrated Digital-Electric Aircraft (IDEA)*. https://core.ac.uk/download/pdf/42846809.pdf (accessed on 14th October 2025).
14. Baldo, L., De Martin, A., Jacazio, G., & Sorli, M. (2025). A systematic literature review on PHM strategies for (hydraulic) primary flight control actuation systems. *Actuators*. https://www.semanticscholar.org/paper/fc04325ba1aaddef54ea503a5a87fc9978b6bbf5 (accessed on 14th October 2025).
15. Bement, L. J., & Schimmel, M. L. (2025). *Advanced Emergency Openings for Commercial Aircraft*. https://core.ac.uk/download/pdf/42842675.pdf (accessed on 14th October 2025).
16. Bergonz, F. H., Bulboaca, M. A., & Derocher, W. L. (2025). *Spacecraft Servicing Demonstration Plan*. https://core.ac.uk/download/pdf/42848239.pdf (accessed on 14th October 2025).
17. Bogaerts, W., Edinger, P., Errando-Herranz, C., Gylfason, K. B., et al. (2020). MEMS-enabled silicon photonic integrated devices and circuits. https://core.ac.uk/download/287939692.pdf (accessed on 14th October 2025).
18. Boukallel, M., Chaillet, N., Grossard, M., Rotinat-Libersa, C., et al. (2009). Mechanical and control-oriented design of a monolithic piezoelectric microgripper using a new topological optimization method. https://core.ac.uk/download/52696285.pdf (accessed on 14th October 2025).
19. Bowles, A., Derrick, A., Jamieson, P., Leithead, W., et al. (1992). Innovative concepts for aerodynamic control of wind turbine rotors. https://core.ac.uk/download/9035308.pdf (accessed on 14th October 2025).
20. Brunton, S. L., Kaiser, E., Nair, A. G., Noack, B. R., et al. (2018). Cluster-based feedback control of turbulent post-stall separated flows. http://arxiv.org/abs/1809.07220 (accessed on 14th October 2025).
21. Cheng, J., Xue, N., Qiu, B., Qin, B., Zhao, Q., Fang, G., Yao, Z., Wang,

W., Zhang, L., & Zhou, H. (2025). Recent design and application advances in micro-electro-mechanical system (MEMS) electromagnetic actuators. *Micromachines, 16*. https://www.semanticscholar.org/paper/f12c30913941508784cfe33c0a2d949e423887b3 (accessed on 14th October 2025).

22. Cheng, M., Wang, Z., Wu, Z., Yuan, S., Cen, M., Wang, J., & Liu, Y. J. (2025). Alignment strategies and morphing mechanisms in liquid crystal elastomers: From fundamentals to advanced applications. *Macromolecular Rapid Communications*. https://www.semanticscholar.org/paper/80a00ab075c01692d60355540b7108ee0622f7bd (accessed on 14th October 2025).

23. Chi, Y., Li, Y., Zhao, Y., Hong, Y., Tang, Y., & Yin, J. (2022). Bistable and multistable actuators for soft robots: Structures, materials, and functionalities. *Advanced Materials, 34*. https://doi.org/10.1002/adma.202110384 (accessed on 14th October 2025).

24. Choi, S. Y., Kim, J., Song, E. H., Park, E., Wu, J., Yoo, J., & Nam, J. M. (2025). Design, mechanisms, and applications of DNA-mediated dynamically reconfigurable plasmonic gold nanostructures. *Small Methods*. https://www.semanticscholar.org/paper/9e399fcf8e65b6d1bc87a46efedf8c492fcc2890 (accessed on 14th October 2025).

25. Cohen, G. C., & Mclees, R. E. (2025). An example of requirements for Advanced Subsonic Civil Transport (ASCT) flight control system using structured techniques. https://core.ac.uk/download/pdf/42816455.pdf (accessed on 14th October 2025).

26. Corpe, S. J. O. (2016). GPS-guided mobile robot platform featuring modular design elements for agricultural applications: A thesis presented in partial fulfilment of the requirements for the degree of Masters of Engineering in Mechatronics at Massey University Turitea Campus, Palmerston North, New Zealand. https://core.ac.uk/download/148643520.pdf (accessed on 14th October 2025).

27. Cyriac, M., Gopakumar, A., Nazeer, N. P., Ds, S., & Warrier, S. J. (2025). Haptic glove guided robotic arm. *International Journal for Multidisciplinary Research*. https://www.semanticscholar.org/paper/5e054d23314b3c389fade110238ae637a9fb9fa0 (accessed on 14th October 2025).

28. de Lange, E. S., Heldens, W. B., Iost Filho, F. H., Kong, L., Li, X., & Dicke, M. (2020). Drones: Innovative technology for use in precision pest management. https://core.ac.uk/download/286361618.pdf (accessed on 14th October 2025).

29. Del Bono, F., Di Trani, N., Demarchi, D., Grattoni, A., & Motto Ros, P. (2025). Active implantable drug delivery systems: Engineering factors, challenges, opportunities. *Lab on a Chip*. https://www.semanticscholar.org/paper/5b9e26825c1b226c657eb3ccbe84080516453bec (accessed on 14th October 2025).
30. Dertien, E. (2006). Dynamic walking with Dribbel. https://core.ac.uk/download/pdf/11461509.pdf (accessed on 14th October 2025).
31. Dodd, L., Fu, R., Luo, J. T., McHale, G., Xu, B., & Wells, G. G. (2020). Integrating microfluidics and biosensing on a single flexible acoustic device using hybrid modes. https://core.ac.uk/download/286716189.pdf (accessed on 14th October 2025).
32. Duffie, N. A., Wiker, S. F., & Zik, J. J. (2025). Test bed experiments for various telerobotic system characteristics and configurations. https://core.ac.uk/download/pdf/42822946.pdf (accessed on 14th October 2025).
33. Ferris, J., Lim, S., Mehta, M., Misuraca, G., Yelin, A., & Williams, C. (2019). Human-centered electric prosthetic (HELP) hand. https://core.ac.uk/download/225379679.pdf (accessed on 14th October 2025).
34. Gaugh, W. J., & Howard, V. W. (2025). Apollo docking test device design study final report. https://core.ac.uk/download/pdf/85255234.pdf (accessed on 14th October 2025).
35. Gill, E., Monna, G. L. E., Scherpen, J. M. A., Leenaerts, B., & Olthuis, W. (2007). MISAT: Designing a series of powerful small satellites based upon micro systems technology. https://core.ac.uk/download/pdf/12941826.pdf (accessed on 14th October 2025).
36. Hassan, H. Z., & Saeed, N. (2024). Advancements and applications of lightweight structures: A comprehensive review. *Deleted Journal, 1*. https://doi.org/10.1007/s44290-024-00049-z (accessed on 14th October 2025).
37. Hr, Y., E N, R., Vutukuri, S. P., & Sharmila, A. (2023). Design of maneuverable robotic arm in pharmaceutical industry. *2023 Innovations in Power and Advanced Computing Technologies (i-PACT)*. https://www.semanticscholar.org/paper/25da14f65a54c2ae49abbfb966725d842111b62b (accessed on 14th October 2025).
38. Koo, B. J. (2014). A 2 degree-of-freedom SOI-MEMS translation stage with closed loop positioning. https://core.ac.uk/download/29175019.pdf (accessed on 14th October 2025).
39. Lipson, H., & Malone, E. (2005). Freeform fabrication of ionomeric polymer-metal composite actuators. https://core.ac.uk/download/287648532.pdf (accessed on 14th October 2025).

40. Luo, M., Su, X., Kazemi, E., Jin, X., & Khayyer, A. (2025). Review of smoothed particle hydrodynamics modeling of fluid flows in porous media with a focus on hydraulic, coastal, and ocean engineering applications. *Physics of Fluids*. https://www.semanticscholar.org/paper/bdb86b2d765868cfe6c7064e138f08d9b7d03709 (accessed on 14th October 2025).

41. Pan, M., Liu, M., Lei, J., Wang, Y., Linghu, C., Bowen, C., & Hsia, K. J. (2025). Bioinspired mechanisms and actuation of soft robotic crawlers. *Advanced Science, 12*. https://www.semanticscholar.org/paper/d26b9a78022f39fff7406e7fa68c50a873c55a17 (accessed on 14th October 2025).

42. Rasheed, A., San, O., & Kvamsdal, T. (2020). Digital twin: Values, challenges and enablers from a modeling perspective. *IEEE Access, 8*, 21980–22012. https://doi.org/10.1109/access.2020.2970143 (accessed on 14th October 2025).

43. Wee, B. J. (2013). Assessment and preliminary model development of shape memory polymers mechanical counter pressure space suits. https://core.ac.uk/download/17029751.pdf (accessed on 14th October 2025).

44. Xiong, J., Chen, J., & Lee, P. S. (2020). Functional fibers and fabrics for soft robotics, wearables, and human–robot interface. *Advanced Materials, 33*. https://doi.org/10.1002/adma.202002640 (accessed on 14th October 2025).

45. Zhang, C., Yang, C., Xu, J., Liu, S., Liu, T., Zhai, X., Xie, X., et al. (2017). A review of smart materials in tactile actuators for information delivery. http://arxiv.org/abs/1708.07077 (accessed on 14th October 2025).

46. Zou, J., Wang, Y., Cheng, J., Cai, W., Li, J., Xu, L., Fan, W., Tan, H., & Li, Y. (2025). One leader of smart materials: Two-way shape memory polymers. *Journal of Applied Polymer Science*. https://www.semanticscholar.org/paper/4d803b1d21a7c6af75bd18c6fabe3dd2306ae69b (accessed on 14th October 2025).

CHAPTER 5

MEMS FABRICATION TECHNIQUES

Contents

INTRODUCTION

The introduction of microelectromechanical systems (MEMS) has transformed various industries by enabling the miniaturization of devices that used to be heavy and cumbersome. The technology brings together mechanical parts, sensors, and electronics in a very tiny scale, paving the way for innovations across the automotive, medical, and telecommunications sectors. Advanced micromachining techniques like deep reactive ion etching (DRIE) and electron-beam lithography have played pivotal roles in facilitating the precision and scalability needed in MEMS fabrication. The enhanced capability to design structures and materials at the micrometer level requires thorough familiarity with the supporting concepts along with diverse processes utilized throughout the fabrication process. As demand for high-performance MEMS devices grows, the search for superior fabrication technologies remains crucial in pursuing further technological development. Wafer bonding and packaging are important in the case of MEMS fabrication, as they ensure device performance and reliability. Wafer bonding enables the integration of heterogenous materials, enhancing the functional characteristics of MEMS devices by creating intricate geometries that are otherwise unimaginable. Additionally, proper packaging solutions play a key role in the protection of sensitive microcomponents from extreme environmental conditions, substantially increasing the life and usage convenience of MEMS. Combining surface micromachining process flows with soft lithography techniques again indicates the diverse methodologies employed by engineers and scientists in this field. Each technique has a particular advantage, and their judicious application is capable of optimizing device performance according to specific operating requirements. Thus, familiarity with various fabrication techniques is a prerequisite for any person who wishes to innovate in the area of MEMS. Furthermore, the evolution of MEMS fabrication techniques continues due to the need to solve practical problems such as noise reduction and improvement of accuracy in sensing systems such as imaging and identification. For instance, where traditional imaging methods fail in certain environments, MEMS technology breakthroughs have brought forth better segmentation methods, which equate to improved quality data obtained for many applications. Studies show that methods like the Circular Hough transform can significantly enhance performance through improved accuracy and reduced error rates, even in noisy environments (Ahmad et al., 2018). Furthermore, MEMS technologies being applied to new fields such as micro fuel cells demonstrate the potential and the versatility of the technology. With the use of conventional microfabrication techniques, scientists were able to create highly efficient miniature-scale fuel cells with the potential to revolutionize portable power sources (Gauthier-Manuel et al., 2006). As we move further into the intricacies of MEMS processing, it is clear that continued innovation will be necessary to achieve the full potential of this cutting-edge technology.

Overview of Fabrication Techniques

The landscape of MEMS fabrication technologies is characterized by the vast variety of various approaches, each tailored to fit specific applications and types of materials. Advanced micromachining processes such as Deep Reactive Ion Etching (DRIE) and Electron Beam (E-beam) lithography are gaining more popularity due to the fact that they are capable of forming complex, high-resolution structures at the microscale. DRIE is particularly valuable for creating deep and narrow etches in silicon and thus enabling complex three-dimensional designs essential to many MEMS applications. In comparison, E-beam lithography possesses unmatched precision and versatility and can be employed to rapidly prototype structures with extremely small feature sizes. These sophisticated techniques not only solidify the foundation of MEMS technology but also open the door to innovation in sensor design and performance, putting devices where they belong for present applications in healthcare, automotive, and consumer electronics. Underlying the MEMS fabrication process are wafer bonding and packaging techniques that enable microstructures to be integrated into functional devices without a hitch. Wafer bonding is the bonding of two silicon wafers or a silicon wafer with some other substrate to create a robust structure that is capable of resisting the required mechanical and thermal properties.

The bonding is done by various techniques like direct bonding, adhesive bonding, and eutectic bonding and each selectively cater to the thermal and mechanical requirements of various devices. Besides, packaging techniques are crucial to ensuring MEMS protection from environmental hazards and ensuring stable operation. They involve material and process selection of cautious materials and methods for encapsulation of the delicate structures with allowances for necessary interactions with the external environment, thus enhancing the reliability and life of MEMS devices in actual use. Surface micromachining and soft lithography form other advances in MEMS device generation. Surface micromachining enables the creation of complex structures through the deposition of thin films on substrates followed by the etching away of layers to reveal the desired geometries. The technique is most attractive during the fabrication of microtransducers, in which the interaction between the material properties can determine device performance. Alternatively, soft lithography employs elastomeric materials to create molds for the replication of microstructures and provides a versatile platform for fabricating polymer-based MEMS devices. Such techniques not only facilitate the creation of devices with specific functions but also facilitate the exploration of new materials and geometries, enhancing the potential of MEMS application in complicated systems. That potential is emphasized in exhibitions of experimental devices incorporating sensors, which demonstrate the promise of MEMS technology, as reflected in research articles at academic conferences (Fluitman et al., 1996) (Figure 5.1).

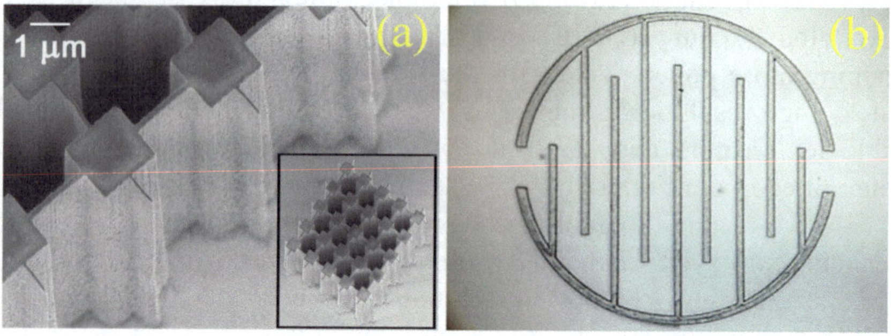

Figure 5.1. Advanced micromachining techniques such as deep reactive ion etching (DRIE) and electron beam (E-beam) lithography have gained prominence due to their ability to produce high-resolution, intricate structures on the microscale.

Source: https://www.mdpi.com/nanomaterials/nanomaterials-12-02754/article_deploy/html/images/nanomaterials-12-02754-g015.png.

5.1. ADVANCED MICROMACHINING

Sophisticated micromachining processes have revolutionized the process of manufacturing complex microstructures involved in designing advanced MEMS devices in the domain of MEMS fabrication processes. Deep Reactive Ion Etching (DRIE) is a pioneering process, enabling the manufacturing of high aspect ratio structures with unprecedented precision. DRIE is a process involving alternating cycles of etching and passivation, enabling deep recesses in silicon substrates with maintaining vertical sidewalls. The efficacy of DRIE is also supplemented by its material compatibility with a variety of materials other than silicon to allow MEMS devices to be coupled with a large number of functionalities. The application of E-beam lithography further enhances the resolution of microscale structures such that patterns can be defined at a nanoscale level. These advances not just improve the structural robustness of the MEMS devices but also provide opportunities for their application in various fields, ranging from biomedical sensors to microfluidics. Wafer bonding has been revealed to be a major application of advanced micromachining that addresses challenges in integrating materials in MEMS design. It makes dissimilar materials integratable to create multilayer structures that are able to take advantage of each layer's individual strength. One such general method is adhesive bonding or thermocompression bonding methods, which give strong interfaces without sacrificing the material's functional properties. Perhaps most importantly, wafers' can be bonded, so complex architectures can be achieved with little dimensional change, thereby enhancing device performance. By employing such methods in the fabrication process, the engineers can utilize the mechanical and electrical characteristics of a number of materials so that there may be new designs in MEMS devices for applications in communication systems as well as micro-electromechanical sensors. Surface micromachining process flows are another significant advancement in the evolution of microfabrication. This involves the layer-by-layer etching and deposition of thin films to build microstructures in layers, enabling the use of three-dimensional structure with two-dimensional patterns. Surface micromachining is particularly desirable due to the fact that it is quite low-cost and straightforward, making it an economically viable means of mass-producing MEMS devices. The integration of soft lithography processes facilitates the fabrication of intricate microfluidic channels and biomimetic structures at reduced costs with potential applications in future biological systems and chemical analysis. Furthermore, polymer MEMS utilize these processes to create light and flexible devices that are better suited for application in wearable devices and portable sensors that include fluctuating environmental conditions (Berenschot et al., 2000). As these techniques mature, they promise to unlock new frontiers within MEMS technology, presenting opportunities for innovation across various industries.

5.1.1. Deep Reactive Ion Etching (DRIE)

The intricate domain of microfabrication relies heavily on advanced techniques that enable the fabrication of high-aspect-ratio structures that are vital in MEMS applications. All these technologies and techniques, DRIE is arguably a pivotal process, best known for its capability to create vertical sidewalls and deep trenches in silicon and other materials. This method uses a sequence of plasma etching and passivation processes to produce etching profiles not possible with conventional wet etching or other forms of dry etching. By switching between etching and passivation cycles, DRIE is able to preserve the integrity of etched features while reducing undercut, which is necessary for applications from microsensors to microfluidic devices. Thus, the precision and efficiency of DRIE have significant impacts on the overall performance and reliability of MEMS devices. Moreover, the application of DRIE has special advantages regarding the speed of fabrication and material compatibility. Through the use of plasma, not only is the etching accelerated but integration of other materials beyond silicon, such as silicon carbide and glass, becomes possible. Such adaptability is a necessity in MEMS fabrication, when device requirements prefer to stipulate a range of material properties and performance. Furthermore, the scalability of DRIE is an advantage for research as well as high-volume manufacturing with support for any batch size and device shape. The complementary nature of high processing rates and material compatibility underscores the pivotal role of DRIEs in the advancement of MEMS technologies and innovation-driven breakthroughs in telecommunication, automobile systems, and medical devices. Current advancements have been focused on maximizing the effectiveness of DRIE through improved system designs and process refinements. For instance, the intersection of MEMS technologies with emerging applications is manifest in projects like EDIFICE, which combine advanced micro-manufacturing techniques that include DRIE for thermal management of components in high-speed electronics (Amon et al., 2003). Such intersection of DRIE with MEMS not only reflects the versatility of the etching technique but also underscores its usefulness in manufacturing devices that can meet expanding needs in heat management and thermal dissipation. As the demand for miniaturized efficient electronics continues to grow, DRIE's role in advancing microfabrication and addressing future technological demands becomes increasingly critical (Kodama et al., 2000). The ongoing advancement of DRIE technologies is an indication of a general movement in MEMS production where innovation takes center stage in guaranteeing efficient and stable microsystems (Figure 5.2).

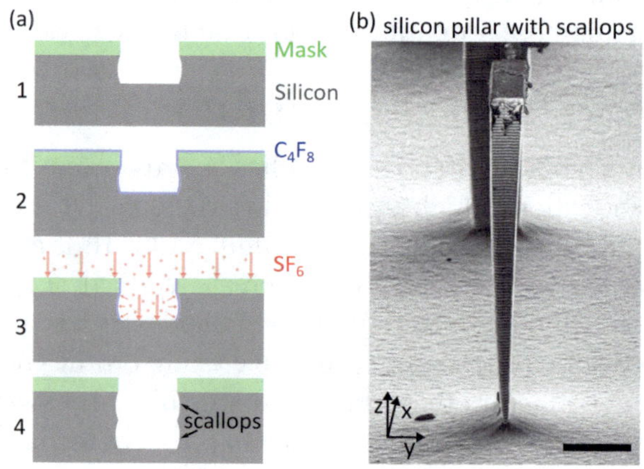

Figure 5.2. Deep reactive ion etching (DRIE) stands out as a critical process, particularly renowned for its capability to produce vertical sidewalls and deep trenches in silicon and other materials.

Source: https://www.mdpi.com/micromachines/micromachines-12-00542/article_deploy/html/images/micromachines-12-00542-g001.png.

5.1.2. Electron Beam Lithography (E-Beam lithography)

The MEMS fabrication field has benefited enormously from the precision and versatility offered by various lithographic techniques, with Electron Beam Lithography (E-beam lithography) standing out as notable among them.

Employing a tightly focused beam of electrons to write intricate patterns onto an electron-sensitive film-coated substrate, E-beam lithography enables one to realize features that are often not within reach of standard photolithography. This capability to form high-resolution patterns makes it ideal in uses such as sensors, actuators, and microstructures. Secondly, E-beam lithography is very effective in forming tailored designs, making it possible for scientists and manufacturers to rapidly prototype systems for specified operational parameters.

Figure 5.3. Electron beam lithography (e-beam lithography).

Source: https://upload.wikimedia.org/wikipedia/commons/thumb/e/e3/EB_litograph.jpg/1200px-EB_litograph.jpg.

Its flexibility not only enhances the design flexibility but also provides the potential in specialized areas of MEMS manufacturing where conventional methods are not precise enough to satisfy the required specifications. In addition to its prototyping characterization role, E-beam lithography also became a critical high-resolution patterning technology of large areas. This feature is particularly applicable in periodic structure applications, for instance those utilized in telecommunications or photonics. It is possible to compare with developments in other lithography technologies, such as the innovative techniques using the Talbot effect, which have benefits like increased depth of focus and high volumes. E-beam lithography stands apart with the possibility of resolving more than 10 nanometers, freeing the industry of the necessity for continuous miniaturization. This capability renders E-beam lithography indispensable in the evolutionary trend of MEMS, particularly with component miniaturization as a key factor to maintaining competitive function and performance in electronic devices. Utilization of E-beam lithography does come, however, with challenges that must be balanced by innovative responses to maintain efficiency. The inherent slowness of E-beam processes compared with other lithographic processes can lead to increased fabrication time, especially with scaling to large area. Researchers are actively pursuing these limitations, including the use of advanced mask technologies that can potentially double throughput. For example, the application of optimized mask designs will reduce the complexity of patterning and boost the rate of E-beam applications. Furthermore, the application of complementary processes, such as the deep reactive ion etching (DRIE) and the electroplating process outlined in recent studies, can provide additional precision to the precise fabrications provided by E-beam lithography, thereby validating its standing in the grand scheme of MEMS fabrication techniques for high quality and efficiency in microstructure production (Ehrfeld et al., 1995; Ekinci et al., 2018) (Figure 5.3).

5.1.3. Comparison of Advanced Micromachining Techniques

The progress made in micromachining technology has led to the development of advanced methods with tremendous improvement in fabrication methods for microelectromechanical systems (MEMS). Among them, deep reactive ion etching (DRIE) and electron beam lithography are two cutting-edge methods. DRIE stands apart in delivering high aspect ratio structures with controlled profiles, which is critical in the majority of MEMS applications. It operates by achieving a balance between the deposition and etching rates to create desired profiles, something that is particularly crucial given the intricacy involved in creating such minute structures. By employing a combination of gases, for example, SF_6O_2 at cryogenic temperatures, the experts are able to obtain maximum etch rates at high selectivity over mask materials, which is best suited for generating intricate structures that are central in advanced microsystems applications (Boer et al., 2002). Furthermore, wafer bonding and packaging technologies are now unavoidable in order to achieve the required integration and functionality of MEMS devices. This module for fabrication allows the accumulation of various materials to enhance device operation and reliability. Some bonding methods, such as anodic, adhesive, and fusion bonding, respond to specific microsystem demands, each offering specific advantages regarding thermal, electrical, and mechanical factors. These techniques make it not only feasible but also possible for complex assemblies to function cohesively within their operating environments. Fundamentally, choice of bonding method can have a significant contribution in deciding the performance and lifetime of MEMS devices, highlighting the importance of employing correct methods in the broader scope of microsystems technology (Boer et al., 2002). Surface micromachining, sometimes used along with soft lithography and polymer MEMS, offers yet another choice to the repertoire of micromachining. It makes it possible to generate large and intricate microstructures on silicon wafer surfaces without involving bulk micromachining methods. Soft lithography, on the other hand, utilizes inexpensive, pliable materials such as elastomers to transfer microstructures, which makes it suitable for rapid prototyping. This integration of techniques allows hybrid devices that take the best from the strengths of rigid and flexible materials, in effect broadening MEMS applications across industries from biomedical devices to microfluidics. These all-contemporary micromachining techniques play their part in the broad mission of enhancing functionality, efficiency, and manufacturability in MEMS fabrication, leading it to be a fast-growing field for future development (Fluitman et al., 1996).

5.2. WAFER BONDING

In microelectromechanical systems (MEMS), the combination of two or more materials is critical in the creation of intended functions and top-level performance characteristics. Wafer bonding is a critical process that is used to facilitate the interface of various substrates, thus allowing the fabrication of complex devices with distinct material characteristics. Wafer bonding aids the stacking of various materials, such as semiconductors, ceramics, and polymers, effectively improving performance. Techniques such as direct bonding, adhesive bonding, and eutectic bonding are usually employed, each with its own advantages depending on the desired application. Direct bonding, for instance, is widely used owing to its compatibility with silicon and other materials, satisfactory low surface roughness, and thus facilitating high-performance devices. Furthermore, developments in wafer bonding have introduced new possibilities for the manufacturing of advanced photonic devices, such as demonstrated in recent success with InGaP-on-insulator substrates. The bonding of these materials through wafer bonding enables high-speed communication and many nonlinear optical effects, which are crucial to sensing and signal processing technologies. Notably, the process outlined in this case exhibits the capability of wafer-scale processing that fabricates thousands of devices in one run. These devices, with their intricate structures like chip-to-fiber couplers and microring resonators, are just right examples of how wafer bonding can lead to breakthroughs in photonic technologies. With the use of high optical nonlinearity and low propagation loss of some materials, the wafer-bonding-inspired techniques greatly enhance microscale device performance, which brings huge promises to possible research directions in photonics technology. Along with the advancements of MEMS technologies, miniaturized and optimized packaging solutions also become more and more necessary. Wafer bonding brings a strategic advantage here, particularly when it is used together with state-of-the-art packaging technologies. Recent research has shown the feasibility of designing wafer-level hermetic structures with very good dimensional constraints—thicknesses of less than 200 μm and areas smaller than 1 mm^2. Not only do these developments allow for micromachined components to be packaged in smaller packages, but also their reliability and performance are enhanced by shielded mechanisms from environmental factors. Techniques like metal patterning for electrical contacts and seal rings permit co-deposition during wafer bonding, making the manufacturing process easier. Taking advantage of these advanced techniques, wafer bonding is a key facilitator of high-throughput MEMS production, which in return paves the way for next-generation electronic devices that are functional yet compact (Huang et al., 2023) (Figure 5.4).

Figure 5.4. Wafer bonding serves as a crucial process that allows for the union of different substrates, enabling the manufacture of complex devices with distinct material properties.

Source: https://atomica.com/wp-content/uploads/2022/03/wafer-bonding.jpg.

5.2.1. Types of Wafer Bonding Methods

One of the primary steps of MEMS fabrication is the high-accuracy process of wafer substrate bonding to enhance the strength and functionality of devices. Various processes of wafer bonding exist, each with unique characteristics applicable for ranging from distinct to selective applications. Mechanical wafer bonding uses physical pressure for bond formation, while adhesive wafer bonding uses polymers or epoxy for maximum possible bond strength. These methods are prevalent since they possess relatively simple implementation and low infrastructure requirements. Thermal bonding, being another prevalent method, employs high temperature to create covalent bonds within the interface and offers considerable mechanical strength, particularly suitable for use in high-temperature applications. Surface treatment or coating is also a significant factor that could influence the strength of the bonding as well as the quality of the resultant MEMS devices. Chemical vapor deposition (CVD) methods are a mature method of wafer bonding where a bond is formed by thin film deposition. The process allows for the integration of numerous materials and hence enhances the functionality of the bonded structures. An investigation study of LPCVD silicon nitride and LPCVD polysilicon CVD sealing materials shows the benefits of such materials as encapsulation materials for micromachined cavities. Specifically, the minimum film thickness required to deliver useful sealing is not fixed, and silicon nitride and polysilicon have been found to be particularly effective sealing materials for specific etch-channel sizes (Liu et al., 1999).

Figure 5.5. Chemical vapor deposition (CVD) methods represent a sophisticated approach to wafer bonding wherein thin films are deposited to form the bond.

Source: https://upload.wikimedia.org/wikipedia/commons/b/b1/PICT0111.JPG.

CVD techniques not only enhance structural integrity but also enable the incorporation of other microcomponents, which is in line with the increasing complexity of requirements of MEMS devices. Contrarily, buried channel technology (BCT) introduces a novel method of fabricating fluidic microstructures in silicon wafers. It involves trench etching to delineate the channels and sealing by depositing an appropriate material to close the trench, obviating the need for complicated assembly and sequential lithographic steps. BCT provides high advantages, whereby it reduces surface topography, thus making it an excellent candidate for secondary processing, such as electronic circuit integration or solid-state sensor fabrication (Berenschot et al., 2000).

In reducing the fabrication process complexity, BCT maximizes efficiency and scalability in MEMS fabrication. Ultimately, choosing the right wafer bonding technology is important in determining the performance and integration of the device and thus needs careful attention in the design and fabrication process (Figure 5.5).

5.2.2. Applications of Wafer Bonding in MEMS

Utilization of wafer bonding techniques in MEMS fabrication has successfully altered the device performance and packaging reliability. The process offers precise wafer surface alignment and bonding, which effectively seals MEMS structures. By bonding the device wafer to a carrier wafer, the MEMS structure is sealed against outside environmental factors, such as humidity and contaminants. This encapsulation is particularly critical, as it maintains the functional integrity of MEMS devices, which are typically operated under various operating conditions. As noted, packaging could account for up to 80% of total device cost, whose significance gives rise to cost-efficient wafer bonding techniques in reducing total manufacturing cost while enhancing the life span of MEMS products (Kirchberger et al., 2007). Furthermore, wafer-level packaging (WLP) processes have streamlined the workflow for MEMS production. WLP minimizes the steps involved in packaging MEMS devices to a minimum by gathering a number of packaging procedures together into one single bonding process prior to wafer dicing. This is important since it allows scaling up production while simultaneously reducing the costs associated with die-level packaging, which can be economically inadvisable (Temel et al., 2020). Implementing WLP minimizes time to market and enhances device performance as it provides better hermeticity and reliability. Wafer bonding is a key process that offers a protected inert environment for delicate MEMS devices, hence accentuating the significance of package quality in the MEMS ecosystem. Among technological advancements in MEMS, applications of wafer bonding come in handy other than surface functionality; they are essential to device miniaturization. With growing demands for smaller and efficient MEMS devices, the ability to bond wafers with precision becomes essential. Techniques such as deep reactive ion etching (DRIE) and surface micromachining complement wafer bonding techniques to provide the potential for the creation of complex and dense MEMS architectures. This synergy enhances the mechanical and electrical properties essential for high-performance applications in various industries, including the automotive and medical industries. Strategically applying wafer bonding in fabricating MEMS ultimately facilitates device miniaturization while fostering innovation and competitiveness in the rapidly evolving MEMS sector (Kirchberger et al., 2007).

5.2.3. Challenges in Wafer Bonding Processes

The complex topography of wafer bonding operations poses a chain of significant challenges that have a direct bearing on the effectiveness of Micromechanical Systems (MEMS) fabrication methods. An essential issue is the thermal limitations of some bonding materials, including Benzocyclobutene (BCB), which requires bonding to be carried out at temperatures greater than 220°C. This demand reduces the application of BCB in temperature-sensitive structures, which are increasingly common in contemporary MEMS devices. As highlighted by recent studies, alternatives that enable low-temperature bonding processes to have been designed. These alternatives aim to enable successful bonding without weakening the structural integrity of sensitive materials, thus overcoming one of the key challenges in achieving optimized wafer bonding in MEMS production. Such low-temperature adhesives are the focus of the research, underlining the ongoing need for new solutions towards overcoming bonding issues in the context of advanced MEMS applications (Rodrigues et al., 2024). Coupled with the challenge of strong wafer bonding, the need for a series of wafer bonding processes in the production of three-dimensional structures often renders the process cumbersome in providing adequate bonding strength. Especially for Al-Ge eutectic bonding situations, it becomes especially crucial to possess sufficient bonding strength in sparsely populated areas during integration of complex MEMS systems. The sparsity of the bonding sites coupled with the requirements of high-temperature processing in the subsequent steps of MEMS fabrications makes this complexity harder to overcome by conventional techniques. Experts have already managed to successfully display methods such as specialized plasma dicing for testing bond strength on small structures, but these methods must continue to be developed to enhance stability and replicating in production settings. The challenge here is not just bonding but also having various MEMS components be able to be counted upon to work in conjunction with one another within the parameters of advanced designs (Wang et al., 2024). Besides, surface condition and interface properties play a critical role in determining the success of wafer bonding processes. Maintenance and acquisition of high-quality surfaces are of prime importance in developing strong adhesive bonds because minor surface contamination or variation of surface roughness can create weak interfaces. In surface micromachining processes, active processes are generally employed for surface activation and cleaning prior to bonding. Techniques such as enhanced plasma treatments have been tried to provide for improved adhesion, but there is still some challenge in reproducing such conditions for multiple fabrication runs. Consequently, providing uniformity in surface properties remains a serious problem that has a direct bearing on the reliability of MEMS devices. Solving these intricacies of wafer bonding processes is vital to further advancing MEMS technologies, which calls for ongoing R&D of surface treatment techniques and bonding processes that can lead to stable device operation.

5.3. SURFACE MICROMACHINING

MEMS has brought to the forefront of attention surface micromachining as a predominant method of fabrication. It involves the manipulation of thin films on a substrate in order to create three-dimensional structures critical to various applications including sensors and actuators. Material deposition frequently begins the process, with lithography being used for the formation of desired shapes. The reason why surface micromachining is particularly valuable is that it is able to form intricate structures with high precision, facilitated first and foremost by improvements of lithography processes such as deep reactive ion etching (DRIE) and e-beam lithography. These improvements enhance the fidelity of dimensions of the MEMS devices without excluding the incorporation of more than one functionality into a single device, being significant advances in seeking miniaturization and enhancement of performance in micro-scale applications. The agility of surface micromachining is also illustrated by its process sequences, which play a central role in maximizing MEMS device performance.

Typical sequences consist of material deposition, lithography, etching, and subsequently microstructure release. Surface micromachining exploits multiple materials like polysilicon, silicon dioxide, and even metals so that one can customize mechanical, thermal, and electrical attributes to suit certain application needs. Further, the technique is also used to combine soft lithography and polymer MEMS and move beyond the silicon-based structures to biodegradable and flexible devices. These advancements not only enhance the ability of surface micromachined devices but also open up new possibilities in biomedical applications and environmental sensors and eventually enhance the potential for impact in applications. Besides technical advantages, surface micromachining also becomes a key part in the overall MEMS manufacturing supply chain by making wafer bonding and package processes possible. These subsequent procedures are essential for device performance and reliability. The precision contributed by surface micromachining makes electronic circuit integration and solid-state sensors integration easy in MEMS devices, enhancing functionality of the devices. The method can also simplify the fabrication procedures by ensuring the ability to create complex structures using a single lithographic mask, such as described in techniques such as buried channel technology (Berenschot et al., 2000). Such efficiency not only reduces the production cost but also increases the potential of producing more reduced and functional MEMS devices, thus making surface micromachining essential in creating MEMS fabrication techniques (Kodama et al., 2000) (Figure 5.6).

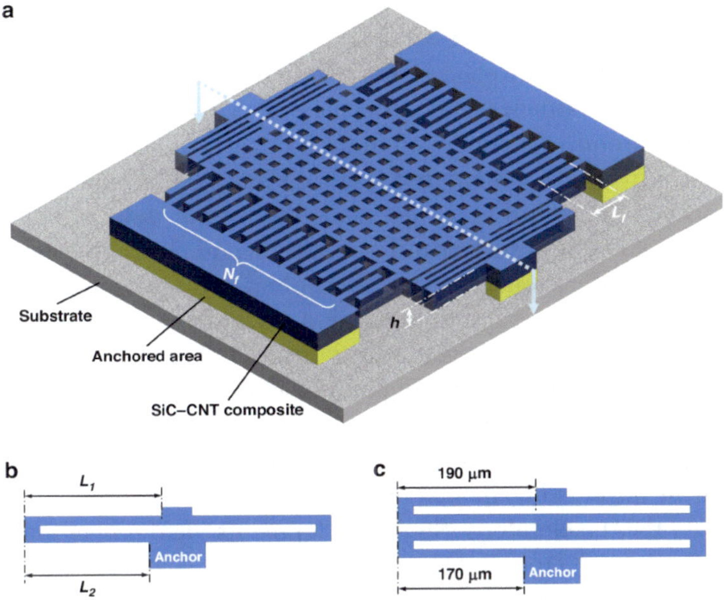

Figure 5.6. Surface micromachining leverages various materials such as polysilicon, silicon dioxide, and even metals, allowing for the tailoring of mechanical, thermal, and electrical properties to meet specific application requirements.

Source: https://media.springernature.com/lw1200/springer-static/image/art%3A10.1038%2Fs41378-024-00672-x/MediaObjects/41378_2024_672_Fig5_HTML.png.

5.3.1. Overview of Surface Micromachining

Surface micromachining is one of the major techniques utilized in the microelectromechanical systems (MEMS) field due to the fact that it is very flexible and cost-effective. Surface micromachining allows for the fabrication of sophisticated microstructures on the wafer surface with thin films deposited by methods such as chemical vapor deposition (CVD) or sputtering. The process typically involves depositing structural materials such as polysilicon or metal and subsequently patterned using photolithography. Surface micromachining is well-suited to the fabrication of high-aspect-ratio features, which are needed in most applications like sensors and actuators. With the integration of greater than one material and functionality into a single substrate, surface micromachining makes a contribution toward the development of MEMS technology and device capability extension. In addition, surface micromachining enables the fabrication of sophisticated devices such as optical phased arrays (OPAs). One such demonstration is that of an OPA fabricated through the PolyMUMPs process, highlighting the practical advantages of surface micromachining. The technique exploits suspended microstructures as optical phase shifters to improve the performance and efficiency of MEMS-based optoelectronic devices. The micromirrors employed in such devices are exceptionally precise and low-sensitivity, with a measured field of view of 4.76° and an impressive actuation speed. These attributes not only render the devices commercially viable but also exhibit how surface micromachining minimizes the entire process of manufacturing, thus rendering the technologies more viable (Mohammad et al., 2021). Besides its usage in the design of actuators and sensors, surface micromachining has also been essential in inductor design for RF MEMS purposes. The technique allows for the introduction of complex geometries required in performance enhancement, such as the solenoid and toroid inductors. Emerging developments emphasize the importance of 3D surface micromachining and its potential in introducing inductors within a single substrate with through-substrate-via (TSV) technology. Such developments yield higher quality factor and power density inductors and lower parasitic capacitance, critical in attaining efficient power electronics and neurotechnologies. As researchers continue to advance the limit of surface micromachining and additive technologies, the potential for further development of even more complex MEMS devices grows even brighter (Le et al., 2021) (Figure 5.7).

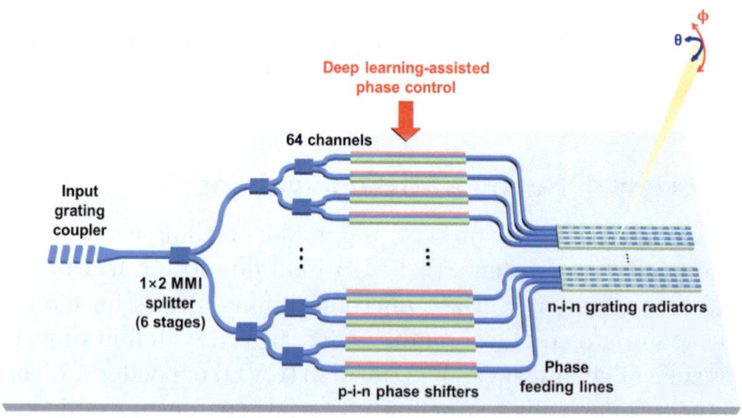

Figure 5.7. Surface micromachining facilitates the development of sophisticated devices such as optical phased arrays (OPAs).

Source: https://media.springernature.com/full/springer-static/image/art%3A10.1038%2Fs41598-023-47004-z/MediaObjects/41598_2023_47004_Fig1_HTML.png.

5.3.2. Process Flows in Surface Micromachining

An integral process of building micro-electromechanical systems (MEMS) is the intricate process flow of surface micromachining, in which intricate structures are attained at the micron scale. The process enables the formation of various components from sensors to actuators that play a pivotal role in current technological progress. Surface micromachining is founded on a sequence of etching and deposition processes in which thin layers of material are deposited onto a substrate, typically silicon, and afterwards the layers get patterned using photolithography techniques. Each layer has a different purpose, whether structuring, insulation, or structural support, finally in forming a functional MEMS device.

Thus, the precision of the process flows is important in ensuring that the resulting structures meet some design specifications and performance criteria. In advanced micromachining, methods such as deep reactive ion etching (DRIE) and electron-beam lithography have become focal in an effort to achieve high aspect ratios and intricate geometries needed by surface micromachining. The effectiveness of DRIE, for instance, is that it can regulate the profile through a delicate balance between deposition and etch rates. It enables micromachining of microstructures with accuracy, optimizing the utilization of material and the reduction of waste in an adaptable design. The intricacy of such high-tech processes is accounted for in methods that optimize etch rate and profile control so that specific design specifications are met without compromising the integrity of the structure of the product (Boer et al., 2002). Thus, accurate calibration of these parameters is the secret to the effective use of advanced surface micromachining techniques. Buried channel technology (BCT) is a beauty of innovation in surface micromachining and most appropriate for the fabrication of fluidic devices. Trench etching is employed in this method, where microchannels and cavities are formed in the silicon substrate through a series of controlled removal and deposition of material. By enabling the fabrication of intricate microstructures within a single wafer with a low level of surface roughness, BCT makes the manufacturing process overall simpler and also facilitates the integration of electronic devices into microfluidics to form new opportunities in MEMS applications. Strikingly, efficiency of BCT is again increased by the capability of BCT to attain intricate designs from a single lithographic mask, which simplifies manufacturing but enhances reliability. These advanced techniques illustrate the versatility of surface micromachining in surmounting the problems confronting microfabrication and therefore highlight its central role in MEMS technology development (Berenschot et al., 2000).

5.3.3. Advantages of Surface Micromachining

The surface micromachining flexibility significantly enhances the functionality of microelectromechanical systems (MEMS) manufacturing. Through the ability to build microstructures layer by layer, this technique enables the integration of complex geometries and functionalities. As a key MEMS technology tool, surface micromachining can be used to produce complex structures from sensors to actuators with high accuracy. The level of control enables the production of devices optimized for specific applications, thereby boosting functionality. In addition, surface micromachuring material can be selected based on its mechanical, thermal, and optical properties to enable the construction of devices that function optimally in varying environments. That it is convenient to integrate varying materials for varying operations represents a major advantage that lays grounds for surface micromachining as one of the cornerstones of today's MEMS fabrication processes. Surface micromachining contrasts with bulk micromachining in being less expensive and having greater yield, primarily due to the broader use of silicon wafers, which are inexpensive and available everywhere. It leverages established semiconductor fabrication processes, hence reducing the cost related to the development and upscaling of the manufacturing process. Furthermore, the effectiveness of surface micromachining processes yields low production times, enabling quick design iterations and prototyping. The ability to produce several layers without requiring enormous material removal is a waste-reducing quality that makes the manufacturing process economically feasible. These advantages not only make surface micromachining appealing to companies but also allow for faster cycles of innovation in the MEMS sector, thereby ultimately benefiting end-consumers through the convenience of employing high-quality and cost-effective products. The enhancement in performance by surface micromachining also extends to more sophisticated uses, namely biomedical applications. For instance, the precision and reliability of this technology allow the fabrication of extremely sophisticated devices such as lab-on-a-chip systems that are capable of conducting a range of chemical or biological assays on a single platform. Furthermore, the fabrication of microfluidic elements, as demonstrated with recent progress, illustrates the diversity of surface micromachining for the fabrication of devices capable of manipulating fluids at the microscale. Applications such as drug delivery and optical trapping are effectively feasible with the aid of micromachined structures, as evidenced from the development of micromachined flow cells for molecular dynamics analysis (Bennink et al., 2001). As the demands for miniaturization and efficiency are on the rise in devices, the advantages of surface micromachining will be instrumental in shaping the future of MEMS technologies (Figure 5.8).

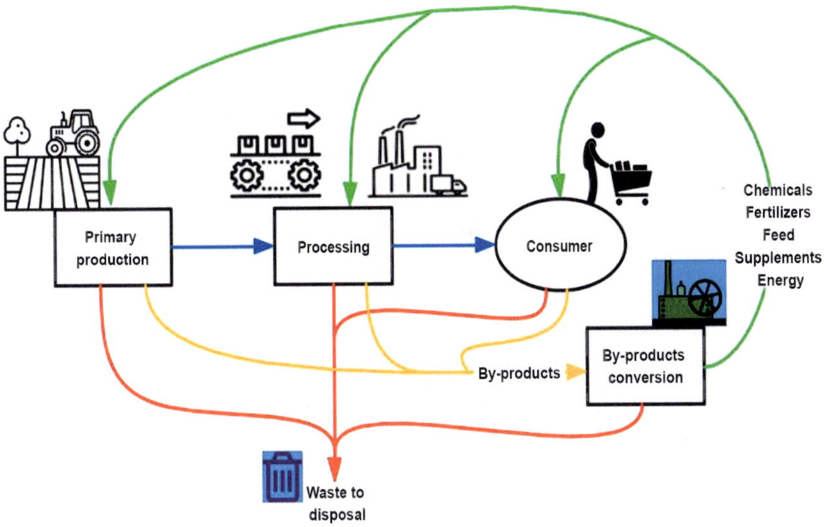

Figure 5.8. The ability to produce multiple layers without requiring extensive material removal minimizes waste, enhancing the economic feasibility of the manufacturing process.

Source: https://www.mdpi.com/energies/energies-16-01752/article_deploy/html/images/energies-16-01752-g001.png.

5.4. SOFT LITHOGRAPHY

The development of fabrication techniques has significantly added to the development of microelectromechanical systems (MEMS), even more so with the development of soft lithography. The process is special in that it can create intricate micro and nanostructures from elastomeric materials like polydimethylsiloxane (PDMS). This is because of the unique properties of PDMS, such as high optical transparency, mechanical deformability, and biocompatibility, thereby making it ideally applicable in a variety of biomedical applications. The method typically involves creating a master template with the desired patterns, which are used to make PDMS replicas. The method not only simplifies the fabrication process but also reduces costs and time invested in prototyping, allowing for rapid development in fields such as microfluidics and biomedicine, where accurate replica of the structure is essential (Inês et al., 2021). Other than its diverse applications, soft lithography has turned out to be a critical tool used in patterning features on the microscale that would allow for biosensor performance enhancement. By using the unique mechanical and electrical properties of nanomaterials in addition to the techniques of soft lithography, researchers have been in a position to develop highly sensitive biosensing platforms. These devices are of vital importance for medical and environmental monitoring, identifying significant biochemical signals and converting them into measurable responses. The use of nanomaterials such as carbon nanotubes (CNTs) and quantum dots (QDs) enables the sensitivity and detection limits to be greatly increased, addressing issues of prime importance in biosensor design, such as signal transduction and miniaturization (Naresh et al., 2021). Soft lithography flexibility and efficiency make it a viable pathway for MEMS technology development with significant implications for real-time analysis and environmental monitoring. Moreover, when combined with other processes of fabrication, soft lithography increases the functionality of MEMS devices. Scalability of soft lithography allows scientists to create complex three-dimensional microstructures that can be bonded and packaged with other MEMS devices, such as microvalves and sensors. This integration allows for deep customizations as well as the formation of multifunctional devices meant for specific applications like drug delivery systems and implants. By allowing for the interfacing of highly dissimilar fabrication techniques, soft lithography enhances system functionality and reliability. As scientists continue to explore new materials and methods, the prospects for soft lithography in the MEMS sector look good, with opportunities for groundbreaking innovation in device fabrication and use (Inês et al., 2021).

5.4.1. Principles of Soft Lithography

The exploration of new fabrication techniques is most crucial in the design of microelectromechanical systems (MEMS), particularly soft lithography. This technique stands out in its ability to replicate complex microstructures with precision and variety, which are most critical in device design meeting the stringent requirements of modern applications. By utilizing elastomeric materials such as polydimethylsiloxane (PDMS), soft lithography is capable of producing high-resolution features that can be easily controlled. Not only is it better for the manufacturing of intricate designs, but it minimizes production cost and time significantly, making it a potential option in MEMS technology. Soft lithography flexibility is best applied in biocompatible uses since it facilitates the integration of biological elements into microfabricated devices and encourages technologies like medical sensors and smart interfaces, which are essential for tactile medicine use (Guglielmelli et al., 2014). In addition to being affordable, the process involved in soft lithography is very efficient in creating high-conformity microstructures and low-temperature processing. Unlike the high-energy exposure and complex masks used in traditional photolithography, soft lithography uses a simple stamp-based process. It begins with the creation of a reusable flexible master mold that can be used to deposit features on various substrates. Such processes are particularly critical in the fabrication of devices like biosensors and microfluidic systems where precise control of fluid dynamics and sensor interaction is required. Medicine applications indicate successful integration of soft lithography with microfabricated tactile sensors that have been promising for application in surgery and robotic interfaces.

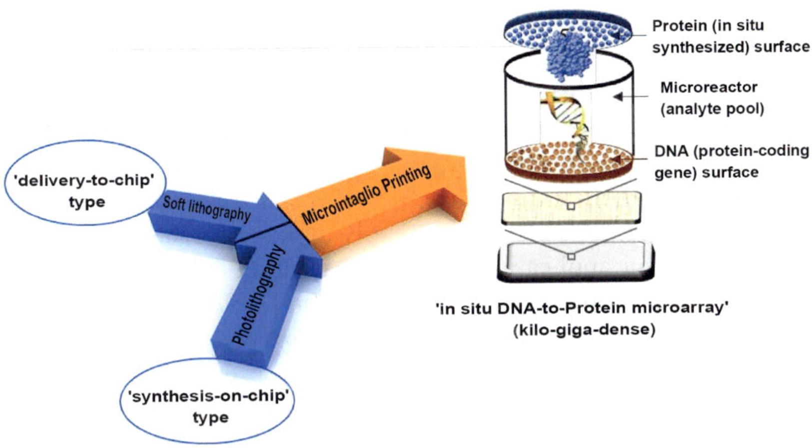

Figure 5.9. Soft lithography.

Source: https://www.mdpi.com/microarrays/microarrays-04-00311/article_deploy/html/images/microarrays-04-00311-g002.png.

Principles behind such methods reflect their significance in ensuring high performance and packaging of multiple functionalities within intricate MEMS devices, an imperative in the fast-changing technology landscape today (Hmurcik et al., 2010). In addition, the future of MEMS processing holds tremendous benefits in the continuous research aimed at developing sophisticated soft lithography techniques. With the development of new materials and improved processing efficiencies through continued research, the soft lithography capabilities will only be reinforced. The possibility of fabricating multi-level structures through soft lithography will unlock the door to intricate devices that will be capable of integrating diverse functionalities on one platform. This is particularly true in the case of tactile sensors, where more metrological qualities such as sensitivity and reliability are most vital for medical applications. As the market demands for MEMS increase across various fields from automobile to biomedical applications, the ideas of soft lithography will always be at the forefront of MEMS technology. The ongoing innovation of this technology reflects its significant input towards the future of smart and miniaturized systems, eventually expanding the area of potential applications (Guglielmelli et al., 2014) (Figure 5.9).

5.4.2. Materials Used in Soft Lithography

The versatility of applications of soft lithography relies very much on the material utilized in the processes. As a general rule, polydimethylsiloxane (PDMS) is among the most frequently used elastomers due to its excellent optical transparency and biocompatibility. PDMS can be very easily manipulated to create high-fidelity microstructures, which are extremely crucial in MEMS applications. Moreover, its inherent flexibility allows for duplication of intricate details from a master model, ensuring accuracy while having the capability to be modified to work on a variety of substrates. Apart from PDMS, other materials such as polyurethane and polyurethane acrylate are also gaining ground, especially for applications that involve improved mechanical strength or required surface properties. The tunability and ability of the materials to adjust broaden the scope of the use of soft lithography in MEMS fabrication technologies for developing innovations in sensor devices and biomedical apparatus. The surface properties of materials in the use of soft lithography are the main factors that determine the outcome of the fabrication process. Surface alteration, functionalization or coating, can be employed to modify adhesion and wettability, which is crucial during photopolymer casting or microstructure replication. Materials like SU-8, a negative photoresist material, are mixed with PDMS in the creation of master molds because of their fine-feature resolution and heat resistance. Fine-feature resolution in mold making enables the replication of complex micro-patterns onto PDMS. However, drying during production creates distortion in structures, an issue described in the latest publications where hybrid approaches based on micro-electrical discharge machining (μ-EDM) have been employed to enhance fidelity and surface finish of parts produced by soft-lithography (El-Sayed et al., 2016). This synergy represents a very worthy direction towards designing precise micro-components for critical applications, e.g., biomedical devices. The applications of soft lithography extend far into the realm of smart devices, especially in the cardiovascular implant field, where the choice of material can dictate device function. Aside from structural precision, the electrical properties of materials used with soft lithography are a prime driver in the performance of implants for monitoring and therapy applications. Material science breakthroughs have seen the development of conductive composites and polymers synthesized for use in conjunction with soft lithography processing, enhancing the efficacy of devices such as stents. Processing of such advanced medical devices is not only based on the mechanical properties of materials but also their biocompatibility and whether or not they can be employed to facilitate communications within the body environment (Bussooa et al., 2018). With advancing technology, the merging of soft lithography with novel materials is sure to come up with more sophisticated options for the therapy of cardiovascular ailments as well as better patient outcomes (Figure 5.10).

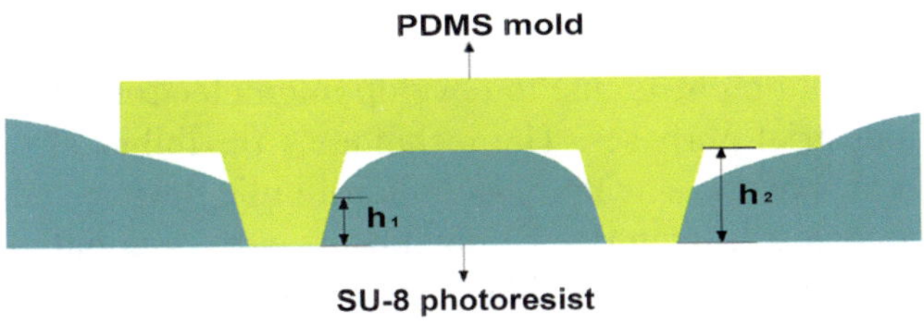

Figure 5.10. Materials, like SU-8, a negative photoresist, are used alongside PDMS for creating master molds owing to their high-resolution features and good thermal stability.

Source: https://ietresearch.onlinelibrary.wiley.com/cms/asset/9facc55c-712f-4d5f-9f6e-1f4b066308c3/mna2bf01764-fig-0007-m.jpg.

5.4.3. Applications of Soft Lithography in MEMS

The increasing demand for miniaturization and precision in various fields of technology has motivated the development and research of novel fabrication techniques for Micro-Electro-Mechanical Systems (MEMS). Among these, soft lithography has emerged as a critical method because it is able to generate intricate micro- and nanoscale features at relatively modest costs. Soft lithography is based on elastomeric materials such as polydimethylsiloxane (PDMS), which enable the high-fidelity transfer of a pattern from a master template. The flexibility of soft lithography allows for the creation of a variety of geometries and features, which is vital in applications ranging from sensors to biomedical devices. It also brings complementary capabilities when combined with other techniques, such as micro electrical discharge machining, enhancing the surface quality of components resulting from it, in addition to overcoming the inherent limitations associated with conventional micro-fabrication processes (El-Sayed et al., 2016). In medicine, soft lithography has its place in the manufacture of advanced medical devices, and it is particularly relevant to cardiovascular therapies. One good example is the smart stent, which illustrates how soft lithography can be utilized to create devices with enhanced electrical properties for monitoring and treating a number of cardiovascular diseases. With the capability to integrate sensors and communication capabilities into the stent structure, soft lithography is facilitating the creation of products that not only maintain vascular access but also enhance patient health through real-time therapy and diagnosis. The ability to craft specialized microstructures is opening up solutions for the therapy of arterial disease, thereby solving some of the most exigent public health problems such as heart attack and stroke (Bussooa et al., 2018). Hence, the integration of advanced MEMS and soft lithography enhances the functionality and performance of medical implants. Soft lithography also has other uses apart from biomedicine to develop microelectromechanical devices for different industrial purposes. The technique's flexibility also finds use in the production of sensors, actuators, and microfluidic devices, where complex fluidic networks and miniaturized components are essential. Its ability to produce high-resolution patterns without the costly expenditures of traditional lithography techniques makes it particularly well-suited for prototyping and low-lot volume production. Furthermore, soft lithography offers the option of using diverse materials, including polymers, which can be tailored to specific mechanical and chemical properties, thereby broadening the area of application in MEMS. As methods improve and innovate continuously, future prospects for soft lithography in MEMS production look good, with new solutions propelling technology in numerous areas (El-Sayed et al., 2016).

5.5. POLYMER MEMS

Microelectromechanical systems (MEMS) fabrication has highlighted the role of polymer materials in fabrication techniques due to their unique properties and low cost. Other advanced micromachining technologies like deep reactive ion etching (DRIE) and electron beam lithography are typically associated with silicon and glass substrates. However, the application of polymers in MEMS extends their functionality since devices can be made not only to be light but also very versatile. Polymers such as polydimethylsiloxane (PDMS) and polymethylmethacrylate (PMMA) have shown to be impressive in soft lithography, possessing the capability of doing rapid prototyping and variable designs that can be created and produced as per requirement. This shift to polymers is a revolutionary approach in MEMS technology that is developing innovations in fields such as biomedical devices and environmental sensors. The polymer MEMS paradigm consistently leverages processes like hot embossing, allowing for the replication of microstructures in a cost-efficient way. The process is based on the utilization of a thermoplastic polymer that melts above its glass transition temperature (T_g) and therefore can be molded into complex geometries. A fascinating expansion of the process involves the deposition of metal layers on to polymer substrates, further enhancing the mechanical and electrical performance of devices. Here, electrostatic switches fabricated using this new approach illustrate the union of traditional methods and novel materials. Metal embedding in hot embossing is a display of MEMS fabrication process versatility, leading to multifunctional polymer MEMS devices that are capable of supporting sensing and actuation on the same platform, as discussed by Chien et al. (2006). Greater application of polymer MEMS is also witnessed through the development of micro fuel cells (μ-FC), which are reflective of transition towards green sources of power for portability. Polymer systems generally vary from traditional fuel cells in that they employ micromachining techniques sensitive to their miniaturized aspect. Scientists have developed μ-FC design and operation significantly by employing MEMS fabrication techniques. For instance, a recently developed proton-exchange membrane of silane-grafted inorganic porous media has proven to be highly effective and applicable. Such innovations represent the broader impact of polymer MEMS on portable electronics and energy solutions, evidencing the ongoing merging of microfabrication and materials science. With the evolution of technology in this field, the necessity of integrating polymer MEMS into other uses cannot fail to be a major driver of coming technologies (Gauthier-Manuel et al., 2006) (Figure 5.11).

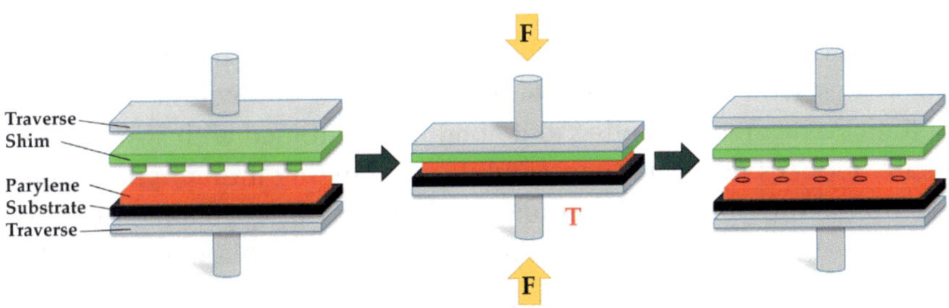

Figure 5.11. The paradigm of polymer MEMS continuously benefits from techniques like hot embossing, which allows for the efficient replication of microstructures.

Source: https://www.mdpi.com/polymers/polymers-16-02218/article_deploy/html/images/polymers-16-02218-g006.png.

5.5.1. Characteristics of Polymer MEMS

The quest for microelectromechanical systems (MEMS) has revealed that polymer materials offer special advantages, especially when the fabrication process such as soft lithography is considered. The polymers are handled more easily at room temperature, and one can produce sophisticated microstructures of high fidelity and low-cost processes. Compared to silicon-based MEMS traditionally requiring precise manufacturing processes such as DRIE and high-precision alignment, polymer MEMS have more flexibility in design. That one can leverage soft lithography techniques offers more flexibility in designing multilayer structures and facilitates applications in various fields such as biomedical devices and sensors. Moreover, polymer MEMS are less dense than silicon MEMS, and this is responsible for reduced system weight and better performance for dynamic applications. Mechanical properties of polymers for use in MEMS applications are significant factors for determining the functionality and reliability of devices. Polymers can be designed to possess desired properties such as elasticity, strength, and resistance to environmental conditions, which are of utmost significance in microactuators and sensors for applications in hostile environments. Polymers' flexibility nature allows for the integration of distinct materials and functionalities on a shared substrate and presents an opportunity to create multifunctional devices. For example, polyimide and polydimethylsiloxane (PDMS) have better mechanical properties and chemical stability and are thus beneficial in applications where durability and performance are needed. Additionally, polymer MEMS can also benefit from the mature micromachining techniques like E-beam lithography with feature precision and enhanced functional integration in microfabricated devices. The fabrication techniques of polymer MEMS also highlight their favorable characteristics with regard to scalability and mass production. Imbuing such technologies as wafer packaging and bonding, combined with soft lithography, drastically reduces fabrication costs while maximizing throughput. As outlined in recent breakthroughs, such approaches as dry etching enable micromolding that gives rise to platforms for embossing and electroplating (Ehrfeld et al., 1995). Such processes enable high aspect ratios, directional flexibility, and high selectivity that is essential in realizing complex patterns and structures in polymer MEMS. Furthermore, the flexibility of polymer material allows for ease of integration with other technologies, including metamaterials and MEMS devices, further diversifying applications in fields such as terahertz spectroscopy and sensor systems, further aiding ongoing developments in MEMS processing technologies (Kodama et al., 2000).

5.5.2. Fabrication Techniques for Polymer MEMS

The emergence of micromachining tools like Deep Reactive Ion Etching (DRIE) and electron-beam lithography has made it easier to manufacture polymer MEMS. The techniques allow the creation of high aspect ratio and complex geometries, which are essential for micro-electromechanical system functioning. DRIE, in particular, allows deep feature etching with excellent vertical profiles and is therefore useful in applications where dimension control is required to be tight. As polymer MEMS find more uses across various industries, the ability to produce such delicate structures using these advanced micromachining technologies grows ever more important. Moreover, breakthrough techniques that combine DRIE with other technologies also contribute to MEMS's flexibility of use, making them more valuable and widening their potential uses from biomedical devices to environmental sensors. Apart from micromachining operations, packaging and wafer bonding are also of utmost importance in ensuring the integrity and reliability of polymer MEMS. Wafer bonding is the technique used to bond two or more wafers together to create a sealed environment, which is required by products that must be shielded from the environment.

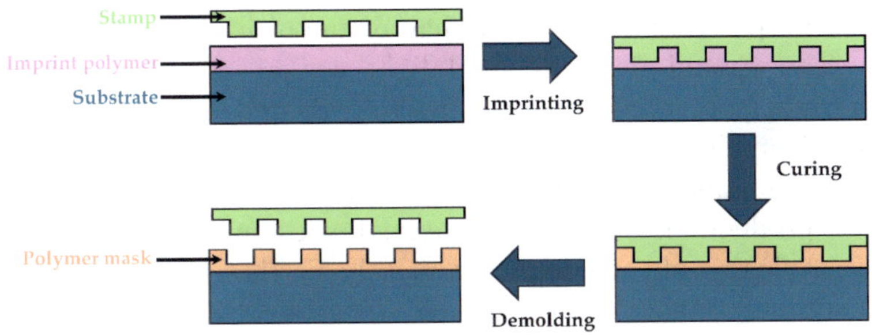

Figure 5.12. As polymer MEMS continue to gain traction in various fields, the ability to fabricate such complex structures with these advanced micromachining technologies becomes increasingly vital.

Source: https://www.mdpi.com/micromachines/micromachines-15-01137/article_deploy/html/images/micromachines-15-01137-g007.png.

The yield of this process relies greatly on adequate consideration of the involved materials because the compatibility between different polymers and substrates can greatly affect the performance of the final device. Moreover, developments in wafer bonding have also made novel architectures like multi-layered structures possible, which are being used more and more to create complex sensor arrays and actuators. As MEMS technology advances, the integration of wafer bonding and packaging technologies that improve performance relative to manufacturability plays an essential role in the realization of commercially competitive devices. Soft lithography has emerged as the prominent technique in polymer MEMS fabrication where features of high resolution can be replicated at relatively low cost. Using elastic materials, such as polydimethylsiloxane (PDMS), soft lithography allows one to create flexible and biocompatible devices, hence the reason for high compatibility for biomedical applications. Easy and efficient pattern transfer is supported by soft lithography, leading to quick prototyping of microstructures without the need for traditional cleanroom facilities. Soft lithography's innate flexibility also facilitates the formation of various functional elements into a single device and therefore enhances overall function. Recent innovations in this field have also shown the potential of soft lithography to be combined with other techniques of fabrication, such as electroplating and embossing, and thus expand its functionality and make it a key player in the future of MEMS fabrication (Ehrfeld et al., 1995; Gauthier-Manuel et al., 2006) (Figure 5.12).

5.5.3. Advantages and Limitations of Polymer MEMS

The study of polymer MEMS has come a long way in microelectromechanical systems, with possible advantages that distinguish them from traditional materials. The natural flexibility and light weight of polymer MEMS are among the most observed strengths, making them suitable for applications related to deformation and toughness, such as wearable technology. In addition, fabrication techniques like soft lithography allow rapid prototyping and inexpensive production. Polymers may ease the design-to-manufacture process since it is simple to mold and duplicate directly. Furthermore, the integration of polymers with microfabrication techniques like the DEEMO process enhances flexibility with the ability to carry out intricate micro-molding and electroplating techniques that result in intricate geometries that traditional materials cannot copy (Ehrfeld et al., 1995). This designer flexibility provides space for innovative applications, e.g., biomedical devices and sensors that utilize polymers unique properties. While the advantages of polymer MEMS are vast, they also have inherent drawbacks to be addressed for broader application. One significant drawback is the mechanical and thermal properties of polymers compared to conventional MEMS materials such as silicon and metals. Polymers have relatively lower strength, lower temperature resistance, and lower long-term stability, factors that may degrade their performance in high-stress applications. Besides, polymers are more susceptible to environmental degradation in the form of UV exposure and chemical attack by solvents, among others, which can lead to long-term reliability issues. Delays in achieving fine resolutions during fabrication, such as in microstereolithography where lateral resolution is limited by layer thickness and settling time, make it that much more difficult to apply polymers to high-accuracy applications (Gibson et al., 2005). Hence, even though polymers possess some merits, their demerits must be counteracted seriously during the design and fabrication of polymer MEMS. The balancing between the merits and demerits of polymer MEMS necessitates ongoing research and innovation of MEMS fabrication techniques. Reducing the material limitations entails utilization of composite materials that combine polymers with other elements for the aim of strengthening them mechanically. Besides, new fabrication techniques like improved microstereolithography and dry etching processes can alleviate some of the limitations revolving around polymer MEMS (Ehrfeld et al., 1995; Gibson et al., 2005). For instance, incorporating numerical simulation methods like Computational Fluid Dynamics can optimize recoating in microstereolithography, providing enhanced control over layer thickness and application flexibility. As studies continue to unravel, the depth of possibility for polymer MEMS in various industries like automotive, healthcare, and consumer electronics may expand exponentially by both leveraging their unique properties and overcoming intrinsic challenges. Such equilibrium is critical in a bid to effectively realize the potential of polymer MEMS in the grand scheme of MEMS fabrication technologies.

5.6. EMERGING MEMS FABRICATION TECHNIQUES

The technological advancement in micromachining processes has significantly transformed the MEMS manufacturing landscape. Of these, Deep Reactive Ion Etching (DRIE) and E-beam lithography have emerged as go-to methods, offering high aspect ratio structures with unparalleled precision. DRIE allows for the creation of intricate three-dimensional geometries, which are significant in applications that demand miniaturization and mechanical performance. Simultaneously, E-beam lithography is the only technique by which nanoscale structures can be defined with great accuracy, the basis for creating next-generation MEMS devices. These techniques are not merely enhancing the resolution of microstructures but creating new opportunities for application in biomedical engineering and consumer electronics, where integration of tiny, sophisticated systems lies at the heart of innovation. As micromachining evolves further, it will dictate the future scalability and function of MEMS technology. Apart from the advanced micromachining, wafer bonding has evolved to become a central aspect of the MEMS fabrication process, particularly where two-dimensional structures need to be encapsulated and functionally integrated. Wafer bonding is a process that involves adhesive or direct adhesion of silicon or glass wafers to close MEMS devices to form a protective but functional layer enhancing performance and reliability. The packaging requirements for MEMS devices can be as much as 80% of the total device cost, which emphasizes the necessity of optimum and standard bonding processes. As the demand for MEMS devices is growing exponentially, particularly in consumer applications, technology advancements in bonding processes are a high-priority need. Standardization and cost reduction of wafer bonding will not only speed up the commercialization process of MEMS devices but also assist in the expansion of the MEMS market, which is projected to exhibit a compound annual growth rate of over 13% over the next several years, contingent upon these developments (Kirchberger et al., 2007). Moreover, surface micromachining and soft lithography are gaining greater popularity for the prospects of realizing MEMS devices with broad material properties and functionalities. Surface micromachining, which involves depositing and patterning thin films to create microstructures, accommodates the integration of a broad spectrum of materials, thus multifunctional devices tailored for specific applications can be achieved. In contrast, soft lithography makes use of elastomeric materials to create accurate patterns and is thus a convenient tool for the fabrication of polymer-based MEMS devices. This simplicity in material choice paves the way for new designs in applications like microfluidics and sensors, which are likely to impact significantly industries like biotechnology and monitoring the environment. As increasingly more research is invested in the capabilities of these fabrication techniques, the potential for MEMS technology to revolutionize space mission applications and more begins to take on a tangible form, ushering in an era of new scientific discovery (Dennehy et al., 2025).

5.6.1. 3D printing in MEMS Fabrication

MEMS fabrication has been revolutionized as a process, particularly with the integration of 3D printing technologies. The process is a solution to some of the difficulties presented by traditional fabrication processes such as photolithography and micromachining, which are both costly and time-consuming. The use of 3D printing means that manufacturers can now fabricate complex geometries and structures that are difficult to produce through conventional processes. 3D printing versatility offers the ability to attain quick prototyping of MEMS devices in a way that scientists are able to produce numerous repetitions of designs and optimize performance in a short time. Additionally, a transition to additive manufacturing methods facilitates material efficiency as it minimizes waste, thereby is in alignment with sustainable manufacturing methods, a key feature in current engineering environments. 3D printing further enables MEMS devices assembly with intricate details, which facilitates enhancing functionality and performance. For example, application of high temperature capable materials or aggressive chemical etching increases the range of uses for MEMS technologies. Additive manufacturing advances have proved highly promising in high-aspect-ratio structure fabrication, which is fundamental to most MEMS applications. The synergistic integration of 3D printing with traditional techniques, such as the combination of photolithography and negative photoresists, as described by Coutu et al. (2011), is an excellent illustration of how the hybrid process can introduce novel routes to creating complementary structures without the attendant extra time delay and cost normally associated with MEMS development. Furthermore, 3D printing has also initiated exploratory forays into novel functional arrangements in the context of MEMS fabrication that combine mechanical and digital processing capabilities. Using the integration of micro-mechanical logic gates, as outlined by Chizari et al. (2019), researchers made it possible to develop intelligent materials that provide computational functionalities in MEMS. All this hints at a pivotal shift in the direction of more autonomous systems that are not solely based on electronic devices but leverage mechanical properties to carry out digital functions. The capability to introduce computational logic into microarchitected metamaterials is a significant milestone, synthesizing form and function and introducing interactive features into MEMS devices with their surroundings. This convergence of disciplines bears witness to the ongoing evolution of MEMS fabrication techniques as spurred on by the new technology of 3D printing (Figure 5.13).

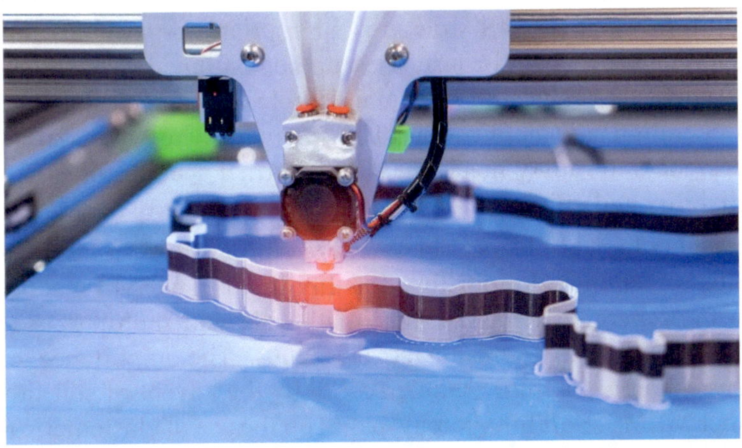

Figure 5.13. By utilizing 3D printing, manufacturers can create complex geometries and structures that are difficult to achieve with conventional techniques.

Source: https://www.3ds.com/assets/invest/styles/banner/public/2023-04/challenges-in-additive-manufacturing.jpg.webp?itok=X6O_kPUg.

5.6.2. Hybrid Fabrication Techniques

The development of microelectromechanical systems (MEMS) has made it imperative to incorporate the use of various different fabrication technologies, resulting in the development of hybrid fabrication processes. By appropriately combining some of the conventional and newer micromachining techniques, including deep reactive ion etching (DRIE) and electron-beam lithography, these processes allow for the expanded functional abilities of MEMS devices. DRIE's high aspect ratio etching makes it possible to produce complex three-dimensional features critical to the manufacture of advanced MEMS components.

Coupled with E-beam lithographs precision in patterning, hybrid techniques can produce features that allow a wide range of applications, from sensors to actuators, and provide better measures of performance. This complementarity of different fabrication routes emphasizes the need for extensive material and method knowledge in MEMS, beckoning the possibilities of innovation in microsystem technology, where sophisticated design and functionality blend, as discussed by Fluitman et al. (1996). Wafer bonding is the center of hybrid fabrication techniques, an important process that enables the integration of disparate materials in order to provide device functionality. Through techniques such as fusion bonding or adhesive bonding, it is possible for scientists to combine semiconductor layers with metals or polymers and create multifunctional MEMS structures appropriate for specific use. Surface micromachining processes, for instance, can be combined with wafer bonding to create devices that consist of rigid and flexible components, expanding their range of operation and application. The redundancy of material properties permits designers to satisfy mechanical and thermal specifications, in order to ensure reliability and efficiency in microsystems.

These technological advances in wafer bonding and packaging are indicative of the miniaturization and multi-functionality trend in MEMS manufacturing, satisfying the emerging demands for small-scale and multipurpose systems in various technologies (Kodama et al., 2000). The importance of hybrid fabrication techniques is further underscored by the advent of soft lithography and polymer MEMS. Soft lithography, based on the use of elastomeric materials in patterning, provides a complementary technique to rigid photolithography and enables one to manufacture MEMS devices with complex geometry quickly by rapid prototyping. In addition, polymer material use provides a new path for developing flexible and lightweight

MEMS devices that are particularly beneficial in situations where conventional materials would be limiting. Having the ability to shape the mechanical characteristics of polymer MEMS significantly enhances their applications in various fields from biomedical tools to flexible electronics and sensors. Thus, the blending of soft lithography with traditional MEMS fabrication methods exemplifies the energetic nature of the discipline, paving the way for innovative approaches in device functionality and design and ultimately extending the boundaries of microsystems technology (Fluitman et al., 1996).

5.6.3. Future Trends in MEMS Fabrication

MEMS manufacturing techniques development means shifting towards more advanced and precise techniques that ensure devices can meet the increasing needs of industries. Deep reactive ion etching (DRIE) is among such developments as a pertinent technique to obtain vertical sidewalls and high aspect ratios necessary in creating intricate microstructures. DRIE not only ensures improved device performance but also facilitates miniaturization for modern applications. In addition, electron beam lithography (E-beam lithography) offers higher resolution, conducive to the creation of complex patterns at the nanoscale. As MEMS devices become more functional, the combination of DRIE and E-beam lithography will lead the way in advancing solutions for issues in precision and efficiency, securing their position in future fabrication processes (Welburn et al., 2025). Wafer bonding and packaging cannot be overstressed in the process of MEMS fabrication. With increasingly complex devices, advanced techniques of wafer bonding enable the seamless integration of various materials with better performance and functionality. Hybrid bonding technologies, which join heterogeneous material at a molecular level, hold the potential to revolutionize the field of packaging in MEMS technology. These processes allow for improved thermal and electrical conductivity, ultimately leading to devices that are more efficient and stable. Furthermore, the packaging should include resistance to environmental factors, and hence it is essential in automotive sensor applications to healthcare devices. Consequently, advancements in wafer bonding have direct influences on possibilities and realization of MEMS devices across a wide range of applications (Basu et al., 2025). The future of MEMS production is also likely to witness the union of soft lithography and polymer MEMS technologies, which provide flexibility and versatility to microdevice design. Through this technique, complex three-dimensional structures are made possible with polymers, which open up more possibilities in biomedicine and wearable technology. The ability to quickly prototype and mass-produce cheap MEMS devices using soft lithography will encourage experimentation and innovation in design. In addition, surface micromachining process flows advancements are also poised to play a key role in propelling the industry, with improved speed of fabrication cycles and higher yield rates. As these methods mature, their synthesis will be poised to open new MEMS application frontiers, addressing integration challenges with biological systems and designing personalized medical devices (Welburn et al., 2025).

5.7. CONCLUSION

The diversity of fabrication processes offered through MEMS technology facilitates the creation and performance of microelectromechanical systems. Techniques such as deep reactive ion etching (DRIE) and electron-beam lithography play a critical role in the formation of high-performance devices' unique geometrical features. They can create intricate patterns on silicon wafers to guarantee that components such as those with micro-scale structures are realized for modern applications. In addition, wafer bonding and advanced packaging methods offer the confidence that such MEMS devices can be interconnected suitably with other microelectronic components so that end-use systems are made efficient. The culmination of these methods leads to added functionality, reliability, and longevity in end-use applications, thus underscoring the importance of continued innovation and enhancement in the context of MEMS manufacturing. Bridging soft lithography and polymer MEMS fabrication techniques with mainstream production methods even further expands the use of MEMS technology.

The ability to utilize flexible materials by using soft lithography increases the scope of applications of MEMS, particularly in biomedicine and environmental sensors. With further development and sophistication of the non-traditional manufacturing processes, there is a possibility for low-cost and high-throughput manufacturing, making MEMS accessible to mass market needs. Besides, these techniques complement traditional silicon-based methods, enabling hybrid approaches to exploit the unique properties of various materials to achieve multifunctional devices. Such integration is a paradigm shift in MEMS design and fabrication, where focus is given to the direction toward the integration of various material systems to enhance performance. The research evidence indicates that advanced MEMS fabrication methods achieve substantial improvement in device performance and reliability. For instance, a study on fracture resistance of temporary restorative materials demonstrates how the use of new approaches like CAD-CAM milling significantly enhances durability compared to conventional methods. Comparative investigation of different fabrication technologies also demonstrates that digital fabrication technologies not only improve marginal and internal fit of endocrown restorations but also offer a pathway to more accurate and responsive systems in MEMS technologies. Successful implementation of such principles shows the ability of new technologies to initiate innovation in mechanical and electronic functions of MEMS, where they have a right to belong in modern-day engineering procedures. Ultimately, a fusion of outdated and modern approaches promises to bring the next generation of MEMS devices to the pinnacle of technological innovations.

5.7.1. Summary of Key Fabrication Techniques

Evolution of Microelectromechanical Systems (MEMS) fabrication techniques has transformed the microtechnology environment radically. Deep Reactive Ion Etching (DRIE) is one of the important techniques, which is better suited for the fabrication of high aspect ratio structures, important in MEMS devices. DRIE entails alternating cycles of etching and passivation to achieve depth without compromising feature integrity. Coupled with Electron-beam lithography, researchers can define complex patterns at nanoscale accuracy, introducing design flexibility. These methods not only offer better control over the dimension characteristics of micro-components but also result in better performance of the device. These newer manufacturing processes open up MEMS applications because they allow engineers to push the boundaries of size and functionality, heighten reliability and efficiency in performance parameters of utmost significance across the industries from telecommunication to biomedical applications. One of the other principal areas of MEMS fabrication is wafer bonding and packaging processes, which are essential to device integrity and functionality.

These processes allow the integration of multifaceted materials, which enhance the functionality of MEMS devices by offering high-end functionalities such as sensing and actuation. Some of these include bonding processes like anodic bonding, which allow for the integration of glass and silicon without gaps, maximizing optical MEMS devices' performance in sensor applications. Also, packaging processes ensure that MEMS devices are protected without diminishing their functional capabilities. The multi-layer approach of packaging not only addresses environmental challenges but also minimizes the chances of contamination, which is essential for accuracy products. With proper wafer bonding and new-age packaging ideas, businesses can significantly increase MEMS technology, guaranteeing durability and functionality for a range of applications, thereby influencing uptake across industries. Surface micromachining and soft lithography are other major fabrication techniques that have expanded the design space for MEMS. Surface micromachining, the deposition and etching of thin films on a substrate, allows intricate structures to be formed on a single chip. It is a layer-by-layer process, and addition of mechanical elements like beams and membranes to electronic circuits means very functional but small devices. Conversely, soft lithography employs elastomeric materials to create pliable molds, and hence polymer-based MEMS can be fabricated in high-throughput fashion. Soft lithography is particularly advantageous in applications requiring low-cost fabrication and material flexibility. The combination of these methods is a significant evolution in MEMS technology because they all, in an integrated fashion, result in the development of cost-effective yet technologically advanced devices that provide new avenues for use in healthcare and environmental monitoring systems.

5.7.2. The Impact of MEMS on Various Industries

MEMS advancements have pushed many industries to new heights by enabling the embedding of miniaturized electrical and mechanical components. Primary manufacturing techniques, such as enhanced micromachining, i.e., Deep Reactive Ion Etching (DRIE) and E-beam lithography, play a key role in achieving the type of precision required to create MEMS. These techniques allow for the production of extremely complex devices used in applications ranging from automobile sensors to medical devices. As these industries increasingly embrace MEMS technology, companies can maximize performance while reducing size and weight, ending with cost-effective solutions that significantly improve performance and reliability in mission-critical applications. Moreover, the impact of MEMS is extended into the consumer electronics sector, such that smartphones and wearable technology employ MEMS sensors in order to deliver functionality. The surface micromachining techniques allow the manufacture of accelerometers and gyroscopes required for devices to detect user movement and interaction. The addition of the MEMS components allows for more interactive purposes, such as navigation and fitness tracking functionalities. As the devices become more prevalent, the demand for high-performance MEMS devices grows, propelling the field and resulting in new advances in materials as well as processes for manufacturing that enable the scalability of MEMS applications to several consumer markets. In addition to the consumer electronics market, MEMS technology has significant influence on the structural health monitoring industry, where it is critical to the integrity of the infrastructure. MEMS fabrication-based sensors can provide real-time information regarding the structural condition, thereby allowing for proactive maintenance strategies. The use of data-driven algorithms, as noted in the literature at present, identifies the importance of MEMS in providing important information regarding the health of structures (Cherta et al., 2020). This aspect not only optimizes safety but also optimizes resource optimization and maintenance planning in areas such as civil engineering and aerospace. With greater demands for sustainable and secure infrastructure, the MEMS technology is set to become an integral part of future development (Albers et al., 2010).

5.7.3. Future Directions for MEMS Technology

Technological advancements in micromachining techniques, such as Deep Reactive Ion Etching (DRIE) and E-beam lithography, are paving the way for next-generation MEMS technology. Such advanced processes allow for the fabrication of intricate structures with micro and nanoscale features, which guarantee better performance characteristics specific to respective applications. As the technologies evolve, they provide for more complex designs and integration of multiple functionalities within one single MEMS device. The innovations in wafer bonding and packaging technology are equally critical, as they ensure MEMS devices are robust, reliable, and functional in dynamic environments. The destiny of MEMS technology hangs in the balance of these innovations, which can potentially stretch the range of embedding MEMS across different sectors from consumer electronics to advanced medical devices. MEMS technology's capabilities extend into the area of structural health monitoring (SHM) in infrastructure such as pavements and bridges. Application of MEMS and nano-electromechanical systems (NEMS) to SHM has significantly greater advantages than traditional wired sensor technologies, including cost reduction and improved long-term performance. As noted in recent studies, commercial off-the-shelf MEMS sensor utilization has been applied in concrete pavements successfully to ensure monitoring of their structural health. Aside from facilitating improved maintenance procedures, it also gives transportation authority's critical data for the effective treatment of distress in pavements at the right time, significantly lowering costs as well as enhancing safety. Thus, integrating sophisticated MEMS technology in civil engineering will lead to more sustainable infrastructure development and management. Another thrilling innovation in MEMS technology is the emergence of stretchable electronics (SE), paving the way for unprecedented opportunities in medical diagnostics and wearable sensor devices. High-density interconnects with ultrastretchability will be needed in next-generation MEMS devices, necessitating unorthodox fabrication methods to achieve micron-scale precision. Free-standing, miniaturized interconnect geometries fabricated via CMOS technology meet these demands by utilizing sophisticated microfabrication methods that leverage inherent material stretchability. This technique not only permits huge elastic deformation but also ensures integration of these structures with standard batch processing in CMOS technology. As studies continue to demonstrate the effectiveness of such techniques, the inclusion of MEMS technology within body-worn electronic devices has the potential to transform healthcare monitoring as a discipline, making diagnostic capabilities more accessible.

REFERENCES

1. Ahmad, A. S., Ahmar, A. S., Asmuni, A., Hishamudin, H., et al. (2018). Iris segmentation. https://core.ac.uk/download/19162806.pdf (accessed on 14th October 2025).
2. Albers, A., Gillner, A., Razali, A., Yabe, A., Brockett, A., Razali, A., Sanchez-Salmeron, A. J., et al. (2010). Micro-manufacturing: Research, technology outcomes and development issues. https://core.ac.uk/download/9023391.pdf (accessed on 14th October 2025).
3. Amon, C. H. (2003). MEMS-based thermal management of high heat flux devices edifice: Embedded droplet impingement for integrated cooling of electronics. https://core.ac.uk/download/4382753.pdf (accessed on 14th October 2025).
4. Basu, P., Verma, J., Abhinav, V., Ratnesh, R. K., Singla, Y., & Kumar, V. (2025). Advancements in lithography techniques and emerging molecular strategies for nanostructure fabrication. *International Journal of Molecular Sciences, 26*. https://www.semanticscholar.org/paper/05904a708b7408d39725678d9bea0244539ca3b6 (accessed on 14th October 2025).
5. Bennink, M. L., Berenschot, J. W., van den Berg, A., de Boer, M. J., et al. (2001). Direct integration of micromachined pipettes in a flow channel for single DNA molecule study by optical tweezers. https://core.ac.uk/download/pdf/11455005.pdf (accessed on 14th October 2025).
6. Berenschot, J. W., van den Berg, A., de Boer, M. J., Burger, G. J., et al. (2000). Micromachining of buried micro channels *in silicon*. https://core.ac.uk/download/pdf/11453503.pdf (accessed on 14th October 2025).
7. Bussooa, A., Mercer, J. R., & Neale, S. (2018). Future of smart cardiovascular implants. https://core.ac.uk/download/161513601.pdf (accessed on 14th October 2025).
8. Cherta, A., Vargas, D. G., C., R., Pedraza, C., Pozo Montero, R., et al. (2020). Damage identification in structural health monitoring: A brief review from its implementation to the use of data-driven applications. https://core.ac.uk/download/288625955.pdf (accessed on 14th October 2025).
9. Chien, C. H., & Yu, H. M. (2006). Fabrication of switches on polymer-based by hot embossing. https://core.ac.uk/download/15488623.pdf (accessed on 14th October 2025).
10. Chizari, S., Hopkins, J. B., Jackson, J. A., Panas, R. M., et al. (2019). Additively manufacturable micro-mechanical logic gates. https://core.ac.uk/download/323078324.pdf (accessed on 14th October 2025).

11. Coutu, R. A., & Ostrow, S. A. (2011). Novel microelectromechanical systems image reversal fabrication process based on robust SU-8 masking layers. https://core.ac.uk/download/213087987.pdf (accessed on 14th October 2025).
12. de Boer, M. J., Elwenspoek, M., Gardeniers, J. G. E., Gilde, M. J., et al. (2002). Guidelines for etching silicon MEMS structures using fluorine high-density plasmas at cryogenic temperatures. https://core.ac.uk/download/pdf/11457799.pdf (accessed on 14th October 2025).
13. Dekker, R., Geers, M. G. D., Hoefnagels, J. P. M., et al. (2017). Ultra-stretchable interconnects for high-density stretchable electronics. http://arxiv.org/abs/1709.04434 (accessed on 14th October 2025).
14. Dennehy, N. (2025). Vision for micro technology space missions. https://core.ac.uk/download/pdf/10513661.pdf (accessed on 14th October 2025).
15. Ehrfeld, W., Elders, J., Elwenspoek, M., et al. (1995). Deemo: A new technology for the fabrication of microstructures. https://core.ac.uk/download/pdf/11454076.pdf (accessed on 14th October 2025).
16. Ekinci, Y., Kazazis, D., & Tseng, L. T. (2018). Improving the resolution and throughput of achromatic Talbot lithography. http://arxiv.org/abs/1810.05406 (accessed on 14th October 2025).
17. El-Sayed, M. A., ElShaer, A., Essa, K., et al. (2016). Manufacturing of metallic micro components using hybrid soft lithography and micro-electrical discharge machining. https://core.ac.uk/download/78900632.pdf (accessed on 14th October 2025).
18. Fluitman, J. (1996). Microsystems technology: Objectives. https://core.ac.uk/download/pdf/11453403.pdf (accessed on 14th October 2025).
19. Gauthier-Manuel, B., & Pichonat, T. (2006). Recent developments in MEMS-based micro fuel cells. https://core.ac.uk/download/15488679.pdf (accessed on 14th October 2025).
20. Gibson, I., & Tan, W. (2005). Numerical study on the recoating process in micro stereolithography. https://core.ac.uk/download/287648599.pdf (accessed on 14th October 2025).
21. Guglielmelli, E., Oddo, C. M., Saccomandi, P., et al. (2014). Microfabricated tactile sensors for biomedical applications: A review. https://core.ac.uk/download/pdf/54936693.pdf (accessed on 14th October 2025).
22. Hmurcik, L. V., Xiong, X., & Zhang, L. (2010). Teaching MEMS curriculum in electrical engineering graduate program. https://core.ac.uk/download/52955950.pdf (accessed on 14th October 2025).
23. Huang, Y. T., Tsai, Y. C., Huang, Y. C., Hu, H. W., & Chen, K. N. (2023).

Ultrathin glass-based wafer-level integration for miniaturized hermetic MEMS application. *IEEE Transactions on Components, Packaging and Manufacturing Technology, 13*, 120–126. https://www.semanticscholar.org/paper/3c9e6be5f0c027dec100ff58a0772b5069af5de2 (accessed on 14th October 2025).

24. Keskin, D. E., Sağlam, G., & Geduk, Ş. E. (2025). Effect of conventional and digital fabrication techniques on marginal and internal fit of lithium disilicate end crowns. *BMC Oral Health, 25*. https://www.semanticscholar.org/paper/09959bdbd096e07e0bc47cb9a1993613da784ca9 (accessed on 14th October 2025).
25. Kirchberger, H., Lindler, P., & Wimpliger, M. (2007). Novel bonding technologies for wafer-level transparent packaging of MOEMS. https://core.ac.uk/download/15495620.pdf (accessed on 14th October 2025).
26. Kodama, C. H., & Coutu, R. A. (2000). THz metamaterial characterization using THz-TDS. https://core.ac.uk/download/213088137.pdf (accessed on 14th October 2025).
27. Kopylov, G. (2015). Overview of sensors suitable for active flow control methods. https://core.ac.uk/download/30294554.pdf (accessed on 14th October 2025).
28. Le, H. T., Haque, R., Ouyang, Z., Lee, S. W., Fried, S., Zhao, D., Qiu, M., et al. (2021). MEMS inductor fabrication and emerging applications in power electronics and neurotechnologies. *Microsystems & Nanoengineering, 7*. https://www.semanticscholar.org/paper/8785f9d29495334a14e6ae8271c44f7c5b6fb815 (accessed on 14th October 2025).
29. Liu, C., & Tai, Y. C. (1999). Sealing of micromachined cavities using chemical vapor deposition methods: Characterization and optimization. https://core.ac.uk/download/4869879.pdf (accessed on 14th October 2025).
30. Lowell, W., Moshref Javadi, A. M., Nguyen, L., & Desai, S. S. (2025). Prospects and trends in biomedical microelectromechanical systems (MEMS) devices: A review. *Biomolecules, 15*. https://www.semanticscholar.org/paper/ce3b8fd63021b9644f1c3f207228625f3e1651b2 (accessed on 14th October 2025).
31. Luo, Y., Abidian, M. R., Ahn, J. H., Akinwande, D., Andrews, A. M., Antonietti, M., Bao, Z., et al. (2023). Technology roadmap for flexible sensors. *ACS Nano, 17*, 5211–5295. https://doi.org/10.1021/acsnano.2c12606 (accessed on 14th October 2025).
32. Miranda, I., Souza, A., Sousa, P., Ribeiro, J., Castanheira, E. M. S., Lima, R., & Minas, G. (2021). Properties and applications of PDMS for

biomedical engineering: A review. *Journal of Functional Biomaterials, 13*(2), 2–2. https://doi.org/10.3390/jfb13010002 (accessed on 14th October 2025).

33. Mohammad, T., He, S., & Ben Mrad, R. (2021). Conventional surface micromachining process for the fabrication of a linear optical phased array based on piston micromirrors. *Journal of Micromechanics and Microengineering, 31*. https://www.semanticscholar.org/paper/0fc39842e092c27e27e6a0784e3d4e5cace8effa (accessed on 14th October 2025).
34. Naresh, V., & Lee, N. (2021). A review on biosensors and recent development of nanostructured materials-enabled biosensors. *Sensors, 21*, 1109–1109. https://doi.org/10.3390/s21041109 (accessed on 14th October 2025).
35. Nikolić, M. V., Milovanović, V., Vasiljević, Ž. Ž., & Stamenković, Z. (2020). Semiconductor gas sensors: Materials, technology, design, and application. *Sensors, 20*, 6694–6694. https://doi.org/10.3390/s20226694 (accessed on 14th October 2025).
36. Rodrigues, A., Andrianov, N., Syed, M. A., Petschnigg, M., Nguyen, T., Dao, T., & Roshanghias, A. (2024). Low-temperature adhesive wafer bonding for film layer transfer. *2024 IEEE 10th Electronics System-Integration Technology Conference (ESTC)*. https://www.semanticscholar.org/paper/3cd846cc663d34145c04aa4565834354c87716d7 (accessed on 14th October 2025).
37. Sankeshwari, B., Kanitkar, A., Pattanshetti, C. V., & Kanitkar, A. (2025). Comparative analysis of 3D printing, computer aided design and milling and conventional fabrication techniques on fracture resistance of interim restorations: A systematic review and meta-analysis. *Journal of Clinical and Diagnostic Research*. https://www.semanticscholar.org/paper/33e0ea9e333844231b6196f66d9d9a2eaa12945f (accessed on 14th October 2025).
38. Temel, O. (2020). Term Project Upload Guide for Non-Thesis Master's Program. https://core.ac.uk/download/480557918.pdf (accessed on 14th October 2025).
39. Thiel, L., Castro, J., Steiner, T., Nguyen, C. L., Pechilis, A., Duan, L., Lewis, N., et al. (2024). Wafer-scale fabrication of InGaP-on-insulator for nonlinear and quantum photonic applications. *Applied Physics Letters*. https://www.semanticscholar.org/paper/ceed0663842d94015a753ba7237346ba3321546b (accessed on 14th October 2025).

CHAPTER 6

MODELING AND SIMULATION OF MEMS DEVICES

Contents

INTRODUCTION

Microelectromechanical Systems (MEMS) has revolutionized a number of fields ranging from telecommunication and automotive industries to biomedical devices. The micro-scale systems with mechanical and electrical integration need sophisticated modeling and simulation techniques to be capable of predicting performance accurately and designing for optimization. Traditional techniques such as Finite Element Analysis (FEA) have proven useful but are known to falter when dealing with the added complexity and nonlinearity associated with MEMS devices. As these systems get increasingly complex, researchers must resort to advanced simulation methods that not only correctly capture the physics but also handle the computational complexities arising from very interactive behaviors. Multiphysics simulation coupling and lumped parameter modeling has been a method by which engineers can model coupled physics in a more realistic manner with computational effectiveness preserved. Against this backdrop, low-order modeling demand has increased in significance, especially in systems with high dimensionality. The curse of dimensionality is one significant impediment that contributes to the difficulty in simulating MEMS because of the enormous quantity of parameters and variability that could affect their functioning. Tensor computation, as has been shown in latest studies, offers a useful starting point in alleviating these issues. By facilitating organization and manipulation of high-dimensionality data, tensor methods can potentially simplify surrogate model extraction and perform uncertainty analysis on design parameters more effectively. The significance of this is the fact that intricate interactions are not only facilitated in the case of the MEMS devices themselves but also in any application scenario in general, which enhances predictive accuracy and reliability. Furthermore, hierarchical uncertainty quantification is presently a central theme in MEMS simulation, useful in assessing stochastic systems and circuits effectively. Computational work in this technique is minimized using advanced strategies, such as spectral techniques, which work at different hierarchical levels of design. These techniques allow the formulation of reliable surrogate models while effectively addressing high parameter dimensionality problems. The efficient simulation of challenging examples, i.e., stochastic oscillators with a large number of random parameters, serves as a real-life demonstration of such principles. As reiterated in the literature, contemporary methods like tensor-train decomposition not only render the process of computation more effective but also provide new avenues for addressing high-dimensional model-related problems, thus reasserting the requirement for continued research into modeling and simulation techniques for MEMS devices (Batselier et al., 2016; Daniel et al., 2014).

Importance of Modeling and Simulation in MEMS Design

In MEMS design, the complexity of the interactions among various physical phenomena calls for an intense practice in modeling and simulation. Tools such as Finite Element Analysis (FEA) feature prominently, enabling designers to simulate stress distribution, thermalization, and dynamic response in MEMS devices. Such predictive capability is essential in the design of the mechanical structures that define the performance characteristics, particularly for the applications with high precision, e.g., sensors and actuators. Second, multiphysics simulations make it possible to simulate several interacting physical laws simultaneously and gain insights into how electrical, thermal, and mechanical properties interact in MEMS. Through the integration of such simulation technologies, prototyping and verification can be significantly reduced in terms of time and cost, thereby culminating in groundbreaking MEMS solutions that push the boundaries of technical innovation. A good example of the use of modeling in MEMS is in the development of extremely sensitive RF filters using extreme throw MEMS actuators.

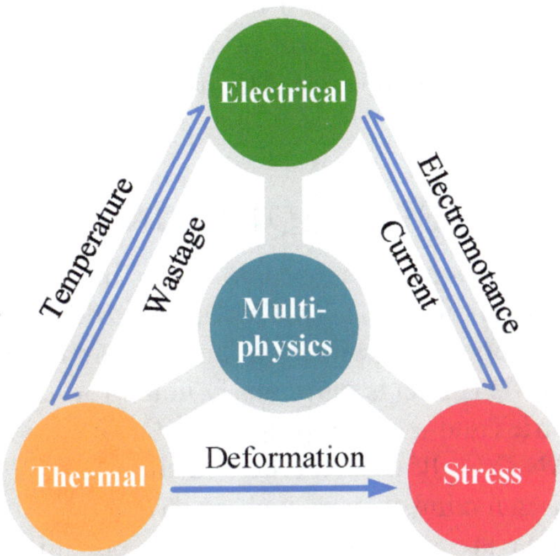

Figure 6.1. Multiphysics simulations allow for the consideration of multiple interacting physical laws, providing insights into how electrical, thermal, and mechanical properties converge in MEMS.

Source: https://www.mdpi.com/symmetry/symmetry-15-00418/article_deploy/html/images/symmetry-15-00418-g001.png.

These actuators deliver large, controlled motion that is critical in filtering over broad frequency bands. Actuator behavior modeling accurately forms the core of understanding how variations in design parameters affect filter performance. Notably, as seen in research, the potential for MEMS filter design to operate optimally in frequencies above 15 GHz relies on precise control of process variables by simulation software. This ensures that the final product not only meets operational requirements but surpasses in a competitive economy. The impact of such modeling techniques goes beyond just functionality; they also promote innovation through boundary-pushing design and fields of application. The relevance of successful modeling and simulation is again substantiated by the manner in which they refine system stability and reliability. Under non-cooperative visual data acquisition situations, i.e., eye-tracking systems, appropriate simulation procedures come into play. For example, noisy image data can play a primary role in making image processing algorithms used in MEMS-based iris recognition systems unreliable, as proposed in the comparison of segmentation algorithms. In this case, simulation software assists in evaluating the performance of different techniques under different noise conditions so that designers may identify the optimum algorithm for real applications. With the use of advanced modeling methods in MEMS design, engineers are able to enhance the stability of their systems, which ultimately results in effective implementation of MEMS technologies in various areas (Figure 6.1).

6.1. MODELING TECHNIQUES IN MEMS

Several simulation techniques are used to study the intricate behaviors of MicroElectroMechanical Systems (MEMS), all with particular strengths depending on the proposed application and requirements. Among them, Finite Element Analysis (FEA) is particularly noted for its capacity to analyze complicated geometries and material properties. Computational packages such as COMSOL and ANSYS allow researchers to simulate MEMS devices under varying conditions, analyzing factors such as stress distribution and dynamic response. Simulations are critical in identifying potential points of failure and enhancing the performance of MEMS. By integrating multiple physical phenomena on a common platform, multiphysics simulation extends the boundaries of traditional modeling methods, enabling engineers to predict how various forces and environment conditions influence MEMS devices. Thus, FEA is an important aid in the modeling process, assisting in the development and validation of new MEMS technology. In addition to FEA, lumped parameter modeling is a quick technique widely employed within MEMS design, particularly in situations where rapid simulations are required. This method simplifies complex systems into basic elements with idealized relationships between them, which allows for shorter computation times than undertaking full-scale FEA. While lumped parameter models omit some of the subtlety of individual components, they capture the significant dynamics and enable making rapid judgment calls on system behavior. This approach is particularly valuable during the preliminary stages of design, where rapid iteration is necessary. Reduced-order modeling also emerged as a viable option, trimming down high-dimensional models to lower forms without sacrificing dominant dynamic characteristics. With the capability to facilitate quicker simulation runs, reduced-order models can lead to improved decision-making processes, ultimately resulting in optimized designs for MEMS devices at no unreasonable computational expense. Current advancements, particularly in hierarchical uncertainty quantification, are revolutionizing the area of MEMS simulation methods to enhance control of high-dimensional systems. The advancements enable retrieval of surrogate models with different levels of complexity for enabling analysis of stochastic systems and circuits that control most MEMS applications. The integration of tensor-train decomposition with ANOVA methods provides greater flexibility in controlling the computational expense of such high-dimensional simulations, balancing the curse of dimensionality that typically plagues MEMS design. For instance, a paper demonstrated the application of this method to simulate a stochastic oscillator with multiple random parameters, achieving excellent results within a short period of time (Daniel et al., 2014). Thus, these novel methodologies provide access to better modeling accuracy and effectiveness, making MEMS devices able to satisfy ever-increasing technology demands without loss of performance or reliability.

6.1.1. Overview of Modeling Methodologies

Among the engaged and intricate landscape of modeling methods, there are numerous methods that play a key role in analyzing and designing MEMS devices. One such technique that is often employed is Finite Element Analysis (FEA), which enables engineers to simulate the intricate physical phenomena involved with MEMS structures under various conditions. With applications such as COMSOL and ANSYS, researchers can model intricate interactions within the device taking into consideration material characteristics, geometry, and forces applied from the outside. This method allows for fine-scale modeling of mechanical stresses, thermal responses, and electrical behavior, giving a complete insight into how the MEMS devices will function in real application. Furthermore, coupling multiphysics simulations can couple these, enabling insights into coupled electro-mechanical dynamics and other effects, which are required to optimize the MEMS design. Aside from FEA, lumped parameter modeling is also a prevalent technique that simplifies the complexity of modeling MEMS devices. Through the discretization of the complexity of models into a collection of parameters, this technique facilitates faster computations and easier interpretation of system behavior. Used mainly in early design phases, lumped models can approximate the fundamental dynamics of a device before conducting more complex analyses. Reduced-order modeling extends this by creating low-order representations of high-dimensional systems, keeping essential dynamics at much lower computational cost. This is valuable where rapid prototype iteration is needed, as it enables engineers to explore many design variations without the time-consuming commitment of full-scale simulation. Lastly, bond graph modeling, particularly in mechatronic devices, is an example of a progressive method of coupling physical systems through energy transfer principles. This method was applied extensively while developing a behind-the-ear (BTE) hearing aid, where simulation models based on bond graphs were defined in an attempt to study interactions between main components like the telecoil and receiver. These methodologies not only yield a more accurate picture of the behavior of systems but also enhance sensitivity analysis, allowing designers to find an optimized component placement to increase device robustness. By uniting the strengths of various modeling paradigms, scientists can construct an overall framework required for the advancement of MEMS technology, finally leading to devices with better efficiency and reliability (Acosta et al., 2011; Atherton et al., 2005) (Figure 6.2).

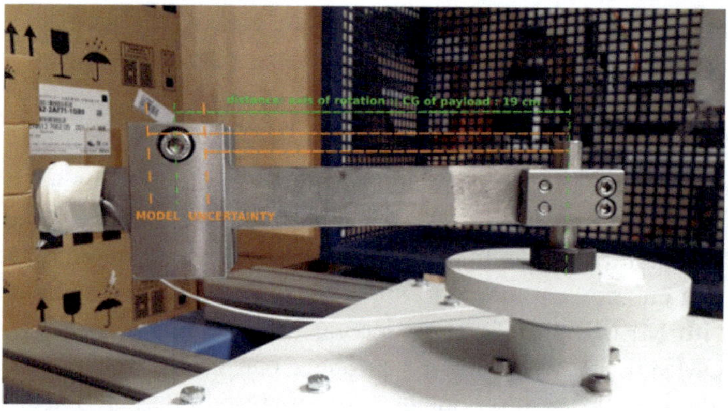

Figure 6.2. Finite element analysis (FEA) stands out as a commonly utilized approach, enabling engineers to simulate the complex physical behaviors of MEMS structures under various conditions.

Source: https://encrypted-tbn0.gstatic.com/images?q=tbn:ANd9GcTvtlmrcUTug20T1RJn7Z6UD7C3k5LejFp5DgbM3UheYsL7x_XY49m6BaPIhmnPZOAwd4A&usqp=CAU.

6.1.2. Comparison of Analytical versus Numerical Methods

MEMS device modeling and simulation increasingly highlight the importance of contrasting analytical and numerical solutions to effective design. Analytical solutions are used in most applications and provide closed-form solutions with inherent benefits of ease of interpretation and simplicity. For instance, perturbation methods, according to research, allow for the development of governing continuum equations tailor-made for micro flows that can give well-defined analytical solutions for low Knudsen numbers. Nevertheless, assumptions needed in analytical methods can limit their use in more complex cases where uniformity and linear assumptions do not hold. Therefore, although analytical methods are precious in offering back-of-the-envelope type insights, they tend to run short under the challenges posed by realistic MEMS applications so that numerical methods would have to be incorporated to outline the important phenomena at micro scales. Numerical methods such as Finite Element Analysis (FEA) and computational fluid dynamics present adaptable modeling environments that can tackle complex geometries and boundary conditions in MEMS applications. These methods allow for general simulations which are able to account for non-linearities and complicated material properties, thus resulting in solution forms that are adjustable to specific design demands. For example, during the performance analysis of micro-accelerometers, numerical simulation is superior to analytical models when it comes to dealing with complexity to meet stringent frequency and sensitivity demands (Wong et al., 2007). However, numerical techniques usually consume a tremendous amount of computational power and are often beset by issues such as convergence and stability, so selecting an appropriate mesh size and solution algorithm is critical. The analytical versus numerical choice in the modeling of MEMS devices is not only one of ability but context and specific requirements. In situations that demand quick assessment and easy to understand analytical knowledge, analytical methods may be adequate for enabling initial design assessment foundations. Yet, when designs get complex or when phenomena are to be analyzed in specific ways, numerical methods unequivocally come first. This complimentary association points to an overarching paradigm in engineering design where hybrid approaches that incorporate both analytical knowledge and robust numerical simulations can allow more effective and creative solutions in the MEMS universe. Therefore, controlling these methods strategically is most important for advancing the objectives of MEMS technology into practical worlds.

6.1.3. Importance of Accurate Modeling for Performance Prediction

Precise modeling is a fundamental cornerstone of micro-electro-mechanical systems (MEMS) design and optimization, particularly in predicting device performance. Integration of various model methods, such as Finite Element Analysis (FEA) and lumped parameter models, allows engineers to model efficiently the complex interactions in MEMS devices. The methods allow for the study of multi-physics phenomena, e.g., mechanical, thermal, and electromagnetic phenomena in RF MEMS switches. By employing simulations that account for all the variables relevant in physics, designers can identify potential performance issues at an early stage in the design process and reduce the chances of costly prototypes and reworking. In addition to that, predictive models enhance the overall understanding of how changes in design impact performance results, thereby enabling more informed decision-making throughout the design cycle. MEMS device optimization calls for a keen understanding of performance parameters, and modeling directly impacts reliability as well as efficiency.

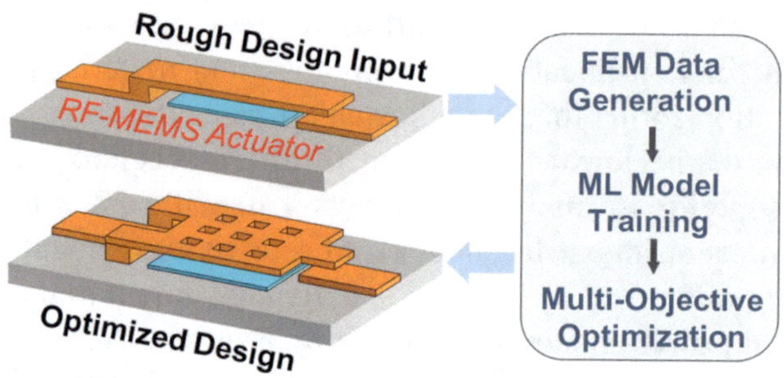

Figure 6.3. Accurate modeling serves as a fundamental pillar in the design and optimization of micro-electro-mechanical systems (MEMS), particularly when predicting device performance.

Source: https://pub.mdpi-res.com/sensors/sensors-23-04001/article_deploy/html/images/sensors-23-04001-ag.png?1681697101.

For the use of RF MEMS, for example, such parameters as capacitive coupling and inductive coupling are significant in determining signal integrity as well as power dissipation. Here, parameterized modeling techniques, as referred to in literature employing the Ansoft HFSS-TM electromagnetic simulator, facilitate an adaptive solution to minimize losses attributed to device packaging and other design aspects (Bartek et al., 2007). Such modeling not only helps in achieving optimal configurations but also plays an important part in checking design choices against experimental outcomes. The resolution of such discrepancies at the model stage can lead to more effective and reliable MEMS device structures meeting stringent performance requirements. Secondly, correct modeling is also important in the fields of longevity and maintenance of MEMS devices. Reliability issues are of utmost concern in the field of applications where switches come close to their limits. Accurate simulation can predict the effect of operational stress on device elements and thus guide designs that are responsive but strong. For instance, use of pulse optimization techniques, founded on empirical modeling, allows engineers to minimize impact forces during actuation, extending the lifespan of the device (Czaplewski et al., 2011). By simulating various actuation methods, designers of MEMS can establish the most effective methods to reduce fatigue so that they may enhance real-world utility and reliability of MEMS. In the end, the commitment to modeling precision is essential to maximize performance and ensure longevity in MEMS applications (Figure 6.3).

6.2. FINITE ELEMENT ANALYSIS (FEA) IN MEMS

The complicated nature of microelectromechanical systems (MEMS) calls for advanced modeling techniques, and Finite Element Analysis (FEA) has emerged as the most prominent technology used in their optimization and design. Breaking down complex geometries into small finite elements enables FEA to solve differential equations that define the behavior of MEMS devices in various physical conditions. The ability to use software like COMSOL and ANSYS allows engineers to model different scenarios such as thermal, mechanical, and electrostatic effects simultaneously. Multiphysics simulation is extremely relevant because MEMS performance often hinges on coupled physical phenomena, such as the manner in which thermal expansion might influence mechanical stress and thus overall sensor and actuator operation. This combined approach not only enhances the predictive accuracy but also facilitates the reduction of design cycles, thus speeding up the development cycle of MEMS devices. Sensitivity and resonant frequency have a significant application in the field of MEMS too, and their significance cannot be overlooked since device performance for all applications is based on them. Utilizing FEA, accurate modeling of these parameters acquires a serious role for effective design. By precisely calibrating material properties and geometric configurations in the finite element model, engineers can make critical assessments of micro-accelerometer resonant frequency and other devices with accuracy. Research indicates that micro-accelerometers can have effective and stable operation only if these prerequisites are fulfilled (Wong et al., 2007). This level of precision enables designers to create MEMS devices that not only meet but also exceed industry standards, thereby improving their performance and reliability upon application. Further, the analysis of mechanical contacts within MEMS devices presents unique problems that can be resolved by FEA. The performance and reliability of MEMS switches, for instance, are significantly reliant on the microcontact between components. By using FEA simulations, the mechanical characteristics of such contacts—e.g., closure time and contact force—can be examined in detail. Experiments show that changes in contact closure time can have a profound impact on switching conditions, including cold-switching and hot-switching situations. Analysis plays a vital role in ensuring durable microswitch designs for use under direct current (DC) and radio frequency (RF) applications (Anwar et al., 2019). By merging advanced FEA methods, engineers can optimize geometries and contact materials to ultimately improve the performance and lifespan of MEMS devices under different conditions of operation.

6.2.1. Introduction to FEA and its Relevance to MEMS

As the complexity of microelectromechanical systems (MEMS) devices continues to increase, design and analysis methods have also followed suit by embracing more advanced simulation techniques. Embracing up-to-date simulation techniques, particularly Finite Element Analysis (FEA), is now a requirement for engineers interested in optimizing functionality in addition to performance. FEA enables complex geometries to be broken down into simpler manageable parts that can be utilized for thorough studies of mechanical responses under varying loads. Such capabilities allow engineers to predict stress distributions, deformations, and resonance frequencies in MEMS structures more accurately, contributing to improved design efficiency. The application of FEA is central when coupling multiphysics simulations because it provides information on the multiscale phenomenon involved in MEMS, ranging from thermal effects to fluid-structure interaction. Its suitability to MEMS is most applicable when applied to research its influence on device optimization and fabrication. Its application to design cantilevers and sensors, for instance, demonstrates the critical importance of precision in the engineering operations being undertaken. Experimental results, as seen in the work of Barrachina et al. (2017), indicate the role of FEA in identifying curvature variation across various post-processing environments. These findings illustrate the way in which parameters like temperature stress and material response can dramatically affect device performance, necessitating large simulations to prevent such risks. This means that the capability of FEA software to model these intricacies before fabrication can save considerable development time, reduce costs, and ultimately produce more reliable MEMS devices. In addition, FEA is a critical tool in the validation and creation of lumped parameter models and reduced-order approximations of MEMS. These models have the tendency to simplify complex systems into a format that allows for quicker analysis, especially at early design stages or where time is of the essence. Adding results derived from detailed FEA simulations can further improve such simplified approaches by calibrating their parameters against more advanced simulations. For example, in the development of a polyimide/SU-8 catheter-tip MEMS pressure sensor, as detailed by Bertsch et al. (2012), FEA allowed researchers to thoughtfully examine the most important parameters affecting device performance and design. Not only does this integration improve the predictive model's credibility but also facilitates the development of MEMS technologies using informed decision-making during the designing phase (Figure 6.4).

Figure 6.4. Embracing advanced simulation techniques, particularly finite element analysis (FEA), has become essential for engineers seeking to optimize both functionality and performance.

Source: https://images.sw.cdn.siemens.com/siemens-disw-assets/public/3K5OUbF4ZUIYqizcU1uj9H/en-US/fe-pre-post-processing-hero-2560x1440.jpg?auto=format,compress&w=1920&q=60.

6.2.2. Popular FEA Software Tools (e.g., COMSOL, ANSYS)

Finite element analysis (FEA) integration has revolutionized the simulation and model process of Microelectromechanical Systems (MEMS) devices. Using advanced FEA tools like COMSOL and ANSYS, engineers can perform multiphysics simulations that facilitate a greater insight into the intricate interaction of MEMS structures. These integrated software suites provide integrated environments for simulating mechanical, thermal, and electrical behavior simultaneously, which is necessary for MEMS design and optimization. For instance, COMSOL's Multiphysics module enables users to couple phenomena—such as heat transfer and structural mechanics—which equates to better predictability. The convenience of use and adaptability of these FEA tools greatly advance MEMS development, with the importance in both academic and commercial settings as they encourage innovation and effectiveness in design cycles. Utilization of FEA tools like ANSYS and COMSOL improves knowledge in MEMS reliability, affecting lifespan and performance in practical applications. For example, stress distribution and thermal gradient simulation in MEMS devices inform design optimization, addressing potential failure modes before physical models are manufactured. Predictive capability is important in applications where technology accuracy is most critical, e.g., telecommunication and biomedical devices. COMSOL facilitates simulation and analysis of complex geometries and the corresponding mechanical properties, addressing problems such as thermal management, mechanical stress, and electrical properties, essential to designs that operate reliably in various conditions. As emphasized in current texts, even the latest progress in system-in-package (SiP) technology relies on FEA tools in attaining improved integration and reliability, positioning software like COMSOL and ANSYS as indispensable tools in current MEMS research and design (Wang et al., 2023). MEMS technology development places more emphasis on reduced-order modeling, as well as traditional FEA methods, for improved simulation efficiency that takes into account the realities of real-life applications. Reduced-order models allow computation at an accelerated rate without compromising enough accuracy, hence finding use in iterative design processes. For example, in design and manufacture of cantilever MEMS sensors for electro-chemical gas sensing, device behavior for diverse operating conditions can be effectively predicted by FEA tools, as elucidated in research studies (Guruprasad, 2020). Moreover, linking lumped parameter modeling to multiphysics simulation provides improved understanding of system dynamics and is thus very important in the optimization of MEMS operation. As industries require faster prototyping and greater reliability, the role of FEA software like ANSYS and COMSOL becomes more central to defining the future of MEMS device design and implementation (Figure 6.5).

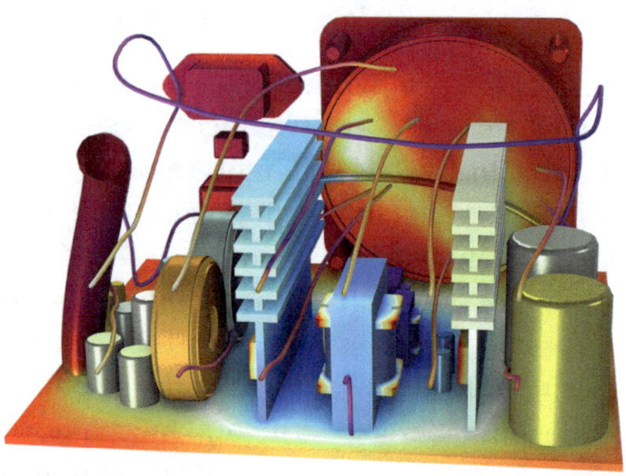

Figure 6.5. By employing advanced FEA tools like COMSOL and ANSYS, engineers can perform multiphysics simulations that allow for a more comprehensive understanding of the intricate interactions within MEMS structures.

Source: https://cdn.comsol.com/product-new/comsol-multiphysics/full/comsol-multiphysics-hero.png.

6.2.3. Case Studies Demonstrating FEA Applications in MEMS

In MEMS applications, the application of Finite Element Analysis (FEA) has been of great help in the prediction of device response under various conditions. FEA case studies exhibit how FEA simulation software like COMSOL and ANSYS have been effectively used to reduce the faults in MEMS design. One particular study demonstrated how curvature characterization of cantilevers fabricated by virtue of complementary metal-oxide semiconductor processes was portrayed. The research brought to the forefront the relationship between cantilever performance and the number of layers stacked in the structure, with different design and processing variations shown to generate stark differences in mechanical performance. Equations that analytically modelled bending performance were employed by researchers to gain invaluable information regarding curvature variation induced by tungsten through design, stressing the significance of FEA to improve MEMS device functions (Barrachina et al., 2017). Other research on MEMS resonators underscores the importance of using FEA in achieving precise specifications such as resonant frequency and sensitivity.

One systematic approach used numerical simulations to analyze micro-accelerometer performance, with attention to how they rely on tailored configurations to meet operational requirements. In a study, parameters were sequentially adjusted—ranging from mass distribution to the variations of geometry—to monitor their effect on resonant frequency and device sensitivity. It is an experimental configuration that is crucial because uniform application of these parameters is deemed to be necessary for offering stable MEMS devices for a broad range of applications (Wong et al., 2007). With the aid of FEA tools, scientists could systematically examine performance, making it possible for prototypes not only according to theoretical prediction but also viable for real-world application. FEA and MEMS technology also show interplay in optimizing complex multi-layered structures. An in-depth case study considered the influence of variations in environmental conditions and design on stacked cantilevers' mechanical reliability. The findings identified the importance of time-dependent stress factors that arise following release can induce exceptional curvatures far in excess of anticipated measurements. This work emphasized the complexity of MEMS design such that an incorrect calculation of the thermal stress profile by a single parameter may have a dominant influence on device behavior. Thus, FEA's role extends beyond mere numerical validation, producing a procedure that promotes analytical understanding of how inherent design factors contribute to overall device performance. This interaction establishes the necessity for advanced simulation software to facilitate the development and improvement of MEMS devices (Barrachina et al., 2017).

6.3. MULTIPHYSICS SIMULATIONS

The development of Micro-Electro-Mechanical Systems (MEMS) has imposed the necessity of encapsulating sophisticated simulation methodologies to facilitate the intense interactions of mechanical, electrical, and thermal fields during device operation. The most effective tool, at least in terms of efficiency, is Finite Element Analysis (FEA), utilized in software such as COMSOL and ANSYS. Both of these packages are essential in carrying out Multiphysics simulations to enable engineers to view and manipulate multiple physical phenomena simultaneously. For example, in transducer modeling of ultrasound, these simulations explain how mechanical vibrations react to different electrical stimuli and thus inform the optimization of device parameters. Such precise understanding is essential to optimizing MEMS designs so their performance can be made to satisfy strict application needs in a wide range of fields. The complexity of interactions highlighted by such simulations highlights their demand for application in MEMS device optimization, development, and design, impacting innovation in various fields, from medical to automotive. As there is an increased need for high-performance MEMS devices, conventional methods of simulation have been inundated by computational inefficiencies. Conventional Multiphysics simulations are computationally expensive and time-consuming, particularly for intricate structures exhibiting nonlinear behaviors. In accordance, researchers have started adopting simplified models, such as reduced-order and lumped parameter modeling techniques, which weigh the accuracy against computational expense. For example, Wang et al. (2024) details how simplified single physics field models can provide results comparable to standard Multiphysics simulations with an eye-popping 80%–90% decrease in simulation time at acceptable error tolerance.

This model progression in modeling strategies is significant as it facilitates accelerated iterations on design loops, which result in quicker prototyping and improved development timelines for next-generation MEMS devices. Multiphysics simulation integration is particularly significant in developing innovative MEMS devices like piezoelectric and capacitive micromachined ultrasound transducers. These technologies have established superior performance characteristics, such as lower manufacturing costs and improved operational efficiency, compared to traditional piezo-ceramic counterparts. The ability to analyze the multi-parameter interactions in these devices—acoustic, mechanical, and also electrical parameters—is an essential aspect of optimizing their design for practical application. Subbiah et al. (2025) emphasizes the need for sophisticated simulation software able to successfully simulate three-dimensional arrays of CMUT and PMUT cells. By applying advanced simulation techniques, engineers are able to comprehend advanced physics of ultrasound transducer functioning, ultimately leading to innovative solutions that maximize the capability of MEMS technology in medical imaging and other areas.

6.3.1. Definition and Significance of Multiphysics Simulations

The intricate nature of micro-electro-mechanical systems (MEMS) demands the utilization of advanced analysis techniques that are capable of capturing their multiplexed behaviors. Among these, multiphysics simulations have emerged as a cornerstone in simulating MEMS devices, since they provide an interfaced model for capturing coupled-electromechanical phenomenon. Unlike conventional modeling practices that handle various physical domains separately, multiphysics simulations allow multi-interacting phenomena to be analyzed concurrently. For example, they enable designers to anticipate the impact of mechanical stress on electrical performance, which is crucial for optimizing MEMS components. Thanks to developments in multiphysics simulations, predictive correctness is significantly enhanced, enabling engineers to forecast device behavior under a wide range of operating conditions, leading to more resilient MEMS products.

Therefore, the significance of such simulations is that they are able to bridge knowledge gaps, resulting in innovations in MEMS technology that would not have been possible otherwise. Multiphysics simulations demand the use of numerous modeling strategies, with Finite Element Analysis (FEA) being one of the prominent ones. COMSOL and ANSYS are some of the software utilities now becoming essential tools for engineers to simulate and analyze complex MEMS structures under various physical conditions. These simulation software programs offer multiphysics capabilities, allowing users to model mechanical, thermal, and fluid dynamics interactions all at once. In addition, these simulations help in the verification of lumped parameter and reduced-order models, which are simplified models for more complicated systems. This integration not only makes the design process efficient but also reduces the time-to-market for emerging MEMS technologies. The ability to visualize and interact with simulated responses gives engineers knowledge that optimizes device performance and reliability at reduced material and manufacturing expense. Growing applicability of multiphysics simulations to MEMS devices also underscores their worth for novel applications, particularly in biomedical and consumer electronics applications. For instance, a simulation of the coupling between MEMS sensors and biological systems must possess an advanced understanding of fluid-structure interactions, thermal behavior, and electromechanical phenomena. To that end, an advance in software capabilities allows researchers and engineers to create more sophisticated models that more accurately reflect real-world conditions. Moreover, emerging applications, like soft robotics, which is based on principles akin to natural muscle in robotics, illustrate the capability of multiphysics simulations to drive innovation. As refined in the development of devices such as HASEL actuators, upcoming simulations will become increasingly important as a future enabler of future smart systems emulating biological functions to advance the versatility and usefulness of MEMS devices into various domains (Ann-Cho et al., 2021; Rothemund et al., 2020) (Figure 6.6).

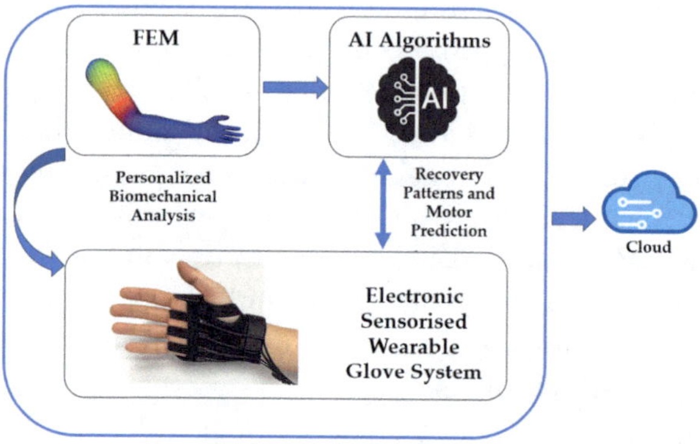

Figure 6.6. As multiphysics simulations evolve, they significantly enhance predictive accuracy, enabling engineers to anticipate device behavior under various operational scenarios, ultimately leading to more reliable MEMS products.

Source: https://pub.mdpi-res.com/electronics/electronics-14-02268/article_deploy/html/images/electronics-14-02268-g001.png?1749038794.

6.3.2. Integration of Different Physical Phenomena in MEMS Modeling

Simulation and modeling of micro-electro-mechanical systems (MEMS) involve an interdisciplinary approach to combine numerous physical phenomena into individual frameworks. The integration is usually done with the help of advanced computational software like Finite Element Analysis (FEA), which allows the consideration of mechanical, thermal, and electromagnetic interactions on a common platform. Software packages like COMSOL and ANSYS have been developed with multiphysics support, allowing researchers to simulate the complex behavior of MEMS devices under various operating conditions. In addition, lumped parameter modeling can be an efficient way to reduce complicated systems to essential parameters without losing reliability in predictions. Minimization of computational needs is also a priority, allowing the development of reduced-order models that have simplified simulations without compromising fidelity. As advances come along, the integration of such tools will allow for accurate design and optimization of future MEMS devices. Incorporating various physical phenomena into modeling MEMS is highly significant in modeling MEMS system behavior in real-world applications. As a case in point, the integration of heat, fluidics, and mechanical stress can be very significant for device operation and duration.

Figure 6.7. Recent developments in quantum technologies (QT) serve as a testament to this trend, as they leverage principles from quantum physics to explore novel applications in sensing and communication.

Source: https://csis-website-prod.s3.amazonaws.com/s3fs-public/2023-05/Adobe-Stock_544067004_cropped.jpg?VersionId=ksgOjDC4hwmmz54jJMnawv1LFL0Dc_79.

This level of complexity requires an integrative modeling strategy where these elements are viewed together in interaction. Space-time modulated digital coding metasurfaces are a good example of integration since electromagnetic wave control with high accuracy is called for in applications ranging from holographic imaging to wireless communications. The behavior of such advanced systems is not only a function of correct modeling but also of smooth interfacing between various physical domains, achievable via advanced simulation techniques. The ability to control physical phenomena in unison highlights the importance of robust MEMS modeling platforms. In addition, the integration of heterogeneous physical phenomena propels the momentum for rapid advancement in MEMS technology, enabling cross-disciplinary research activities. Recent advances in Quantum Technologies (QT) are a testimony to this trend since they utilize the principles of quantum physics to explore novel applications in sensing and communication. Development of QT has also exhibited significant interaction with MEMS, particularly with the advent of quantum sensors based on highly calibrated mechanical systems. This cross-pollination of ideas is a one-of-a-kind advantage for MEMS researchers, who can draw upon quantum theory to advance device functionality. As the technology continues to mature, an efficient research community will be essential to solving challenges and unlocking new possibilities, ultimately leading to revolutionary innovations that bridge conventional MEMS and quantum-enabled technologies (Antonio et al., 2018; Zhang et al., 2018) (Figure 6.7).

6.3.3. Examples of Multiphysics Simulations in MEMS Applications

In MEMS technology, multiphysics simulation application has brought new doors for device performance enhancement and reliability improvement. Finite Element Analysis (FEA) methods, particularly those supported by software like COMSOL and ANSYS, enable researchers to explore complex physical phenomena governing MEMS behavior. By bringing mechanical, thermal, and fluid dynamics together in a single modeling framework, designers are able to predict the effect of design parameter changes on overall behavior. The co-simulation is crucial in optimizing parameters such as piezoelectric responses or structural integrity, ultimately leading to high-performance devices. The knowledge acquired by these multiphysics simulations is central in achieving a better insight into material behaviors and interactions and hence enhancing the design and manufacturing process of MEMS applications. A typical illustration of such integration is piezoelectric microcantilever simulation and optical characterization, which is produced by aluminum nitride (AlN). Utilization of the devices based on Complementary Metal Oxide Semiconductor (CMOS) technology not only facilitates standard manufacturing processes but also cuts down the overall cost of production. This CMOS standard compatibility is attractive, especially due to the increasing needs for efficient, cost-effective MEMS solutions for various industries. The challenge of ascertaining the reliability and performance of the cantilevers was overcome through a mix of optical techniques and multiphysics modeling. Especially, resonance frequencies and vibration modes were analyzed with laser Doppler vibrometry supplemented by finite element model simulations confirming the presence of distinct flexural and torsional modes. The satisfactory correspondence between simulated and measured data confirms the applicability of using multiphysics simulations within MEMS design approaches (Calle et al., 2015). Further research into multiphysics modeling can also be achieved in RF MEMS capacitive switches, but now through the application of a coupled 3D model that analyzes fluid-structure interaction (FSI). These advanced modeling techniques take into account the intricacies that accompany time-dependent viscous damping, which plays a significant role in the operation dynamics of the switch. The application of intricate three-dimensional geometries allows for a better assessment of the damping characteristics, which are typically oversimplified by conventional models. The short-time simulations using this model ultimately aim to determine an equivalent damping coefficient relevant to time to be used as a vital input for reduced-order models that follow. By addressing the complex aerodynamic phenomenon behind the actuation process, this approach emphasizes the significance of multiphysics simulations towards better performance and reliability of MEMS devices towards achieving revolutionary applications within the discipline.

6.4. LUMPED PARAMETER MODELING

In the design of MEMS devices, a multitude of modeling methods are employed in order to project performance accurately and facilitate design iteration. One method that excels at breaking down complex physical systems into manageable components is lumped parameter modeling. This method models MEMS components as separate entities, relating their electrical characteristics with simplified forms similar to electronic circuits. Through the transformation of physical motion into electrical quantities, engineers are able to rapidly model and examine MEMS device performance without the computational cost of more advanced methods like Finite Element Analysis (FEA). This efficiency is particularly useful at the early stages of design where rapid iteration is necessary in order to maximize specifications and compare design choices. Thus, lumped parameter modeling is a critical bridge between theory and practice in MEMS design. By incorporating lumped parameter models in the design workflow, not only are simulations made faster, but device response prediction accuracy is also enhanced. By including salient features, such as capacitance and resistance components, engineers are capable of predicting accurately the performance of MEMS devices under varying operating conditions. For example, in a wafer-level encapsulated (WLE) RF-MEMS switch, small-signal model development supports speedy simulations that are centered on electrical properties. Such models are particularly helpful in estimating RF performance parameters like isolation and insertion loss, which are crucial in ensuring reliable operation under radio frequency applications. The successful performance measures uphold the fact that the lumped parameter modeling method closely approximates empirical data, upholding its efficacy and precision in MEMS application (Göritz et al., 2017). Nevertheless, despite its strengths, lumped parameter modeling comes with unique challenges towards model accuracy and stability. Though reduced models may make simulations easier, they do not necessarily capture the intricate dynamics of actual MEMS devices, especially when non-linear responses or environmental variations are taken into consideration. The resulting models should comply with basic properties like causality and passivity so that they are truly representative of physical realities. Inaccuracies will ensue and predictive failure in the engineering design process if these basic properties cannot be generated. Therefore, a detailed understanding of these limitations is essential for engineers to work towards maximizing lumped parameter models in their potential when utilized in MEMS design, with experience dictating the need for constant development and comparison to experimental data (Achar et al., 2007) (Figure 6.8).

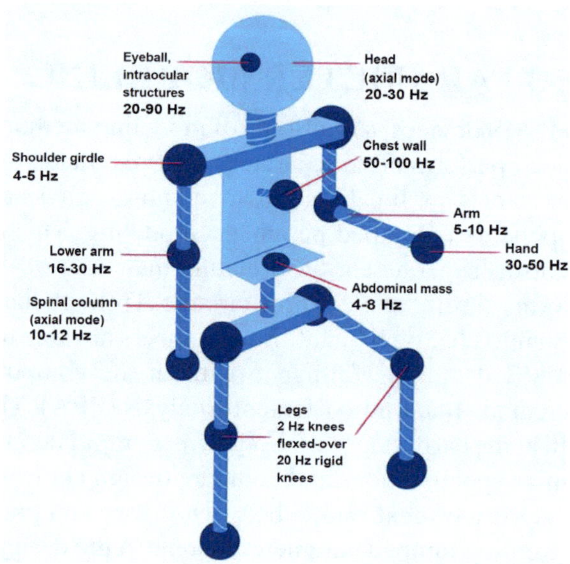

Figure 6.8. Lumped parameter modeling holds a unique position due to its ability to simplify complex physical systems into manageable components.

Source: https://cdn.comsol.com/wordpress/sites/1/2019/05/human-body-model-lumped-approach.png.

6.4.1. Explanation of Lumped Parameter Modeling Concepts

Knowledge of lumped parameter modeling is also important in the context of MEMS modeling and simulation because it reduces complex systems into smaller elements that are easier to work with. This method enables the modeling of a physical system in terms of discrete parameters like resistances, capacitances, and inductances, which can correspond to actual components. By compressing physical interaction to these parameters, engineers can simulate various behaviors without having to navigate computationally burdensome methods like Finite Element Analysis (FEA). Furthermore, lumped parameter models will also help facilitate reduced-order modeling, enabling faster simulation with conservation of key system dynamics. This simplification technique is particularly valuable during early design when quick iteration is important in inducing innovation. Thus, the application of lumped parameter theory supports efficient and effective design procedures in the field of MEMS technologies. In practice, the application of lumped parameter modeling should consider the size and nature of the MEMS devices in question. For instance, the validity of using lumped parameters could be sacrificed under conditions where physical dimensions approach the mean free path of the molecules or electromagnetic effects due to size are involved. However, these limitations are usually overshadowed by the advantages of diminished analysis, especially in electrical interconnect systems that include high frequency signaling, as discussed in recent models that emphasize stability, causality, and passivity in electrical interconnect models (Achar et al., 2007). These underlying properties ensure that as lumped models are utilized in simulation, they are physically consistent and predictive of actual performance. Adherence to these guidelines improves the credibility of simulations, which is essential in making reliable MEMS devices. Advances in wireless sensors emphasize the engineering importance of lumped parameter modeling in day-to-day life. For instance, new approaches like isospectral party time reciprocal scaling (PTX) have optimized model design to enhance microelectromechanical systems, leading to increased sensitivity and resolution (Alù et al., 2018). By applying lumped parameter ideas to tunable resonator design, extrinsic quality factors can be enhanced by simplifying circuit profiles. Reduced models, therefore, open the door for engineers to accelerate the design process while ensuring that the final sensors will operate effectively under rough conditions. Thus, lumped parameter modeling not only supports theoretical studies but also finds expression in concrete contributions and hence forms the basis of the ever-growing dynamism in MEMS device technology.

6.4.2. Advantages and Limitations of Lumped Models

In MEMS device simulation, selecting the most suitable analysis method is critical in terms of efficiency and accuracy. Among all the available methods, lumped models stand out because they are simple and can provide quick estimates of device performance. Lumped models treat the system as a collection of isolated elements, capturing basic dynamics without delving into the details of spatial dependencies. For instance, in the design of an electrostatically actuated micro peristaltic pump, a lumped-parameter model was utilized to perform dynamic analysis with accurate prediction of trends in accordance with experimental observations. In this case, the lumped model justified its utility in quantifying the peristaltic pumping operation with additional focus on its advantages in rapid evaluation of system performance and informing iterative design. These characteristics emphasize the use of lumped modeling in the early phases of MEMS device development. While the overwhelming advantages, lumped models are not without their shortcomings that can never be overlooked. One such significant drawback is that they portray constant properties all over the system and can produce overidealizations, particularly in scenarios of highly sensitive applications where localized effects are of importance. This limitation is particularly strong when working with micro devices of complicated geometry or more than one material property. Further, lumped models cannot represent the interaction between different components very well, particularly in devices requiring very strict coordination between different operating states. As an example, in discussing reconfigurable split ring resonators (SRRs) with RF-MEMS switches, one can see that lumped modeling might overlook important aspects of electrical and mechanical interactions that are crucial for obtaining filter optimality. Thus, depending only on lumped models may result in suboptimal designs and demands additional accurate simulations, e.g., Finite Element Analysis, to acquire full insight into performance nuances. To enhance the accuracy in the modeling of MEMS devices, researchers prefer to adopt a combination of lumped and detailed modeling techniques. By integrating lumped models with sophisticated approaches like Finite Element Analysis (FEA) and multiphysics simulations, engineers can avail each other's benefits and offset their drawbacks. This synergistic strategy allows the initial rough estimates provided by lumped models to then be used in guiding and directing further detailed examinations through FEA for confirmation and calibration purposes. Betterment of modeling techniques is essential to maximize devices such as microfluidic systems and RF-MEMS switches, where precision is paramount. From this viewpoint, experiments have demonstrated that a combined strategy can successfully provide prototypes bridging the gap from theoretical model to real application, thus unlocking the door to more solid insights and innovations with MEMS technologies. Incorporating these modeling strategies is part of the development progress of model fidelity and device performance with MEMS technology.

6.4.3. Applications of Lumped Parameter Models in MEMS Design

Under microelectromechanical systems (MEMS) engineering, the effective application of lumped parameter models has greatly enhanced the efficiency and reliability of device performance simulation. In contrast to other advanced simulations that require enormous computational capacities, lumped parameter models minimize the physical complexities into manageable parts that facilitate rapid analysis. This method proves most beneficial in sensing and predicting the dynamic responses of MEMS devices to external inputs. For instance, an electromagnetic (EM) model and a small-signal lumped-element model of a wafer-level encapsulated radio frequency MEMS switch illustrate how such models can accurately forecast RF performance and allow for fast simulations in integrated circuit design environments. The inclusion of such models in a 0.13 μm SiGe BiCMOS process technology demonstrates not just the utility of lumped parameter models but also their suitability in achieving high-performance predictions in RF applications, which attained a high capacitance Con/Coff ratio of 11.1 and achieved isolation measurements of 51.6 dB at strategically critical frequencies (Göritz et al., 2017).

Additionally, lumped parameter model creation is approached in power generation applications in MEMS. Particularly, the need for self-powered devices, such as sensor systems of personal area networks, requires research into micro-power generator integration based on ambient energy. One notable innovation proposed a novel approach to energy conversion from mechanical to electrical via charge transportation between two parallel capacitors, showcasing the ubiquity of lumped parameter modeling across various areas of MEMS. With the application of large-signal and small-signal models, designers can now measure output power as a function of mechanical input, thus facilitating precise tuning of design parameters for maximum performance. This dual modeling approach not only facilitates ease of design but is primarily concerned with minimizing device footprint with enhanced power generation efficiency, key to providing long-life battery-operated devices (Baert et al., 2001).

MEMS technology development is increasingly demanding an integration of lumped parameter models with multiphysics simulations, thus bridging unrelated fields such as mechanical, electrical, and thermal phenomena. By combining lumped parameter techniques with finite element analysis (FEA), engineers can create reduced-order models that retain the fidelity of complex interactions without significantly raising computational burden. This hybrid technique allows for faster innovation cycles, which facilitate rapid prototyping and testing of MEMS devices. Therefore, with designers now realizing the advantages of lumped parameter modeling alongside top-level simulation techniques, the next generation of MEMS will be engineered to provide more efficient, high-performance devices and those with increased versatility for use in consumer electronics to biomedical systems. These advances are paramount in positioning MEMS technologies at the leading edge of modern-day engineering solutions.

6.5. REDUCED-ORDER MODELING

The fundamental nature of the complexity in modeling Micro-Electro-Mechanical Systems (MEMS) significantly brings into being obstacles that can complicate computational efficiency and reduce the rate at which rapid design iteration can be achieved. Any time the simulation space dimension is increased, typical modeling techniques become intractable, leading to computational times that are far above requirements and resource allocation problems. This issue, known as the curse of dimensionality, raises serious issues about the suitability of existing algorithms in dealing with high-dimensional parameter spaces efficiently. To address such issues, reduced-order modeling (ROM) stands out as an option. By simplifying the system description while maintaining the key dynamics, ROM allows for economical simulation of MEMS devices. Not only does it reduce computation, but it also gives designers an easier method to deal with the intricate interactions in MEMS. In practice, lower-order modeling techniques have the tendency to leverage advanced mathematical paradigms like proper orthogonal decomposition (POD) and balanced truncation to create efficient, yet accurate, models of the system's behavior. The techniques come to fit especially in conditions where finite element analysis (FEA) and multiphysics simulations generate vast amounts of data that could be cumbersome to handle and utilize effectively. Consequently, ROM can have the potential to significantly reduce the computational burden typically associated with such sophisticated instruments. For instance, in stochastic circuit simulations involving high-dimensional subsystems, the integration of ROM can have the potential to make rapid estimates with the accuracy of results maintained for design verification purposes. Through application of techniques such as tensor decomposition, researchers can construct surrogate models that capture system behaviors with very much fewer degrees of freedom (Batselier et al., 2016). Reduced-order modeling is not just about mere computational savings; it also facilitates increased design space exploration and optimization processes. Engineers can more readily include uncertainty quantification in the design process by developing reduced models that preserve essential dynamic behavior. Hierarchical uncertainty quantification techniques, for example, in the context of MEMS, can significantly generalize simplification of the simulation of a complex system with many random parameters. By utilizing an ANOVA-based stochastic simulator, it is possible to efficiently achieve surrogate models at a design hierarchy lower and avoid dominating dimensionality challenges (Daniel et al., 2014). Such methodological complementarity underlines the crucial role that reduced-order modeling plays not only in enhancing computational efficiency but also in ensuring reliability and accuracy in design and simulation activities of MEMS devices.

6.5.1. Definition and Purpose of Reduced-Order Modeling

Microelectromechanical systems (MEMS) complexities need very strict analytical techniques to accurately forecast their behaviors under various conditions. Among these approaches, reduced-order modeling (ROM) is a key methodology, which effectively simplifies complex models without compromising significant accuracy. ROM is applied for the purposes of capturing the primary MEMS device dynamics in such a way that effective simulations can be conducted, which do not demand the computational expense of full-scale numerical techniques such as finite element analysis (FEA). This reduction is particularly useful in cases where rapid prototyping and real-time analysis are required, enabling the engineer to make quick decisions. By employing reduced-order models, researchers can maintain the fidelity of the most critical phenomena while significantly reducing simulation time and resource consumption and thereby accelerating the design and optimization process inherent within MEMS technology. Reduced-order modeling is applied in MEMS device design particularly for multiphysics simulations. The complexity of co-interacting physical effects—e.g., mechanical deformation, fluid flow, and heat conduction—makes it nearly impossible to form complete models. Reduced-order modeling succeeds with such challenges by enabling the decomposition of high-dimensional systems into lower-dimensional representations that capture essential features.

Thus, such models allow for the examination of parameter spaces that can potentially maximize device performance. For instance, sensor design or actuator operation contexts, ROM allows analysis across various designs while computational efficiency is preserved, hence facilitating support to iterative refinement processes characteristic of MEMS development. Furthermore, the application of reduced-order modeling is also crucial not only for performance enhancements but for the design of real-time monitoring systems for MEMS applications. Similar to the application of data-driven approaches in structural health monitoring, the utilization of low-complexity models is able to yield effective identification and characterization of damages or operating anomalies (Cherta et al., 2020). The integration of ROM with predictive analytic software, such as that used in compliance and treatment modality adherence analysis in health research, shows how models of lesser order can effectively blend vast quantities of data while retaining key variables (Acosta et al., 2011). This integration is particularly crucial when continuous monitoring of MEMS devices is needed, since it enables the models created to provide information on the operational integrity of devices in real-application environments, eventually for improved reliability and performance outcomes (Figure 6.9).

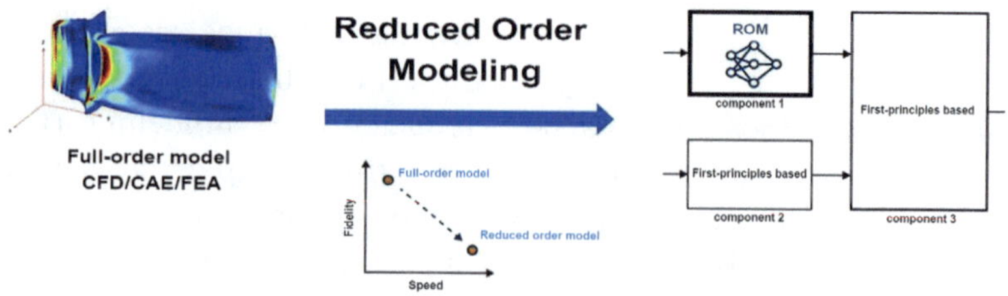

Figure 6.9. Reduced-order modeling (ROM) stands as a pivotal approach, effectively simplifying complex models without sacrificing critical accuracy.

Source: https://ch.mathworks.com/discovery/reduced-order-modeling/_jcr_content/imageParsys/thumbnail_1940094120.adapt.full.medium.jpg/1740983307324.jpg.

6.5.2. Techniques for Creating Reduced-Order Models

The formulation of techniques for simulating reduced-order models significantly enhances the efficiency of simulations in Micro-Electro-Mechanical Systems (MEMS). Such techniques aim to simplify intricate models in a manner where significant dynamic characteristics are preserved, which is particularly valuable within the scope of MEMS devices where intricate interactions and behaviors must be simulated. Finite Element Analysis (FEA) is generally applied in the modeling of MEMS, allowing for complex, multiphysics simulations to take many phenomena into account such as stress, thermal effects, and fluid dynamics. However, high-cost computations with full-order models typically exclude their practical application in real-time situations. The construction of reduced-order models allows engineers to perform rapid analyses without accuracy loss, ensuring that critical insights into device performance can be realized at design time. These models are thus vital in speeding up the optimization of MEMS devices while not sacrificing the precision of the intricate physics. Lumped parameter modeling is an effective process employed in reduced-order modeling that scales down the physical model by representing the distributed parameters through discrete components. This can be suitable for MEMS devices because it scales down the model to a more manageable size which can be analyzed more efficiently. In practice, it works by modelling the device as a network of masses, springs, and dampers, something that can be easily manipulated and understood to describe dynamic behaviors. Such a technique can be used as the basis to model more complex models, facilitating efficient computation even of highly nonlinear systems. Moreover, embedding this strategy within software applications like COMSOL or ANSYS enhances accessibility, in that researchers and engineers can focus on refining design and iteration without being inordinately encumbered by computational issues. Accordingly, lumped parameter models emphasize the importance of abstraction in modeling. Aside from lumped parameter models, the use of intelligent technologies, such as the recently coined term reconfigurable intelligent surfaces (RISs), presents new avenues for ongoing reduction in computational needs in MEMS simulations. As noted in recent research, RISs are characterized by their ability to accurately control the transmission properties of electromagnetic waves. This technology is similar to the aspirations of reduced-order modeling in the pursuit of minimalist designs that need to use less resource to operate efficiently. The integration of traditional modeling approaches such as FEA and new approaches such as RISs has the potential to make huge advancements in simulating intricate MEMS environments. By taking up these new approaches, researchers can help foster the development of smart radio environments conducive to MEMS use, the ultimate beneficiary being the interaction between the physical world and computational models. The convergence is the promise of sophisticated modeling techniques in establishing the next generation of MEMS technology.

6.5.3. Benefits of Reduced-Order Models in Simulation Efficiency

In the rapidly expanding field of Microelectromechanical Systems (MEMS), efficient simulation of device behavior is crucial to propel design and operation. Traditional modeling techniques, such as Finite Element Analysis (FEA), are heavy in computational resources due to the high dimensional problem-solving nature. This type of challenge arises from numerous spatial and temporal influences, nonlinearities in device response, and numerous design parameters involved in MEMS usage. This kind of complexity typically results in a computational burden that restricts simulation speed and efficiency, making it inefficient for engineers to sample design space. Reduced-order models come as a strong alternative that can bring considerable dimensionality reduction in these problems with the retention of key dynamic properties. Reducing the curse of dimensionality, reduced-order models make it possible to speed up simulations without compromising accuracy, opening doors to fast iterative design cycles. The integration of reduced-order models makes it possible for engineers to decouple simulations, thus improving workflow efficiency in MEMS design. By applying approaches such as lumped parameter modeling and model order reduction, engineers can simplify complex systems into mathematically manageable representations. Not only does this reduce computation times, but real-time analysis is also facilitated, which is of most value while in the design and optimization stages. Reduced-order models enable engineers to quickly assess and iterate between design alternatives, thus improving productivity in the overall context. Therefore, the application of these models reduces both the time and economic expense needed for MEMS design as well as enhances the ability to simulate more diverse operating conditions, essentially leading to more innovative solutions in optimizing device performance. Furthermore, the application of reduced-order models optimizes the use of multiphysics simulations, where different physical phenomena dynamically couple with one another. For instance, in the case of problems with complex thermal and mechanical couplings, reduced-order modeling allows efficient coupling of heterogeneous physical domains without the computationally intensive weight that high-detail FEA simulations normally carry. Advances in reduced-order modeling highlight its significance in attacking high-dimensional problems with characteristics prevalent throughout much of MEMS's many applications, including power-saving interconnect designs, as highlighted throughout relevant literature. Through the interaction between computational speed and accuracy, reduced-order models not just enhance simulation performance but also enable engineers to design responsive designs that can meet the adaptive demands of the modern MEMS technology (Amoroso et al., 2017; Batselier et al., 2016) (Figure 6.10).

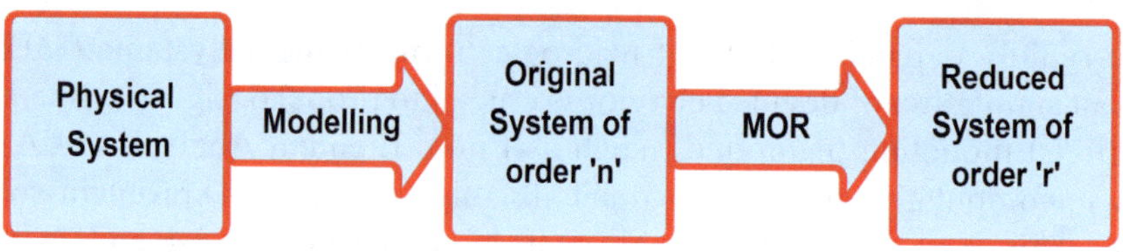

Figure 6.10. Reduced-order models emerge as a powerful solution, allowing for a significant reduction in the dimensionality of these problems while maintaining essential dynamic characteristics.

Source: https://media.springernature.com/lw685/springer-static/image/art%3A10.1007%2Fs40435-022-00985-7/MediaObjects/40435_2022_985_Fig1_HTML.png.

6.6. CONCLUSION

The use of sophisticated modeling and simulation techniques has profoundly impacted the design and analysis of Microelectromechanical Systems (MEMS) devices. Through the use of Finite Element Analysis (FEA) and other multiphysics simulations, designers gain the potential to forecast device behavior under real-life conditions. Not only does this ease the design process but also enhances MEMS product reliability. For instance, analytical models using lumped parameter modeling can be useful in determining the dynamic response of MEMS structures and reduced-order models cut computation time in half without loss of accuracy. Both of these methods assist in improving device performance and minimizing production costs to achieve more functional and innovative MEMS technologies. An in-depth examination of specific applications confirms that FEA has a key role to play in the optimization of the performance of MEMS devices, i.e., electrostatically actuated micro-switches. Past work has established that precise contact force predictions are dependent on the understanding of the collapse voltage-pull-in voltage relationship, as in research where these forces were rigorously investigated both analytically and through advanced simulations like CoventorWare (Coutu et al., 2004). By establishing this correspondence, micro-switches of improved working characteristics are constructed, thereby decreasing contact resistance and optimizing overall efficiency. The ability to simulate electrode dimensions further demonstrates the flexibility of these modeling techniques, offering data that can be utilized to make radical device performance improvements. Besides, ongoing innovations in simulation and modeling are not limited to mechanical performance but also in fields such as image processing for MEMS-based systems, bearing witness to the multidisciplinary nature of MEMS research. For example, in studies investigating segmentation methods for deteriorated eye image data, the application of robust simulation techniques greatly improves recognition systems by eliminating environmental noise (Ahmad et al., 2018). This convergence of image processing and MEMS technology shows the full breadth of simulation tool capability, enabling researchers to address multidimensional problems in a broad variety of applications. Lastly, continued advancements of modeling and simulation techniques can help speed MEMS technology, fuel innovation, and enhance device performance in an increasingly complex technology environment.

6.6.1. Summary of Key Points Discussed

The intricate nature of MEMS devices demands advanced modeling and simulation techniques to correctly predict their performance in a range of operating conditions. The method of Finite Element Analysis (FEA) is widely used because it has a built-in facility to solve intricate geometric as well as physical effects and provides data on the distribution of stress, deformation, and thermal properties within MEMS structures. Software such as COMSOL and ANSYS offers an entire platform to conduct such analyses. There is also required multiphysics simulation, which allows scientists to visualize how different physical principles—thermal, mechanical, and electromagnetic forces—interact with one another in MEMS devices. This integrated approach to modeling not only leads to precise predictions but also deeper insight into the limits of the technology's performance, paving the way for the design of more sophisticated and effective MEMS systems. One of the newer trends in MEMS device simulation is the application of lumped parameter models, which approximate large systems as smaller, manageable elements while preserving the important dynamics. These models can accelerate analysis and design parameter optimization without the computer cost of complete FEA. By leveraging reduced-order model techniques, engineers can make simulations much more efficient, leading to quicker design cycles and improvement. These advancements find particular importance in real-time performance monitoring applications such as sensors and actuators in consumer appliances and industrial applications. The use of these modeling strategies speaks volumes about embracing more flexible engineering paradigms, a step towards committing to the requirement of flexibility in the design process of MEMS. The need for efficient methods to overcome uncertainties in MEMS design has driven new solutions, such as hierarchical uncertainty estimation methods. As the complexity of the circuit and systems increases, stochastic simulation becomes increasingly challenging, particularly with high-dimensional parameters. Novel developments, such as the ANOVA-based stochastic circuit/MEMS simulator, highlight the utility of such models for overcoming the curse of dimensionality. This approach uses tensor-train decomposition and Gauss quadrature points to make the simulation process simpler. To illustrate, for a stochastic oscillator with MEMS capacitors, the simulation times were reduced considerably, thereby enabling real-time analyses. These are important methods because they do not only promote efficiency but also boost simulations' credibility such that designers can confidently base the performance predictions of their MEMS devices under varying circumstances, tidily in concert with the forthcoming trend of MEMS technology development (Beacken et al., 2018; Daniel et al., 2014).

6.6.2. Future Trends in MEMS Modeling and Simulation

Improvements in computational power and numerical methods are creating windows of opportunity for emerging trends in MEMS device simulation and modeling. The complexity of MEMS systems demands advanced simulation capability, particularly by means of technologies such as Finite Element Analysis (FEA) and Multiphysics simulation. Software packages for FEA such as COMSOL and ANSYS enable comprehensive analysis through allowing engineers to predict the behavior of a device under varying conditions. In such a situation, lumped parameter models are a useful tool for rapid design cycles with a reduced description that can encompass the key physical phenomena without the computational effort of more detailed models. With growing computational power, reduced-order modeling is becoming increasingly prevalent as it effectively separates simulation from detail and thus significantly enhances the design process of MEMS devices. New technologies are also transforming the simulation platform for MEMS devices, particularly by incorporating machine learning (ML) and artificial intelligence (AI) techniques. These techniques bring advanced prediction capabilities, enabling models to learn from large data and refine designs based on performance specifications. The incorporation of AI-driven algorithms not only makes the simulation process faster but also enhances parameter extraction techniques that facilitate real-time optimizations. Further, the marriage of ML with traditional modeling approaches facilitates the discovery of intricate design patterns and interdependencies that are not easily discernible in conventional studies. This movement towards AI-powered simulations can lead to revolutionary advancements in MEMS design, eventually presenting devices with improved performance and decreased time-to-market. Thermal and mechanical simulation domains are increasingly becoming significant in the modeling of MEMS in the future, particularly as devices are being pushed into more and more extreme operating conditions. Sealing methods employed within micro-cavities, for instance, as shown by Liu et al. (1999), demonstrate the significance of robust thermal simulations involving material behaviors under various deposition schemes. Improved modeling techniques will allow accurate simulation of mechanical and thermal load interactions, providing precious data on the reliability and lifetime of devices. As the industry migrates towards miniaturization and integration with other MEMS technology, such as RF MEMS packages discussed by Bartek et al. (2007), more sophisticated hybrid simulation environments will be required. These trends not only promise to enhance the fidelity of MEMS simulations but also ensure that products are achieving stringent performance requirements in increasingly demanding applications.

6.6.3. Final Thoughts on the Impact of Modeling on MEMS Technology

Advances in modeling techniques have been crucial for advancing Microelectromechanical Systems (MEMS) technology, enhancing device performance and reliability to a great extent. Techniques like Finite Element Analysis (FEA) have provided such detailed information on the mechanical, thermal, and fluid dynamics behavior of MEMS devices. Of greatest significance, computer platforms like COMSOL and ANSYS allow engineers to conduct multiphysics simulations, allowing analysis of complex interactions between MEMS structures. These interactions often involve not only mechanical stresses but also thermal effects and electrostatic forces impacting the device's function and durability. These kinds of holistic modeling allow designers to optimize device geometry and material selection, which leads to breakthroughs that can meet the stringent demands of new applications with reduced risks of failure. Lastly, the integration of sophisticated modeling techniques marks an era of revolution in MEMS design with improvements in performance that were previously unattainable. In addition, lumped parameter modeling provides a leaner analytical vehicle for MEMS application, permitting quicker design cycles. As they decompose complex systems into tractable subsystems, designers can rapidly evaluate the impact of discontinuous design changes without using profuse computing facilities. This method is especially useful in the early stages of design when rapid prototyping and feedback loops are essential for maximizing device performance. Reduced-order modeling further enhances this by allowing the retention of critical dynamics at the expense of lower computational costs. These methods therefore accelerate the design process, enabling faster innovation cycles and reduced time-to-market for novel MEMS products. The increased reliance on such modeling methodologies emphasizes the pivotal role that simulation plays in the MEMS context, leading the technology toward more efficiency and emergence of novel applications. Not the least, ongoing innovations in modeling technologies, particularly in the context of MEMS, help show the critical role of interdisciplinarity. For instance, the use of features like carbon nanotube interconnects in MEMS devices prompts us to reconsider existing models to facilitate the formation of new materials and their intrinsic properties (Amoroso et al., 2017). Similarly, adaptive optics application in MEMS can be improved by leveraging cutting-edge modeling techniques to focus system performances, such as with top-of-the-line telescope tools. These multi-disciplinary approaches render the model relevant and sensitive to issues accompanying next-generation MEMS devices. In time, with emerging advances in modeling, they shall not only be a driving force for engineering excellence but also an integral linkage to energize cross-disciplinary breakthrough, charting the path of future MEMS technology in revolutionary directions.

REFERENCES

1. Achar, R., Canavero, F., Grivet-Talocia, S., et al. (2007). Stability, causality, and passivity in electrical interconnect models. https://core.ac.uk/download/pdf/11395534.pdf (accessed on 14th October 2025).
2. Acín, A., Bloch, I., Buhrman, H., Calarco, T., Eichler, C., Eisert, J., Estève, D., et al. (2018). The quantum technologies roadmap: A European community view. *New Journal of Physics, 20*, 080201–080201. https://doi.org/10.1088/1367-2630/aad1ea (accessed on 14th October 2025).
3. Acosta, E. P., Holden-Wiltse, J., Huang, Y., et al. (2011). A dynamic Bayesian nonlinear mixed-effects model of HIV response incorporating medication adherence, drug resistance and covariates. http://arxiv.org/abs/1104.3464 (accessed on 14th October 2025).
4. Ahmad, A. S., Ahmar, A. S., et al. (2018). Iris segmentation. https://core.ac.uk/download/19162806.pdf (accessed on 14th October 2025).
5. Alù, A., Chen, P. Y., Cheng, M., et al. (2018). Generalized parity-time symmetry condition for enhanced sensor telemetry. http://arxiv.org/abs/1812.08296 (accessed on 14th October 2025).
6. Amoroso, S., Asenov, A., Brown, A., et al. (2017). A survey of carbon nanotube interconnects for energy efficient integrated circuits. https://core.ac.uk/download/82958608.pdf (accessed on 14th October 2025).
7. Anwar, F., Coutu, R. A., & Mahanta, P. (2019). Novel test fixture for characterizing MEMS switch microcontact reliability and performance. https://core.ac.uk/download/228028443.pdf (accessed on 14th October 2025).
8. Atherton, M. A., Bates, R. A., Kaymak, E., et al. (2005). Experimental and bond graph-based sensitivity calculations for micro-scale robust engineering design. https://core.ac.uk/download/13641780.pdf (accessed on 14th October 2025).
9. Baert, K., Borghs, G., Puers, R., et al. (2001). Power extraction from ambient vibration. https://core.ac.uk/download/55824876.pdf (accessed on 14th October 2025).
10. Barrachina, L., Fernández Martínez, D., Madrenas Boadas, J., et al. (2017). Curvature of BEOL cantilevers in CMOS-MEMS processes. https://core.ac.uk/download/157810913.pdf (accessed on 14th October 2025).
11. Bartek, M., Gaddi, R., Gnudi, A., et al. (2007). A fully parameterized FEM model for electromagnetic optimization of an RF MEMS wafer level package. https://core.ac.uk/download/15495639.pdf (accessed on 14th October 2025).

12. Batselier, K., Daniel, L., Liu, H., et al. (2016). Tensor computation: A new framework for high-dimensional problems in EDA. https://core.ac.uk/download/95558233.pdf (accessed on 14th October 2025).
13. Beacken, M., Bishop, D. J., Chang, J., et al. (2018). High performance, continuously tunable microwave filters using MEMS devices with very large, controlled, out-of-plane actuation. http://arxiv.org/abs/1804.08658 (accessed on 14th October 2025).
14. Bertsch, A., Forchelet, D., Hasenkamp, W., et al. (2012). Polyimide/SU-8 catheter-tip MEMS gauge pressure sensor. https://core.ac.uk/download/18158609.pdf (accessed on 14th October 2025).
15. Bogaerts, W., Edinger, P., Errando-Herranz, C., et al. (2020). MEMS-enabled silicon photonic integrated devices and circuits. https://core.ac.uk/download/287939692.pdf (accessed on 14th October 2025).
16. Bonache Albacete, J., & Martín, F. (2014). Application of RF-MEMS-based split ring resonators (SRRs) to the implementation of reconfigurable stopband filters: A review. https://core.ac.uk/download/132084746.pdf (accessed on 14th October 2025).
17. Calle Gómez, F., Fuentes Iriarte, G., Lebedev, V., et al. (2015). Electro-optical characterization of IC compatible microcantilevers. http://oa.upm.es/40909/1/INVE_MEM_2015_223830.pdf (accessed on 14th October 2025).
18. Cherta, A. D., Gomez Vargas, R. C., Pedraza, C., et al. (2020). Damage identification in structural health monitoring: A brief review from its implementation to the use of data-driven applications. https://core.ac.uk/download/288625955.pdf (accessed on 14th October 2025).
19. Cho, A. N., Jin, Y., An, Y., Kim, J., Choi, Y. S., Lee, J. S., Kim, J. H., et al. (2021). Microfluidic device with brain extracellular matrix promotes structural and functional maturation of human brain organoids. *Nature Communications, 12.* https://doi.org/10.1038/s41467-021-24775-5 (accessed on 14th October 2025).
20. Coutu, R. A., Jr., Kladitis, P. E., Reid, J. R., et al. (2004). A comparison of micro-switch analytic, finite element, and experimental results. https://core.ac.uk/download/213087969.pdf (accessed on 14th October 2025).
21. Czaplewski, D., Hyman, D., Karatzidis, D., Rebeiz, G., Newman, H., Sumali, H., McKillop, J., et al. (2011). RF-MEMS switch actuation pulse optimization using Taguchi's method. https://core.ac.uk/download/337977.pdf (accessed on 14th October 2025).
22. Daniel, L., Karniadakis, G. E., Oseledets, I. V., Yang, X., et al. (2014). Enabling high-dimensional hierarchical uncertainty quantification

by ANOVA and tensor-train decomposition. https://core.ac.uk/download/78065180.pdf (accessed on 14th October 2025).

23. Di Renzo, M., Zappone, A., Debbah, M., Alouini, M. S., Yuen, C., de Rosny, J., & Tretyakov, S. (2020). Smart radio environments empowered by reconfigurable intelligent surfaces: How it works, state of research, and the road ahead. *IEEE Journal on Selected Areas in Communications, 38*, 2450–2525. https://doi.org/10.1109/jsac.2020.3007211 (accessed on 14th October 2025).
24. Ehwald, K. E., Gürbüz, Y., Kaynak, M., et al. (2010). Characterization of an embedded RF-MEMS switch. https://core.ac.uk/download/pdf/11742165.pdf (accessed on 14th October 2025).
25. Euclid Collaboration, Scognamiglio, D., Schrabback, T., Tewes, M., Gillis, B., Hoekstra, H., Huff, E., Marggraf, O., et al. (2025). Euclid preparation LX. The use of HST images as input for weak-lensing image simulations. https://www.semanticscholar.org/paper/aab36fa0857315e56921ad1b7b8be1a93af05981 (accessed on 14th October 2025).
26. Fan, Z., Li, Z., Tang, R., Wu, G., Yang, S., Tu, L., Wang, Y., et al. (2024). Quality factor optimization and thermal noise analysis of silicon MEMS resonators. *2024 IEEE SENSORS*. https://www.semanticscholar.org/paper/3f60baeb7c9b2880e948c8120cfedd8877b1c8f9 (accessed on 14th October 2025).
27. Gill, E., Monna, G. L. E., Scherpen, J. M. A., et al. (2007). MISAT: Designing a series of powerful small satellites based upon micro systems technology. https://core.ac.uk/download/pdf/12941826.pdf (accessed on 14th October 2025).
28. Göritz, A., Kaynak, M., Mai, A., Tillack, B., et al. (2017). Electromagnetic and small-signal modeling of an encapsulated RF-MEMS switch for D-band applications. https://core.ac.uk/download/187203021.pdf (accessed on 14th October 2025).
29. Guruprasad, B. (2020). Design and fabrication of cantilever MEMS sensor model for electro-chemical gas sensor. *International Journal of Engineering Research and, V9*. https://doi.org/10.17577/ijertv9is070262 (accessed on 14th October 2025).
30. Kerkhoff, H. G. (2007). Testing microelectronic biofluidic systems. https://core.ac.uk/download/pdf/11468013.pdf (accessed on 14th October 2025).
31. Lin, Q., Shih, J., Tai, Y. C., Xie, J., et al. (2004). Surface micromachined electrostatically actuated micro peristaltic pump. https://core.ac.uk/download/4872581.pdf (accessed on 14th October 2025).

32. Liu, C., & Tai, Y. C. (1999). Sealing of micromachined cavities using chemical vapor deposition methods: Characterization and optimization. https://core.ac.uk/download/4869879.pdf (accessed on 14th October 2025).
33. Rothemund, P., Kellaris, N., Mitchell, S. K., Acome, E., & Keplinger, C. (2020). HASEL artificial muscles for a new generation of lifelike robots—Recent progress and future opportunities. *Advanced Materials, 33*. https://doi.org/10.1002/adma.202003375 (accessed on 14th October 2025).
34. Subbiah, N., Banker, J., Mohanraj, S. P., Dominik, B., Landgraf, E., Pregl, S., Khouya, B., et al. (2025). Modeling and simulation of CMUT and PMUT ultrasound transducers and package using a Multiphysics approach. *2025 26th International Conference on Thermal, Mechanical and Multi-Physics Simulation and Experiments in Microelectronics and Microsystems (EuroSimE)*. https://www.semanticscholar.org/paper/b29a393b0cccfeefb638c2017c3e504d1c2d89fd (accessed on 14th October 2025).
35. Wang, H., Ma, J., Yang, Y., Gong, M., & Wang, Q. (2023). A review of system-in-package technologies: Application and reliability of advanced packaging. *Micromachines, 14*, 1149–1149. https://doi.org/10.3390/mi14061149 (accessed on 14th October 2025).
36. Wang, L., Liu, H., & Liu, X. (2024). Advancing simulation efficiency for MEMS electrostatic actuators. *Journal of Physics: Conference Series, 2740*. https://www.semanticscholar.org/paper/e4eaebb9e458f40c6ae1960256c180b413ff5824 (accessed on 14th October 2025).
37. Wong, W. C. (2007). Analysis of the suspension beam in accelerometer for stiffness constant and resonant frequency by using analytical and numerical investigation. https://core.ac.uk/download/32600339.pdf (accessed on 14th October 2025).
38. Wu, Q., Zhang, S., Zheng, B., You, C., & Zhang, R. (2021). Intelligent reflecting surface-aided wireless communications: A tutorial. *IEEE Transactions on Communications, 69*, 3313–3351. https://doi.org/10.1109/tcomm.2021.3051897 (accessed on 14th October 2025).
39. Zhang, L., Chen, X., Liu, S., Zhang, Q., Zhao, J., Dai, J. Y., Bai, G. D., et al. (2018). Space-time-coding digital meta surfaces. *Nature Communications, 9*. https://doi.org/10.1038/s41467-018-06802-0 (accessed on 14th October 2025).

CHAPTER 7

MEMS SENSORS FOR PHYSICAL QUANTITIES

Contents

INTRODUCTION

The rapid advancement of microelectromechanical systems (MEMS) has significantly enhanced the sensors' performance in diverse applications. This technology has led to the availability of highly sensitive, compact, and low-cost sensors, which are revolutionizing sectors such as automotive, aerospace, and healthcare. Gyroscopes and accelerometers, normally packaged as inertial measurement units (IMUs), are evidence of this milestone. MEMS sensors play a crucial role in applications where precise sensing of motion and orientation is required, such as in smartphones and drones. The miniaturization process offered by MEMS technology makes the sensors easily implanted in confined spaces, thus making them a critical component of today's technology advancements. Greater reliance upon such systems necessitates an extensive understanding of how they operate and the individual advantages they lend themselves to in measuring physical parameters, creating new avenues of application that could not be accessed with traditional sensor technology. Pressure sensors and flow sensors are two important areas within the domain of MEMS technology, illustrating the flexibility and versatility of the methodology. These sensors have important uses across a range of industries, including medicine, transportation, and environmental monitoring. They allow for accurate measurement of atmospheric and fluid pressure and provide real-time information required for system control and safety. MEMS pressure sensors can monitor vital signs in medicine, and flow sensors can provide proper intravenous fluid supply in hospitals. Similarly, MEMS technology enables monitoring of the environment through sensing changes in the quality of fluid flow and air in ecosystems. Due to the high demand for sensitive and precise sensor networks, researchers continue to explore the promising use of MEMS pressure and flow sensors, projecting their significance in achieving a sustainable and intelligent world. Inertial navigation systems (INS) and environmental monitoring MEMS also demonstrate the revolutionary impact of MEMS sensors on knowledge of physical phenomena. INS employs accelerometers and gyroscopes to determine the position and orientation of an object with respect to an external reference, providing vital information for aircraft and vehicle navigation. The integration of MEMS technology into these systems has significantly improved their reliability and accuracy, making them capable of support for autonomous vehicle navigation technologies. Furthermore, the capability of MEMS sensors to monitor environmental conditions is essential for support against rising global challenges such as climate change and resource management. The development of miniature low-power MEMS sensors enables large-scale data collection and analysis, which enables real-time assessment of environmental factors. With these new technologies, society is more capable of responding to shifts in physical quantity, possessing an enhanced knowledge of our evolving world, and creating solutions to its most important challenges.

Importance of Sensors in Modern Technology

In the modern era of technology, small sensors have revolutionized industries by enabling precise measurement and data collection on a scale never seen before. Accelerometers and gyroscopes, or inertial measurement units (IMUs), are the core of applications ranging from mobile phones to outer space. They monitor variations in motion and orientation, providing critical information for navigation and stabilization systems. Their compact nature makes it possible for them to be incorporated into small devices without compromising performance, thereby driving wearable technology and autonomous technology. In addition to measurement intent, these sensors are significant because they play a fundamental role in user experience and safety, particularly in the automotive and robotics sectors, where real-time response to data is critical for efficient operations and decision-making protocols. Apart from motion sensing, the role of pressure and flow sensors in modern-day technology cannot be overstated. The sensors provide critical information to a wide range of applications, ranging from industrial automation to environmental monitoring and medical applications.

Figure 7.1. Accelerometers and gyroscopes, commonly referred to as inertial measurement units (IMUs), are instrumental in applications ranging from smartphones to aerospace engineering.

Source: https://live.staticflickr.com/65535/48326791917_4fc9d7b53a_b.jpg.

Pressure sensors, for instance, are indispensable in hydraulic systems, whereas flow sensors are at the hub of fluid management and energy efficiency programs. The ability to make accurate measurements of physical quantities allows for better control and optimization of processes, which is essential to making operations more efficient and safer. Moreover, as businesses push for better energy efficiency and sustainability, these sensors complement by being able to capture the data that can be used to drive patterns of energy consumption and facilitate smart grid technology integration in accordance with the needs of today's efficient utilization of resources. MEMS sensors are particularly significant in environmental monitoring, a critical area in the contemporary era of climate change and the environment. MEMS micro-electromechanical systems provide real-time measurement of various environmental parameters such as air quality, temperature, and humidity. Widespread application of MEMS technology has enabled data collection at high resolution levels, hence enabling more informed decision-making on matters of public health and environmental preservation. Furthermore, the flexibility provided through MEMS technologies allows for portable sensor networks to be installed in different environments and monitor responses to the problems brought about by urbanization and industrial contamination. Through increasing our ability to observe environmental conditions in an integrated way, scientists are able to more effectively respond to the grand challenges presented by current ecological realities, as outlined in research that identifies bottlenecks and recommends holistic strategies for adopting sensor technology (Alzubaidi et al., 2023; Luo et al., 2023) (Figure 7.1).

Overview of Physical Quantities Measured by MEMS Sensors

The numerous applications of MEMS sensors enable precise measurement of a variety of physical parameters critical to modern technology. A prominent category includes accelerometers and gyroscopes, commonly used in inertial measurement units (IMUs). These sensors can measure the acceleration and angular speed of an object and find application in navigation systems and robotics. Miniaturization of their dimensions makes it possible to integrate them into consumer electronics, where they form an essential component of functions such as screen orientation and motion sensing. Moreover, advances in MEMS technology have enhanced these sensors sensitivity and precision, among other aspects, and thus broadened their application across various sectors. As they continue to evolve, the potential for IMUs to facilitate high-level motion tracking across a wide variety of applications—everything from self-driving vehicles to augmented reality—becomes increasingly relevant, underscoring the central role of MEMS in modern technology. A similarly functionally significant type of MEMS sensor is pressure and flow sensors, offering critically valuable functionality in numerous industrial and environmental applications. These pressure change and fluid dynamics sensors provide real-time data that is essential for system control and optimization across applications from the automotive markets to air conditioning systems. MEMS technology integration has brought size and power minimization, reduction in response time, and increased reliability. For instance, MEMS pressure sensors are used comprehensively in medical devices for blood pressure measurement or in intelligent irrigation systems for water efficiency. Furthermore, MEMS sensors aid in precise flow measurement that is crucial for chemical processing and water management. Their ongoing development incites innovative approaches, resulting in increased efficiency and sustainability across diverse areas and highlighting their central position in addressing challenges of the modern era. MEMS sensors are also significant contributors to inertial navigation systems (INS) and environmental sensing. INS utilize data obtained from accelerometers and gyroscopes to determine an object's location and orientation in relation to itself without any external reference. This is beneficial in environments where the GPS signal may be weak or unavailable, i.e., underwater navigation and space exploration. Along with environmental surveillance, MEMS sensors find application in gathering data on atmospheric conditions, temperature variations, and air quality. Due to their low power requirements and miniaturization, they are ideal for use in inaccessible or remote regions, leading to increased data acquisition for climate monitoring and pollution tracking. The flexibility and generic nature of the MEMS technology in such uses showcase the wide-ranging impact of such sensors in providing enhanced insights into navigational as well as environmental motion, thus creating innovative avenues in different industries.

7.1. ACCELEROMETERS

The growth in advanced technology has been considerably influenced by miniaturized sensors with the ability to measure physical values with precision. Amongst these, accelerometers have emerged as integral parts of diverse devices ranging from consumer goods to advanced navigation systems. As part of Inertial Measurement Units (IMUs), accelerometers are used along with gyroscopes to detect and analyze motion and orientation. Their ability to sense acceleration forces allows for varied applications, such as for smartphones, where they allow screen rotation and gesture detection. Additionally, for robotics and aerospace, these sensors are instrumental in providing real-time feedback on velocity change or position alteration, thus facilitating accurate enhancement in the guidance system and motion control systems. This bridging of accelerometers with the gyroscopic data increases their functional worth even more across platforms requiring robust motion sensing ability. In MEMS sensors employed to sense environmental parameters, accelerometers also play a crucial role in inertial navigation systems (INS). The latter utilize the precise measurements provided by accelerometers for position, orientation, and velocity estimation without reference requirements. This independence places INS in the role of being a viable solution, particularly in negative environments where GPS is weak or absent, i.e., underwater or cityscapes with much interference. Research has attempted to look for methods of optimizing sensor calibration in an attempt to increase the accuracy of INS drastically. Improvements in sensor calibration methodologies, such as ones that use a shared sensor model to normalize inertial measurement into usable applicable physical quantities, demonstrate these advances. The performance improvement of such calibration techniques greatly improves the overall effectiveness of accelerometers in providing credible data in dynamic conditions, thereby addressing significant issues faced in navigation applications across diverse fields (Yang et al., 2014). Additionally, MEMS sensor development extends to environmental monitoring, with accelerometers being central to effective determination of environmental variability and phenomena. Their integration with data acquisition systems provides tools for the measurement of a range of conditions, including seismology and structural integrity monitoring.

For instance, accelerometers can serve as pre-impact earthquake sensors by detecting vibrations and movements in the earth's crust. With increasing demand for sustainable infrastructure in urbanization, utilizing MEMS technology to check structural integrity in real-time is becoming imperative at a rapid pace. Through the examination of the data collected by accelerometers using advanced algorithms, engineers and scientists can identify approaching failures and initiate preventive measures, thus enhancing safety protocols. The universality of accelerometers in navigation and environmental uses is used to illustrate their intrinsic value in driving the progress of MEMS sensor technology for physical parameters toward achieving precision and reliability in applications (Figure 7.2).

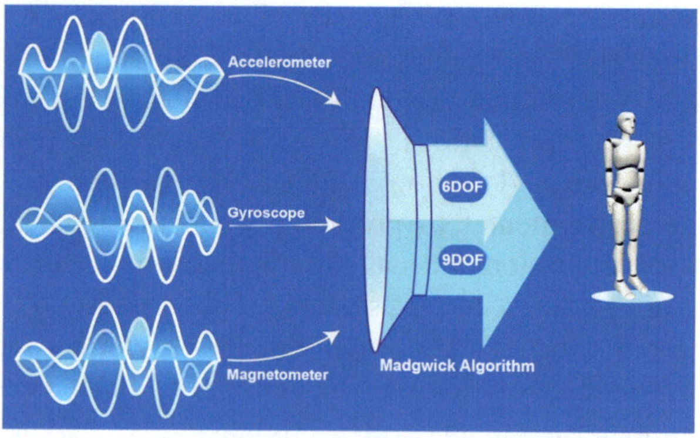

Figure 7.2. As part of inertial measurement units (IMUs), accelerometers work in tandem with gyroscopes to detect and analyze motion and orientation.

Source: https://qsense-motion.com/wp-content/uploads/2024/11/Data-visualization-DOF.png.

7.1.1. Working Principle of Accelerometers

The dynamic field of microelectromechanical systems (MEMS) has revolutionized the measurement of physical parameters, and the accelerometers are a case in point. Sensors that form the core of inertial measurement units (IMUs) detect movement or orientation changes in various ways. The most conventional type of MEMS accelerometers uses a proof mass that travels due to acceleration forces. This weight is converted to an electric signal, from which accurate acceleration can be read. With advances in manufacturing techniques, such as surface micromachining, new accelerometers have been designed to model biological systems, as evidenced by progress in state-of-the-art models mimicking the cricket clavate hair and the haltere in the fly, enhancing sensitivity and performance in use (Boer et al., 2015). Aside from applications in consumer electronics and automotive safety systems, accelerometers are at the heart of inertial navigation system (INS) building.

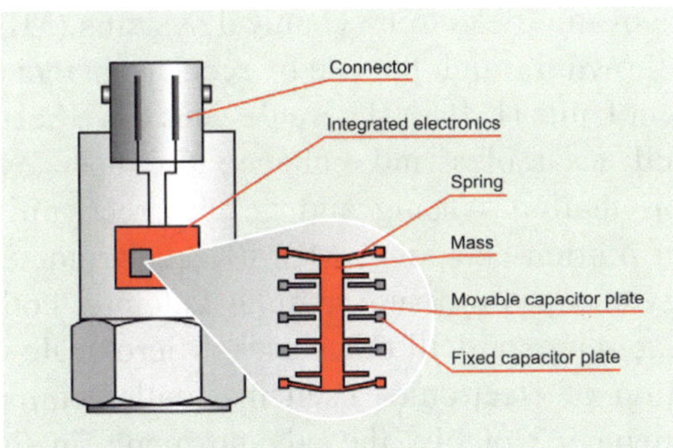

Figure 7.3. MEMS vibration sensors.

Source: https://power-mi.com/sites/default/files/images/blog/MEMS%20sensor%20for%20condition%20monitoring.jpg.

These tap the accelerometer's ability to measure acceleration changes to calculate velocity and position over time. Unlike traditional navigation methods based on external references, INS is free from external reference dependency and hence retains its outstanding performance in situations where GPS signals are disrupted or weak. Through the integration of information from multiple sensors, including gyroscopes, accelerometers provide an unobstructed perspective of an object in motion. The relationship between these sensors provides true tracking and navigation, offering a broad array of advantages in use ranging from aerospace to robotics, pointing to the utility of MEMS technology in developing navigation systems (Dong et al., 2010). Moreover, accelerometers have an extremely crucial function in environmental monitoring, which is one of the fastest emerging areas of MEMS applications. In other fields, like seismic monitoring and structural integrity assessment, accelerometers are precious sources of information that are inputs to decision-making. Based on the detection of minute vibrations or shifts in position, the sensors are beneficial for the evaluation of situations requiring immediate attention, such as natural disasters or structural collapses. Miniaturization of the instruments without affecting sensitivity presents prospects for large-scale deployment within numerous sensitive environments. The complementarity between scalability and accuracy offered by MEMS accelerometers solidifies their status as building blocks of modern sensing technology, justifying their use to further advance scientific investigation and daily use in the physical world (Boer et al., 2015) (Figure 7.3).

7.1.2. Applications in Consumer Electronics

The innovative role of microelectromechanical systems (MEMS) in consumer electronics can be shown through the use of accelerometers and gyroscopes in Inertial Measurement Units (IMUs). These devices have been used extensively in smartphones and wearables and enhance the user experience through features such as orientation sensing and gesture recognition. Along with a full complement of motion-detecting, MEMS accelerometers measure linear acceleration, and gyroscopes measure angular velocity, both being critical to fitness monitoring, gaming, and all other uses. Micro-scale design contributes to the miniaturization of electronics used in smaller configurations without sacrifice in performance. Notably, the advancements in the fabrication and optimization process have improved the responsiveness and dependability of such sensors, facilitating their versatility towards various dynamic environments and applications, highlighting the key role played by MEMS technology in contemporary consumer electronics (Hsieh et al., 1995). In addition to motion detection, MEMS sensors also have key functionalities in pressure sensing and flow measurement in electronic devices. These sensors play a major role in influencing user interaction and functionality in smartphones, smart home devices, and automotive applications. For instance, MEMS pressure sensors monitor ambient conditions to optimize pressure-sensitive device performance, like drone altitude compensation or air flow detection in heating, ventilation, and air conditioning systems. Besides, the combination of pressure sensing with smart technology facilitates real-time data analysis, energy efficiency, and comfort for users. As a result of MEMS technology advancements, its capability for high precision and low power makes it an important building block in future consumer electronics, which can facilitate a harmonious conversation between users and their devices (da Gomes et al., 2014). But yet another possible use of MEMS sensors is in the environmental monitoring field. MEMS sensors are employed in a variety of consumer electronic devices to sense air quality, temperature, humidity, and noise levels, thereby providing a better user experience through real-time information provision. One such example is in MEMS-based sound sensors for noise pollution localization, which should be capable of identifying sources of sound with accuracy in urban settings. Technologies such as the SoundCompass use arrays of MEMS microphones to perform noise mapping with high accuracy, producing critical information that supports public health initiatives (da Gomes et al., 2014). This integration not only adds more functionality to consumer electronics but also to sustainability in that it enables users to make decisions regarding their surroundings. With the increasing demand for eco-friendly consumer choices, MEMS sensors are poised to become a crucial part of expanding environmental awareness with smart devices (Figure 7.4).

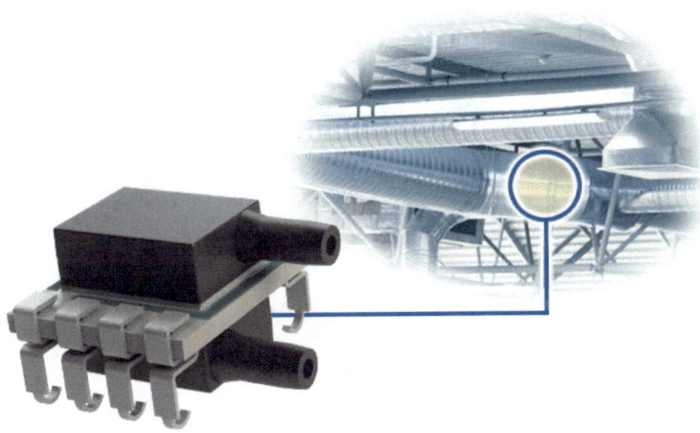

Figure 7.4. MEMS pressure sensors monitor atmospheric conditions to ensure optimal performance of pressure-sensitive applications, such as altitude adjustment in drones or air flow monitoring in HVAC systems.

Source: https://meritsensor.com/wp-content/uploads/LP_HVAC__705x436.png.

7.1.3. Role in Automotive Safety Systems

As the automotive industry continues to enhance safety features, the adoption of micro-electromechanical systems (MEMS) has been instrumental in the development of sophisticated safety systems. Through accelerometers and gyroscopes, typically packaged within inertial measurement units (IMUs), vehicles are capable of accurately tracking motion change and orientation. Such functionality enables sophisticated features such as electronic stability control and anti-lock braking systems, which form the basis for managing emergency maneuvers or harsh weather conditions. Furthermore, the sensitivity and accuracy of MEMS accelerometers are crucial in airbag deployment systems to make sure such safety features deploy in response to proper measurements of acceleration when there is impact. The application of such sensors extends beyond the immediate actual safety features and into predictive analysis, where data can be utilized for real-time vehicle dynamics analysis, thereby further enhancing safety for both drivers and passengers. In addition to IMUs, MEMS technology plays a crucial role in automotive pressure and flow sensors that are integral components of most car systems.

Figure 7.5. Inertial navigation systems (INS) utilizing MEMS technology have revolutionized automotive safety through enhanced situational awareness and navigation capabilities.

Source: https://guidenav.com/wp-content/uploads/2024/08/%E6%88%AA%E5%B1%8F2024-08-06-%E4%B8%8B%E5%8D%883.45.44-e1722930436804-1024x683.png.webp.

They monitor important parameters such as tire pressure, oil pressure, and coolant flow to ensure vehicles are within safety margins. Real-time output of MEMS pressure sensors allows for early failure detection and reduces the likelihood of accidents due to mechanical failure. Further, the innovation of these MEMS sensors has also brought with it increased precision and reaction times, which are essential in safety-oriented domains. By providing sustained and consistent measurements, these sensors contribute toward a culture of safe driving as well as minimize the scope for accidents caused by mechanical irregularities. In essence, the coming together of MEMS technology and automotive engineering has underscored the importance of safety systems designed to protect lives and maintain vehicular integrity. Moreover, MEMS-based inertial navigation systems have also revolutionized the safety of cars by providing greater situational awareness and navigation functions. The systems collect data from multiple MEMS sensors and provide true positional information even in situations when GPS signals may be poor or unreliable, for example, heavy urban canyons or tunnel environments. High-strength INS helps promote advanced driver-assistance systems (ADAS) that support lane keeping, collision avoidance, and automated parking features. Simultaneously, environmental monitoring with MEMS offers an additional safety aspect in terms of analyzing external parameters such as air conditions and temperature fluctuation. By integrating them into vehicle systems, the manufacturers can optimize operational performance based on real environmental conditions. Collectively, these advances reveal not only how MEMS sensors are enhancing the typical car safety features but also how they are opening the door to the next generation of smart, adaptive cars, on their way towards safer roads for all (Figure 7.5).

7.2. GYROSCOPES

The intricately involved world of microelectromechanical systems (MEMS) has revolutionized the development of gyroscopes, which are at the heart of a wide array of applications ranging from inertial navigation to mobile phone orientation. As sensors of angular velocity, gyroscopes are components of inertial measurement units (IMUs) that operate commonly in synergy with accelerometers to provide integrated motion monitoring. Emerging technologies have enabled the creation of MEMS gyroscopes, meant to offer more performance through miniaturization and sophisticated engineering. Derived from bio-inspired models, the fly's halteres for example, such MEMS devices create substantial bandwidth and response time enhancements without the loss of accuracy. The combination of these technologies is a leap of galactic proportions in the functionality of gyroscopes to address the requirements of modern applications in robotics, consumer electronics, and numerous other fields. The development of gyroscope technology in the field of inertial navigation systems (INS) cannot be overstressed. Traditional navigation methods have relied heavily on external references, but gyroscopes provide a means for calculating orientation and position based solely on internal measurements. By measuring the angular momentum and applying Coriolis forces, MEMS gyroscopes enable continuous measurement of orientation changes, thus playing an enormous role in navigational performance. The biomimetic technologies that were developed from nature, particularly the complex mechanisms found in living organisms, have yielded gyroscopes with high precision and robustness. Specifically, the fabrication of fly-like haltere gyroscopes is a demonstration of the potential of incorporating biological principles in technology, as exemplified by their resonance frequency of around 550 Hz, which indicates not only a design focused on performance but also on daily practicality for diverse applications (Brookhuis et al., 2014). Along with it, MEMS technology advancement has facilitated improvement in environmental monitoring with gyroscopes as sensors in various systems. The gyroscopes facilitate accurate measurement of movement in instruments that monitor environmental parameters, used in meteorology and pollution monitoring. Understanding the dynamics of airflow, pressure change, and rotation helps these monitoring systems function appropriately. With ongoing research, frequent comparisons of MEMS gyroscopes and their biological equivalents reveal how biomimicry is able to create models not only mimicking nature but even outperforming in technological uses. This research contributes towards a greater understanding of biological systems and engineered sensors, a sign of the potential for future developments in MEMS gyroscopes that will be capable of continuing to enhance the accuracy and reliability of environmental monitoring systems (Boer et al., 2015).

7.2.1. Functionality of MEMS Gyroscopes

Inertial Measurement Units (IMUs), which are usually comprised of both gyroscopes and accelerometers, have revolutionized navigation and motion tracking with their compact and efficient design. MEMS gyroscopes, in particular, have a critical role to play in such systems as they provide angular velocity data essential for calculating orientation. Such a capability is immensely beneficial across diverse applications like aerospace, automotive, and consumer electronics where accurate navigation and stabilization are critical. MEMS gyroscopes are able to sense rotation with the assistance of the Coriolis effect, and this makes it possible to monitor the object's orientation in three-dimensional space continuously. The technology can potentially solve problems across applications ranging from drone flight control to virtual reality, where performance directly correlates with end-user satisfaction and the requirement for accurate motion sensing. Therefore, MEMS gyroscopes are at the forefront of technological innovation, enabling efficiency and enhancing user experience across a broad spectrum of modern applications. Within the MEMS sensor family, pressure and flow sensors play a complementary yet differentiated function which complements the performance of systems incorporating MEMS gyroscopes. For measurement of pressure and fluid dynamics variation, these sensors provide vital environmental feedback that is cross-checked with data obtained from MEMS gyroscopes. This integration improves performance in products such as automotive stability control, where body movement as well as external condition need to be accurately interpreted to deliver maximum safety measures. The interrelationship of these sensors allows for the whole analysis of environmental parameters alongside rotational parameters, paving the way to advanced algorithms for inertial navigation systems (INS).

Figure 7.6. *MEMS gyroscopes, serve a critical role within these systems by providing angular velocity data essential for determining orientation.*

Source: https://guidenav.com/wp-content/uploads/2025/03/MG-Collection-1.webp.

Thus, the development of next-generation MEMS gyroscopes in association with pressure and flow sensors is a seminal step toward more robust and dynamic solutions for navigation for industries. MEMS gyroscopes have a significant function in environmental monitoring, particularly in data collection and interpretation, in applications such as detecting earthquakes and environmental monitoring. Through the integration of these gyroscopes into ground motion-monitoring systems, researchers can acquire more accurate and timely data about seismic activity, vital to disaster mitigation and urban planning efforts. Miniaturization of MEMS technology allows for the application of sensors in far-flung and hard-to-reach settings to increase environmental monitoring capacity. This is becoming more critical as climate change and urbanization create new challenges to infrastructure resilience. The ability to integrate detailed information regarding movement and the environment underscores the multidisciplinary aspect of MEMS gyroscopes, which becomes essential in the provision of advanced research and practical implementations towards addressing the issues that the world is experiencing today (Fluitman et al., 1996; Lampo et al., 2012) (Figure 7.6).

7.2.2. Use in Navigation and Stabilization

The subject of guidance and stabilization has seen unprecedented advancements with the integration of Micro-Electro-Mechanical Systems (MEMS) sensors, i.e., gyroscopes and accelerometers, merged together to form Inertial Measurement Units (IMUs). They play a vital role in enhancing navigation accuracy for a vast range of applications, ranging from consumer products to space systems. With linear acceleration and angular velocity measurement, IMUs enable real-time processing of information required for stability and orientation during dynamic maneuvers. As demonstrated in different UAV deployments, from research-oriented projects conducted by the Hungarian Academy of Sciences to conventional practice, MEMS technology facilitates control mechanisms with precision that are important for applications like aerial surveillance and environmental monitoring. Therefore, the ongoing growth of MEMS sensors improves navigational accuracy as well as facilitates complex maneuverability in wide-ranging applications, emphasizing their significance in modern technological settings (Gill et al., 2007). Pressure and flow sensors are also essential in the stabilization of platforms due to their capability to sense environmental states that are capable of affecting navigation. For instance, MEMS pressure sensors are a critical component of aircraft and vehicle altimetry, providing vital feedback used for enhancing altitude control and safety. Coupled with inertial navigation systems (INS), these sensors improve the accuracy of position estimation by compensating for drift errors inherent in standalone inertial navigation. This combined system, employing both MEMS accelerometers and pressure, contributes to higher system reliability. As part of missions such as the MISAT mission, miniaturized sensor systems are gaining attention and paving the path towards future satellite missions with real-time stabilization as essential (Gill et al., 2007). The interaction among MEMS sensors not only enhances navigation but also facilitates advancements in other fields of science applications (Gill et al., 2007). Finally, the new field of environmental monitoring has also been transformed employing MEMS technology, suggesting its application potential for navigation and stabilization applications. Environmental sensors enable tracking of atmospheric conditions and thereby affect navigational calculations for meteorology and climate studies. MEMS sensors can detect very subtle changes in temperature, humidity, and gas mixtures that can become vital for the stabilization algorithm of UAVs and other autonomous systems. Through embedding environmental data within the navigation frames, systems can reset their trajectories and self-recover with respect to real-world atmospheric conditions. Not only does this new approach enhance the operational effectiveness of autonomous systems, but it also permits deeper understanding into environmental dynamics. Thus, the use of MEMS technology in environmental monitoring establishes its role in enhancing navigation methods, thereby informing the pivotal role of these sensors in contemporary technological culture (Bauer et al., 2007) (Figure 7.7).

Figure 7.7. By measuring linear acceleration and angular velocities, IMUs enable real-time data processing that is vital for maintaining stability and orientation in dynamic environments.

Source: https://www.jouav.com/wp-content/uploads/2023/11/inertial-measurement-unit.jpg.

7.2.3. Integration with Accelerometers in IMUs (Inertial Measurement Units)

The integration of accelerometers in Inertial Measurement Units (IMUs) significantly enhances the precision and reliability of inertial navigation systems, which are very important in numerous applications, including robotics, automotive systems, and aeronautics. Accelerometers are utilized to measure specific force, thus allowing for continuous tracking of the motion and orientation of an object. Integration with gyroscopes, which track angular velocity, offers a comprehensive realization of linear and rotational dynamics. As per recent studies, the integration of the sensors improves data fidelity by means of functional redundancies that help counteract the impact of single sensor faults (Faragher et al., 2013). Furthermore, the technology has been boosted by improvements in micro-electro-mechanical systems (MEMS), and this has resulted in miniaturized, inexpensive IMUs being used in huge quantities in a wide range of applications while maintaining high levels of performance. The fusion of MEMS accelerometers with gyroscopes has facilitated opportunities for solutions in which there is a requirement of good navigation solutions, such as pedestrian dead-reckoning and simultaneous localization and mapping (SLAM). By synergistic utilization of these sensors, data fusion techniques can improve navigation performance, particularly under poor conditions where traditional global navigation satellite systems (GNSS) normally fail. Research has shown that by joining accelerometers and gyros, their inherent errors as separate sensors—such as drift in gyros or acceleration bias in accelerometers—can be nicely balanced out (Barber et al., 2013). The cross-support provides higher accurate trajectory estimates, which is crucial in applications ranging from autonomous vehicles to indoor navigation where GPS signals may be subtle or absent. Further, scalability in integrating MEMS accelerometers with other technologies broadens the application of IMUs to areas outside navigation. For instance, pressure and flow sensors in the same system can be integrated to provide useful environmental data, enabling application in environmental monitoring and industrial processes. Such hybrid systems can potentially offer insight into atmospheric conditions as they monitor movement and orientation simultaneously, offering a multi-dimensional approach to data acquisition and analysis (Faragher et al., 2013). Such integration is one of the dominant sensor technology trends—diverging towards multifunctional systems that leverage the strengths of several MEMS devices to get more information out of physical quantities and improve overall system performance across a broad spectrum of applications.

7.3. INERTIAL NAVIGATION SYSTEMS (INS)

Among the significant advances in the field of navigation technologies is the development of Inertial Navigation Systems (INS), which widely utilize sensor technologies such as micro-electro-mechanical systems (MEMS) accelerometers and gyroscopes. The inertial measurement units (IMUs) are crucial to defining the position and orientation of a body by quantifying data on its motion in real-time. The integration of MEMS technology has enabled the easy deployment of small, light, and low-cost IMUs, which are now accessible for general usage like personal navigation, automotive, and aviation. Since the sensors are capable of sensing acceleration and angular velocity, displacement and attitude can be calculated in real time. The main challenge is to eliminate the inherent measurement drift with time, typically requiring recalibration periodically to be precise (Bertuletti et al., 2017). Hence, development and optimization of MEMS-based INS become imperative to augment navigational dependability in various applications. MEMS INS also gain significance due to their capability to operate in various kinds of environments, allowing facilitated navigation in conditions where global navigation satellite systems (GNSS) may fail. The

IMUs can be utilized for supporting pedestrian dead-reckoning solutions, and thus they are vital for urban navigation, particularly in adverse satellite visibility conditions. Most encouraging is their potential to enhance the performance of existing technologies. The integration of accelerometers and gyroscopes with other sensors, such as pressure and flow sensors, provides insight into motion dynamics as well as environmental parameters. This multi-fold integration also seeks improved navigation accuracy but additionally addresses functionalities in environmental monitoring, further extending the utility of MEMS sensors in diverse applications (Bertuletti et al., 2017). Such innovations point to the revolutionary capability of INS technology in revolutionizing navigation paradigms, presenting new solutions. Moreover, the exploration of new configurations and calibration schemes maximizes the effectiveness of MEMS INS systems and minimizes error propagation. Researchers have focused on developing sophisticated algorithms that take advantage of the merits of different types of sensors to optimize navigation performance. For instance, integrating complementary data obtained from in-plane and out-of-plane MEMS sensors can remove biases and improve positional estimate accuracy. Additionally, novel methods from MEMS sensor arrays reveal promising results in mitigating common-mode error and optimizing overall measurement fidelity. Current research into less complex static calibration methods may further democratize INS technology by simplifying the process of users calibrating sensors without costly machinery (Faragher et al., 2013). As a whole, the development of MEMS-based Inertial Navigation Systems heralds a new era, where advanced navigation aids become more resilient, accurate, and available to a broad range of applications and users.

7.3.1. Components of INS using MEMS Technology

MEMS technology integration accelerated the revolution of inertial navigation systems (INS) primarily through advancements achieved in accelerometers and gyroscopes. These inertial measurement units (IMUs) consist of tiny MEMS accelerometers and gyroscopes, providing crucial information regarding orientation and motion tracking. MEMS accelerometers gauge linear acceleration along various axes, while MEMS gyroscopes gauge angular velocity to enable precise positioning in three-dimensional space. The precision of these IMUs may be significantly enhanced by calibrating against common-mode error or exploiting variations in different sensor orientations and measurement ranges (Faragher et al., 2013). Additionally, by exploiting methods such as differencing between outputs from aligned sensors, scientists are able to remove temperature-induced bias and improve overall accuracy, which is particularly critical in mission-critical application areas such as autonomous navigation and aerospace systems. In addition to accelerometers and gyroscopes, MEMS technology also encompasses a variety of pressure and flow sensors that are at the heart of environmental monitoring and control systems. They enable the measurement of atmospheric pressure, elevation changes, and fluid dynamics in a wide variety of applications. Pressure sensors based on MEMS, for example, find widespread use in weather forecasting and aeronautics where precise measurement of altitude is needed. Furthermore, integrated with accelerometers and gyroscopes, MEMS pressure sensors are part of the multi-sensor fusion common in modern INS, providing a more informed view of environmental conditions. Such fusion is more precise since it makes up for the limitations inherent in each sensor type, ultimately creating robust navigation solutions essential for land and aerial vehicles (Bertuletti et al., 2017).

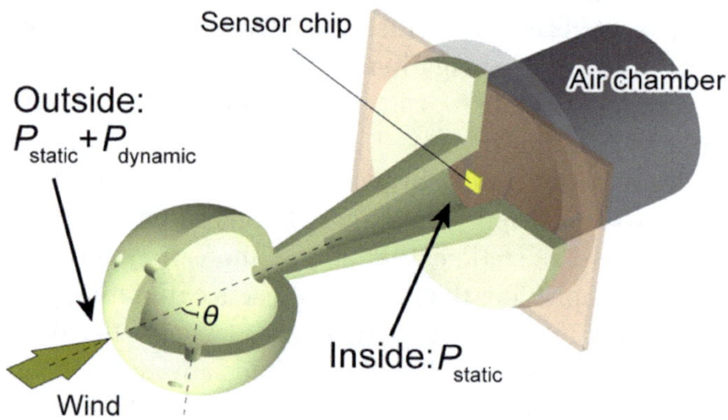

Figure 7.8. MEMS-based pressure sensors are widely used in weather forecasting and aerospace, where precise altitude measurements are crucial.

Source: https://www.mdpi.com/micromachines/micromachines-14-01941/article_deploy/html/images/micromachines-14-01941-g001.png.

To enhance the maximum potential of MEMS technology within inertial navigation systems, continuous research and development are critical. Research into the synergistic effects of combining MEMS sensors with allied technologies, such as magneto-inertial measurement units (MIMUs), lead to wider application areas in human movement analysis and robotics. With the integration of MEMS IMUs with infrared and other proximity sensors, richer data capture is made possible, particularly in dynamic situations where the motion of individuals must be tracked closely. By fusing information from multiple sensors, scientists are able to comprehend intricate physical quantities, which opens up innovation in medical fields and automated navigation systems. The unique advantages provided by MEMS technology illuminate its substantial contribution to the advancement of the capabilities of inertial navigation systems and their usage in many real-world situations (Bertuletti et al., 2017; Faragher et al., 2013) (Figure 7.8).

7.3.2. Advantages of MEMS-Based INS Over Traditional Systems

The development of inertial navigation systems (INS) has undergone a paradigm shift with the advent of MEMS-based technology, which has greater accuracy and reliability compared to its traditional counterparts. One significant advantage is miniaturization of MEMS devices; They are significantly smaller, making integration into limited space much easier, especially for application in aerospace and automotive. Miniaturization in this case is done without hampering performance; instead, MEMS-based inertial measurement units (IMUs) usually use advanced algorithms to make precise measurements. Studies depict how MEMS IMUs combined with global navigation satellite systems (GNSS) yield improved performance under dynamic conditions, thus ensuring maximum navigational feedback in unmanned aerial vehicles (UAVs) (Bartel et al., 2012). Moreover, the reduced weight of such sensors provides the capability to implement lightweight systems, which can enhance mobility and efficiency in numerous applications, from hand-held devices to large drones. Pressure and flow sensors benefit from MEMS technology as well, which allows them to be more sensitive and responsive in numerous environments. Such MEMS-based sensors are effective across a wide range of parameters and are appropriate for implementation in industrial monitoring and environmental measurement applications.

Their greater sensitivity and accuracy are a result of miniaturized sensors that amplify measurable phenomena without introducing substantive error. The use of MEMS sensors also yields lower power dissipation profiles, thus appropriate for energy-constrained applications. This attribute is critical in maintaining long operation cycles during remote or handheld devices. Consequently, MEMS-based flow and pressure sensors play key roles in modern environmental monitoring systems, supporting real-time data collection and analysis that can dynamically adapt to changing conditions. MEMS-based inertial navigation systems provide better fault tolerance than their traditional counterparts, with improved reliability in mission-critical applications. These MEMS devices, though challenging to be as accurate as conventional IMUs, have the potential to trigger redundancy actions through onboard processing abilities. On the occurrence of discrepancy in data streams from multiple inertial sensors, systems detect potential faults automatically and respond with compensatory actions, thus providing navigation integrity (Brady et al., 2012). Not only is this self-monitoring feature increasing overall system robustness, but it is also extending the operating lifespan of MEMS-based INS in adverse conditions. Lastly, continuing innovation in MEMS technology promises a new generation of navigation technologies where the benefits of miniaturization, precision, and fault tolerance come together to redefine performance requirements in several industries (Figure 7.9).

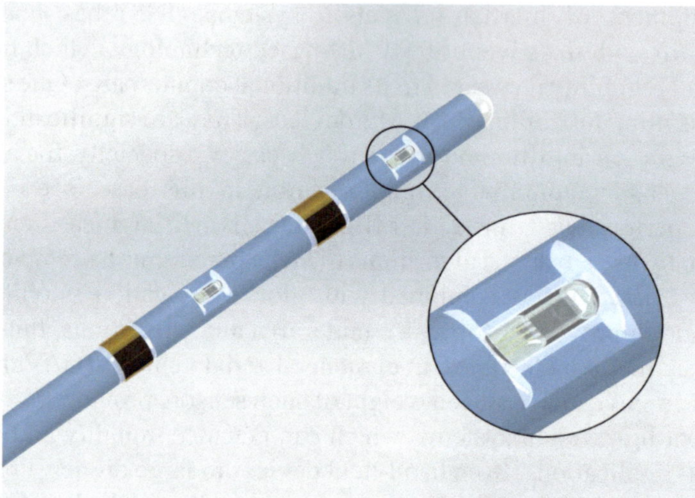

Figure 7.9. Pressure and flow sensors also benefit from MEMS technology, which allows for increased sensitivity and responsiveness in various environments.

Source: https://eu-images.contentstack.com/v3/assets/blt0bbd1b20253587c0/blt7fed70d4475e3a5e/65b83bb66d245e040a85eac6/dual_sensor_integrated_catheter_with_electrodes_(finished_look).jpg_(1).png?width=1280&auto=webp&quality=80&disable=upscale.

7.3.3. Applications in Aerospace and Robotics

For obtaining precision and reliability in application of aerospace and robotics, MEMS sensors are now a foundation technology. Among them, accelerometers and gyroscopes, generally integrated as Inertial Measurement Units (IMUs), are critical data for tracking motion and orientation. These sensors enable autonomous flight for drones and aircraft through the detection of changes in velocity and angular location. The merging of these measurements plays a crucial role to stabilization algorithms, which play a critical part in the safe flight of aerospace craft. Moreover, miniaturization by MEMS technology not only reduces the total mass of the systems but also reduces power usage, which is ideal for long-duration space missions where payload ounces matter. Thus, the transition to MEMS-based IMUs revolutionized the aerospace industry, improving functions and ensuring stringent performance requirements. Pressure sensors and flow sensors in robotics and aerospace cannot be overstressed. MEMS pressure sensors are nowadays central to monitoring altitudes in aircraft, ensuring safety and efficiency. With provision of real-time data on atmospheric pressure change, such sensors help maintain flight conditions at their best level. Apart from pressure measurement, MEMS flow sensors also play a significant role in sensing fluid dynamics in robot systems. The sensors provide feedback signals required for controlling fluid systems, enhancing efficiency in robotic systems such as autonomous underwater vehicles (AUVs) and programmable robots navigating through varying environments. As more robotics rely on the accuracy of environmental variable control, MEMS pressure and flow sensors remain essential to increasing responsiveness and flexibility across a range of operating conditions, solidifying their value in modern engineering applications. Inertial navigation systems (INS) also heavily rely on MEMS technology, integrating different sensors to find position and velocity with no external references. This is particularly beneficial in aerospace applications, where GPS signals are occluded or unavailable. By integrating information from MEMS gyroscopes and accelerometers, INS is capable of producing accurate navigation solutions required for airborne and ground vehicles. Furthermore, MEMS environmental sensors like humidity and temperature sensors are increasingly employed in aerospace to ensure systems operate within the limits of safe parameters. These MEMS devices assist in the monitoring of onboard environmental conditions, contributing to improved system performance and reliability. This application of MEMS technology both for environmental sensing as well as navigation underscores its ubiquitous use and relevance in robotics and aviation, culminating in revolutionary advancements in the field.

7.4. PRESSURE SENSORS

In the field of microelectromechanical systems (MEMS), pressure sensors are paramount, particularly in many applications needing precise measurement of pressure in a wide range of environments. MEMS pressure sensors benefit from miniaturized dimensions to achieve high sensitivity and accuracy, thus being indispensable in a wide range of fields, including biomedical devices, automotive systems, and industrial automation. Their miniaturization means that they can be built into advanced systems, thus enhancing overall performance without consuming much space. Their creation has led to advancements like wireless microdevices with ultrasensitive response, significantly improving measurement quality and accuracy. Advanced technologies, such as isospectral party time reciprocal scaling (PTX) symmetry, have also increased their performance by increasing the extrinsic quality factor (Q factor), which increases sensor sensitivity and enables more precise pressure measurements in diverse working conditions (Alù et al., 2018). Finding the applications of pressure sensors marks their basic use in environmental monitoring, as they aid in measuring atmospheric pressure variation with weather patterns and climatic conditions. Pressure sensors can be used in remote and harsh environments, yielding real-time data that aids us in understanding ecologic changes. Also, in the medical field, MEMS pressure sensors are integrated into devices that track vital signs so that there can be constant monitoring of parameters such as respiratory health and blood pressure.

Figure 7.10. MEMS pressure sensors leverage miniaturized components to achieve high sensitivity and accuracy, making them indispensable in a wide range of fields, including biomedical devices, automotive systems, and industrial automation.

Source: https://product.minebeamitsumi.com/en/tech/mems/case_studies/img/pressure_sensor_img_001.jpg.

They are well-positioned for the application of wearable medical technology, as there is growing focus on customized healthcare solutions. Due to the advancements of sensor technology, not only are scientists improving accuracy within the data but also expanding possibilities for new applications in all areas, solidifying the role of pressure sensors in MEMS design even further (Oliver et al., 2009). Pressure sensors incorporated within inertial navigation systems (INS) are merely a demonstration of the multifunctionality of them within MEMS structures. Pressure sensors complement accelerometers and gyroscopes in INS to enhance navigation accuracy, chiefly where GPS signals are absent or poor, such as underwater or indoors. The sensors constantly observe pressure variations relative to sea level or atmospheric pressure and provide essential data used to ascertain an object's direction and trajectory. Besides, in conjunction with pressure sensor functions alongside environmental monitoring sensors, pressure sensors enable in-depth working environment analysis of an object, thus facilitating the development of autonomous vehicles and drones. The implementation of MEMS technology coupled with intelligent materials and systems not only lessens design complexities but also optimizes navigation system responsiveness and flexibility (Alù et al., 2018; Oliver et al., 2009) (Figure 7.10).

7.4.1. Mechanism of MEMS Pressure Sensors

The intricate design of MEMS pressure sensors enables them to operate consistently across different applications including industrial automation and environmental monitoring. Central to their operation is the micro-electromechanical system (MEMS) which converts pressure variation into electrical signals. This conversion is achieved primarily by the deformation of a diaphragm in reaction to pressure variation. When exposed to varying pressures, the diaphragm deforms, generating a capacitance or resistance variation that is detected electronically. Accuracy of such sensors has been enhanced by advances in manufacturing techniques, allowing extremely sensitive structures that precisely track infinitesimal pressure variations. Such ability is applied in virtually any field, from medical to aerospace, where pressure measurement is critical to ensure safety and performance. The operation mechanism of MEMS pressure sensors is also optimized by the application of new materials and designs. The development in silicon-based microfabrication has introduced pressure sensors with higher durability and thermal stability. These developments make it possible to utilize sensors to provide accurate measurements under harsh environments, which is particularly important in applications like aerospace and automotive. Also, the pairing of other MEMS sensors, such as accelerometers and gyroscopes, with pressure sensors has paved the way to multifunctional applications. Such synergy allows the creation of fault-tolerant inertial navigation systems (INS) that are capable of combining pressure data with motion sensing to provide situational awareness. As the wider perspective of microsystems technology highlights, versatility and efficiency in MEMS pressure sensors are central to their growing usage in different fields (Fluitman et al., 1996). Variety in usage demands continuous innovation in mechanisms of MEMS pressure sensors. Miniaturization while maintaining high performance has encouraged researchers to explore other modes of sensing, such as piezoelectric and piezoresistive sensing. These techniques leverage the intrinsic material properties to enhance sensitivity, and thus overall pressure measurement reliability. Additionally, the integration of MEMS pressure sensors into larger networks for environmental monitoring facilitates real-time data acquisition. The advancements not only enable the exact monitoring of atmospheric conditions but also the proactive response to environmental parameter fluctuations. The ongoing effort to create MEMS devices to be even superior guarantees that they will remain at the forefront of sensor technology, adapting to an increasingly large catalog of applications and addressing critical matters of data accuracy and sensor longevity (Ahmad et al., 2018) (Figure 7.11).

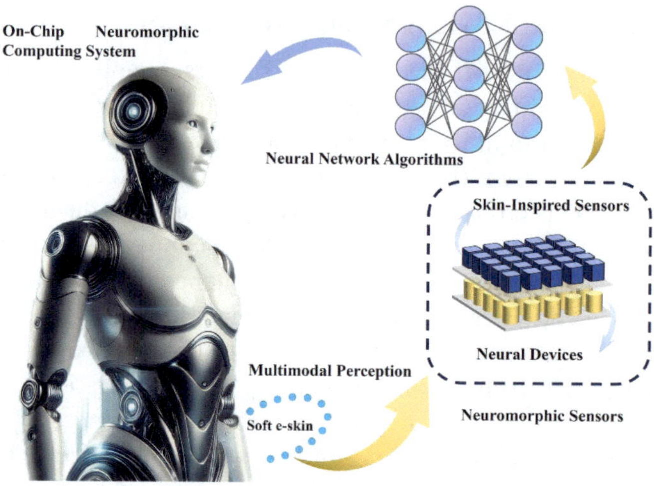

Figure 7.11. Innovations in silicon-based microfabrication have led to the development of pressure sensors that exhibit enhanced durability and thermal stability.

Source: https://image.oaes.cc/57b8d54c-3197-448c-9697-c65d59cc610b/ss4077-coverimg.jpg.

7.4.2. Applications in Medical Devices

Medical diagnostic devices have been significantly improved with the integration of inertial measurement units (IMUs) into them, which are made up of MEMS accelerometers and gyroscopes. By enabling precise tracking of movement and orientation, they have gained entry into rehabilitation medicine, assisting patients in tracking their bodily activity and progress in treatment. For instance, accelerometers are able to effectively identify tiny changes in patients' movement patterns and hence present clinicians with accurate information about their rehabilitation process. Moreover, the real-time information generated by these sensors assist with more tailored rehabilitation through the ability to alter treatment plans in accordance with empirical data. These applications highlight the focal position of MEMS-based IMUs in advancing patient care, not just to demonstrate their technological relevance but also to have the potential to advance health outcomes within various medical settings. Pressure and flow sensors that have stemmed from MEMS technologies also revolutionize medical applications, particularly in vital sign monitoring and drug delivery systems. These sensors can potentially offer high reliability and sensitivity in measuring body fluids, which is crucial in making rapid medical intervention. MEMS pressure sensors, for example, are commonly used in devices like portable blood pressure meters and respiratory devices because they can offer accurate measurements in small size and at low cost. Additionally, such sensors are an essential component of intelligent infusion systems offering accurate medication administration, significantly minimizing the risk of human error. Thus, the use of MEMS-based flow and pressure sensors contributes not just to the efficiency of medical devices but also safeguards patients and offers a high degree of care rendered. MEMS-based inertial navigation systems (INS) provide another instance of the technological development in medical device applications, i.e., surgical navigation and robotic surgery. Providing precise location data without external references, these systems provide enhanced precision for minimally invasive surgery. This level of precision helps surgeons navigate complex anatomical structures, which ultimately means less time required for surgeries and improved patient outcomes. Moreover, MEMS-driven INS can be applied to various therapeutic approaches, such as in physical therapy, where patient positioning and motion are accurately measured to optimize recovery plans. Continued development of MEMS for environmental monitoring applications is also promising for medical applications because such sensors are capable of quantifying environmental conditions that can affect patients' health and therefore facilitate active management of health through real-time data acquisition and analysis.

7.4.3. Role in Industrial Automation and Monitoring

The integration of MEMS sensors into industrial automation revolutionized processes by offering more efficient and reliable monitoring of physical parameters. Some of the most significant elements include the gyroscopes and accelerometers, typically integrated into inertial measurement units (IMUs), which are tasked with delivering vital information for motion tracking and orientation control. The sensors play a critical role in applications ranging from robotics to automated production lines, where detection of movement with precise accuracy enhances overall efficiency. MEMS-based IMUs' role extends well beyond the acquisition of data; they, in fact, command automated system response, enhancing the precision and safety of operations. Application of these technologies can significantly lower downtime and operating costs, thereby making production processes run more smoothly in an increasingly competitive industrial environment. Pressure sensors and flow sensors also play a significant contribution to industrial applications through real-time monitoring and process control. These MEMS sensors are essential for maintaining fluid system integrity, and pressure and flow fluctuations result in catastrophic system failure. With advanced microfabrication techniques, the sensors have high sensitivity and response rates, allowing real-time adjustments to automated systems. Additionally, use of the sensors enables predictive maintenance through data-driven decision-making because continuous monitoring optimizes preventive strategies. The ability to identify anomalies in fluid systems prior to when they are full-blown issues shows the ways through which MEMS pressure and flow sensors enhance productivity and safety within industrial processes, hence solidifying their growing pertinence to automation.

Figure 7.12. The role of MEMS-based IMUs extends beyond mere data collection; they actively influence automated systems responsiveness, thereby improving operational safety and accuracy.

Source: https://img.electronicdesign.com/files/base/ebm/electronicdesign/image/2019/04/electronicdesign_19712_imu_promo.png?auto=format,compress&fit=crop&q=45&h=356&height=356&w=640&width=640.

The innovation of inertial navigation systems (INS) powered by MEMS technology shows the paradigmatic impact of these sensors on advising industrial mobility solutions. Through the combination of MEMS accelerometers and gyroscopes, INS are being used more frequently as part of the autonomous car and drone navigation used in various industrial settings. These systems offer precise location positioning, therefore the significance to logistics, delivery, and inventory management. Furthermore, advances in communications capability and analytics capacity, as illustrated by today's movements in the Internet of Things (IoT), enable INS to learn and improve, refining their navigation algorithms over time. This learning capability not only amplifies operational performance but also reduces significantly the risk of human error in mission-critical applications. With the continued pressure for automatization, the union of MEMS technology with AI-based analytics promises an era of industrial monitoring and operational excellence (Baaziz et al., 2015) (Figure 7.12).

7.5. FLOW SENSORS

In considering the intricate world of MEMS sensors of physical parameters, it is significant to observe the various applications that flow sensors possess. Flow sensors play a significant role in applications such as automotive, aerospace, and industrial processes where accurate measurement and control of fluid flow is required. The downscaling of flow sensors employing micro-electromechanical systems (MEMS) technology has made it possible for them to be more accurate and sensitive. Therefore, advances in MEMS flow sensors have also been focused on making them more robust and reliable, especially in harsh environments because they are used in wireless sensor networks for security monitoring (Kerkhoff et al., 2009). Such a capability is even more relevant in the case of the operational demands placed on sensors used in safety applications, for which a number of physical parameters are observed simultaneously. The MEMS flow sensors' design commonly integrates the use of temperature sensors for enhanced measurement accuracy. Recent work, as presented at the 3rd Micro and Nano Flows Conference, includes new designs that use conventional silicon planar fabrication techniques and new layouts to accommodate diverse microstructures (Brandner et al., 2011). Such developments show very promising prospects for MEMS flow sensors, particularly in applications to study rarefied gases wherein the normal measurement techniques may fall short. With the calibration of such sensors intensively verified, preliminary results show immense potential for the fine-tuning of sensor design for better performance characteristics. This continuous development also places emphasis on the technological advancement of MEMS as well as emphasizes the pressing need for specific sensor solutions within niche markets. Additionally, the integration of flow sensors into big systems such as inertial navigation systems (INS) is a prime example of how crucial they are for enhancing overall performance and precision. In flow-dominant applications, MEMS flow sensors are complemented by other sensors, such as accelerometers and gyroscopes, to provide an overall view of operational conditions. Such synergy is essential in autonomous vehicles to aerospace engineering applications, where every parameter has to be diligently monitored to ensure optimum performance as well as safety. As research continues to evolve in the field of MEMS for environmental monitoring, focus on flow sensors will also continue, making them critical components towards the achievement of precision in fluid measurement and control.

7.5.1. Types of MEMS Flow Sensors

In microelectromechanical systems (MEMS), various flow sensors based on varied designs have been designed to address particular measurement requirements. Pressure and flow sensors are one of the prominent categories among these and are increasingly being used for precise fluid dynamics analysis in various applications. These predominantly operate on mechanical or thermal principles for quantification of fluid velocity and pressure without the need for large-scale elements. Thermal flow sensors, for instance, measure flow rate using temperature gradients induced by a heat element inserted into the stream of flow. Mechanical flow sensors, on the other hand, such as diaphragm or pitot tube configurations, depend on physical displacement resulting from pressure fluctuations in the fluid. These differences enable tailored sensor solutions to meet specific system performance requirements in different domains of aerospace, automotive, and biomedical applications. Inertial navigation systems (INS) combine gyroscopes and accelerometers, and they constitute another key class of MEMS flow sensors. Although they are most commonly identified with motion tracking, they are also very important in flow measurement, particularly in dynamic conditions where fluid behavior is subject to motion in the environment. For example, an IMU installed with MEMS accelerometers can detect slight changes in velocity and orientation that affect fluid flow, thus offering real-time information on the pattern of flow. Sensor miniaturization makes it possible to employ such sensors in small spaces without the loss of high-resolution data collection, thereby making it possible to apply MEMS technology in advanced systems such as robotics and self-driving cars. Thus, these sensors are priceless towards making fluid behavior analysis more accurate in settings ranging from industrial production to consumer goods. Environmental monitoring is one essential application area where MEMS flow sensors play a central part in sustainability and safety initiatives. Sensors provide valuable feedback on air and water quality through measurement of various flow parameters, which can be correlated with pollutants or anomalies within these mediums. Moreover, the growth of micro-machined tunable resonators, as studied currently, will further improve the functionality of MEMS flow sensors by improving sensitivity and resolution in detecting environmental changes. The integration of such technologies as PT-symmetry will also make these sensors more optimum, leading to new generations of devices exponentially reacting to varying fluid environments. This development in MEMS flow sensor technology not only facilitates comprehensive environmental analysis but also facilitates proactive intervention against environmental problems, underscoring the pivotal function such sensors have in modern-day ecological and industrial environments (Figure 7.13).

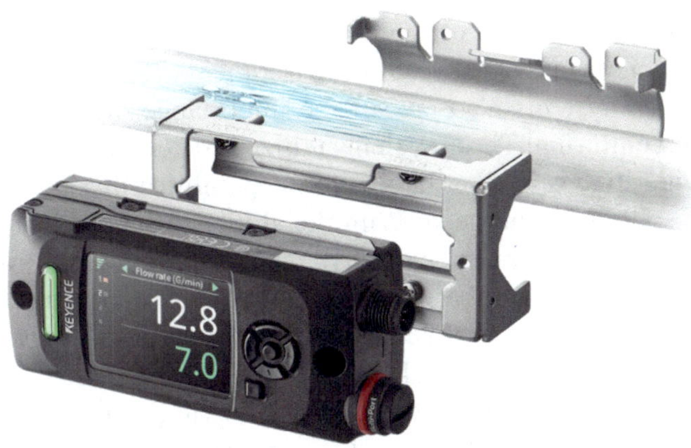

Figure 7.13. Various types of flow sensors exist, each engineered to fulfill different measurement requirements.

Source: https://www.keyence.com/Images/type_keyvisual_fd-h_2015136.png.

7.5.2. Importance in Environmental Monitoring

MEMS sensors integration into environmental monitoring systems has increased with the miniaturized structure, low power consumption, and high sensitivity. Accelerometers and gyroscopes, commonly integrated in inertial measurement units (IMUs), have maintained promising potential in monitoring environmental change. For instance, their responsiveness to variations in land and atmospheric positions can be utilized to track seismic motion and infrastructure shifts, thereby serving as an early warning mechanism for calamities. Depending on the manner in which environmental factors change, these sensors can provide real-time data, making possible interventions in time. The consistency and adaptability of such systems react suitably to harsh conditions, which is vital in ensuring perpetual consistency and working effectiveness in surveillance work, where human presence could be scarce or hazardous (Kerkhoff et al., 2009). Pressure sensors and flow sensors are also crucial in assessing the health of ecosystems as well as controlling environmental systems. In fluid dynamics, such MEMS technologies can measure atmospheric pressure, water quality, and flow rates of various water bodies accurately. Such data becomes essential for environmental scientists while analyzing climate change impacts, management of freshwater resources, and pollution monitoring. The broad application of MEMS pressure and flow sensors enables ongoing monitoring of the environment, providing the scope for policymakers and researchers to make decisions based on readily accessible and accurate information. Therefore, the sensors are the culmination of the necessity for the use of state-of-the-art MEMS technology for the protection and conservation of natural environments (Kerkhoff et al., 2009). Inertial navigation systems also talk of the importance of MEMS sensors in environmental monitoring. These systems utilize accelerometers and gyroscope inputs to provide high-precision navigation capabilities that are calibrated to earth and atmospheric phenomena. With the ability to achieve robust tracking and positioning, INS supports extensive area environmental surveys and geospatial mapping. Among the many applications fuelled by MEMS-based INS are simonifying climate change effects, tracking wildlife migration routes, and estimating land-use changes. Furthermore, these technologies facilitate interdisciplinarity in environmental study, allowing for more integrative solutions to present ecological issues. Ongoing system advancement will be essential to opening the door to novel environmental monitoring approaches with the ability to adapt to evolving needs in humankind's drive toward sustainability (Figure 7.14).

Figure 7.14. Inertial navigation systems (INS) further highlight the importance of MEMS sensors in environmental oversight.

Source: https://www.datocms-assets.com/53444/1661501351-what-is-inertial-navigation-systems.jpg?auto=format&w=1200.

7.5.3. Use in HVAC (Heating, Ventilation, and Air Conditioning) Systems

With the growing demand for energy-efficient HVAC equipment, new technologies to simplify control and monitoring become unavoidable. MEMS sensors in the shape of gyroscopes and accelerometers are significant in the improvement of the dynamics of HVAC systems. By applying inertial measurement units (IMUs), HVAC equipment achieves greater precision in detecting position and movement, which becomes necessary in adjusting temperature settings and air flow automatically. These sensors can be utilized to gather instantaneous information, allowing HVAC systems to respond dynamically to fluctuating environmental conditions as well as user preferences. This capability not only optimizes energy use but also enhances comfort levels inside, illustrating how MEMS technology may enhance conventional heating and cooling practices to current standards of efficiency and sustainability. Pressure and flow sensors form another key application of MEMS technology to HVAC systems. By providing accurate measurements of airflow and pressure inside duct systems, the sensors allow HVAC systems to run at optimal capacity and enhance energy efficiency.

MEMS pressure sensors with high-tech capabilities are capable of sensing very small changes in pressure, allowing for corrections before wastage of energy and maintaining constant air supply. Such a feature is extremely important in complex systems managing various loads and weather conditions. Besides, the application of MEMS flow sensors encourages more effective maintenance plans since they can notify when filters or ducts should be cleaned, ultimately leading to equipment longevity and reducing operational costs. Thus, the role of MEMS pressure and flow sensors is crucial to enhancing the reliability and efficiency of HVAC systems. Environmental monitoring has also become more critical in the context of HVAC systems, especially with greater focus on indoor air quality. MEMS technology can be used to design small and highly sensitive sensors for sensing various environmental parameters such as humidity, temperature, and particulate matter. The sensors can be implemented in HVAC systems to ensure continuous monitoring and control of air quality, ensuring a healthy environment for the occupants. By utilizing real-time indoor and outdoor weather data and indoor air composition data from MEMS environmental sensors, the HVAC can automatically react to alterations in indoor air composition and exterior weather through modulating ventilation rates. Such capability not only promotes occupant health but also facilitates adherence to regulatory standards on indoor air quality. Hence, the role of MEMS sensors in environmental monitoring is an indicator of their importance in developing intelligent, more effective HVAC systems that are health and sustainability oriented (Figure 7.15).

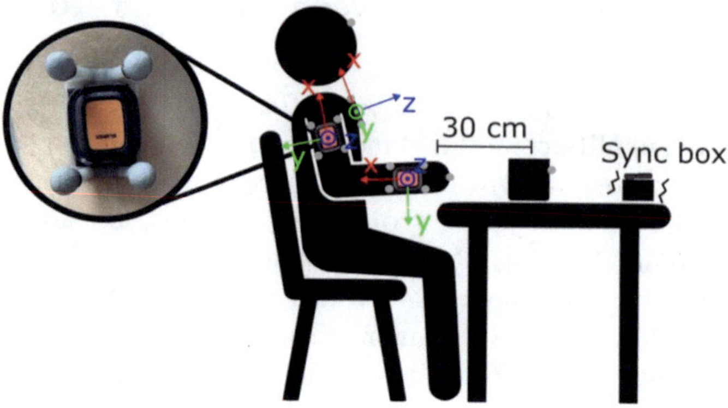

Figure 7.15. By employing inertial measurement units (IMUs), HVAC units can achieve superior precision in positional and motion detection, which is vital for automated adjustments in temperature regulation and air flow.

Source: https://images-provider.frontiersin.org/api/ipx/w=1200&f=png/https://www.frontiersin.org/files/Articles/1359776/fdgth-06-1359776-HTML/image_m/fdgth-06-1359776-g001.jpg.

7.6. MEMS FOR ENVIRONMENTAL MONITORING

MEMS technology has revolutionized environmental monitoring as it became possible to develop sophisticated sensors capable of measuring all sorts of physical parameters with unbelievable accuracy. Engineers and scientists can create devices not only of small size but also low power, which is critical for long-term use in harsh or distant places, thanks to microelectromechanical systems (MEMS). Such miniaturization has enabled the integration of different functionalities, such as gyroscopes and accelerometers, to assess movement and rotation in environmental conditions. The advancement of MEMS technology in inertial navigation systems (INS) characterizes its significance; such systems aid in the understanding of atmospheric phenomena by giving accurate information on movement and orientation, which is required to monitor the state of weather patterns as well as climate change impacts. Thus, MEMS sensors play a crucial role in developing smart systems based on the growing demand for environmental data. With environmental problems arising, pressure and flow sensors developed with MEMS technology cannot be overstated enough. They play a critical role in monitoring variables such as air quality, fluid flow, and other important atmospheric components. With pollution and climate change worries escalating, MEMS pressure and flow sensors are central to measuring and quantifying emissions to aid in regulatory compliance and policy formulation.

By maintaining high stability and sensitivity, the sensors help monitor pollutants in real time, thereby enhancing the capability of environmental monitoring networks. In addition, developments in MEMS technology have led to the creation of more durable sensors that can function under varying environmental conditions, allowing uninterrupted data acquisition. In doing so, pressure and flow sensors based on MEMS are a witness to the mingling of sophisticated technology and pressing ecological imperatives, as they contribute substantively toward the cause of environmental sustainability. Application of MEMS technology in environmental monitoring systems also pushes the boundaries of smart materials and systems. For instance, advances in electroactive polymers and bio-responsive hydrogels using MEMS/NEMS devices hold the potential to revolutionize the interaction with environmental data acquisition processes. These technologies provide the potential for interactive systems that enhance user engagement and sensitivity towards environmental change. As detailed in recent studies, the advancement of materials in MEMS systems provides the capability to create intelligent sensors that dynamically respond to environmental change, hence enhancing decision-making processes. These advancements not only optimize the performance of MEMS sensors but also foretell far-reaching implications for ambient living environments and intelligent ecosystems. By leveraging these attributes, environmental monitoring MEMS sensors have the power to have a significant influence on how society responds to ecological challenges while facilitating the integration of green technologies into daily life.

7.6.1. Types of Environmental Parameters Measured

In environmental parameter measurement, gyroscopes and accelerometers play a vital role in the application of Inertial Measurement Units (IMUs) for robotics and navigation. The sensors provide extremely critical data on motion dynamics, enabling devices to detect orientation and acceleration with high precision. The reliability and stability of the bias and noise properties of IMUs play an important role in their accuracy and competence in filtering and data acquisition for navigation. For instance, the above discussion by Suvorkin et al. (2024) examines variations in performance levels between different grades of MEMS-type IMUs, using parameters like Allan deviation to evaluate their bias stability and accuracy over time. These test performances are of critical significance, particularly for application in applications that require strict navigation accuracy, thereby making it the need to understand about these sensors and their efficiencies in measurement of environmental parameters. Pressure and flow sensors are another critical group of sensors, which allow one to measure and regulate fluid systems in the natural as well as engineered environment. These MEMS sensors are employed widely in various industries including environmental monitoring, automobiles, and HVAC applications to measure parameters such as atmospheric pressure, liquid flow rate, and gas composition. Such sensors' accuracy is very critical since minimal errors can cause significant inefficiency in the process or even represent safety hazards. Moreover, the operation of MEMS pressure sensors is enhanced with their integration into smart systems with the ability to process real-time information, as seen through the explication of state-of-the-art environmental monitoring technologies. All the mentioned sensors as a group yield an improved perception of environmental parameters, which is critical for making fact-based decisions and proactive planning management. Apart from individual sensors, Inertial Navigation Systems (INS) entail sophisticated integration of various types of sensors, including MEMS accelerometers and gyroscopes, to offer comprehensive environmental sensing. INS operates by estimating the position and velocity of a moving vehicle from initial conditions and sensor measurements over time, with applications in aviation and naval navigation. Significantly, MEMS environmental monitoring devices have been miniaturized, low-cost alternatives to traditional sensors that can be deployed more broadly in a range of environments. Ongoing research on personal monitoring devices, as addressed by Víctor et al. (2024), illustrates how MEMS sensors are being designed specifically for environmental monitoring applications, including air quality monitoring. This innovation unleashes the potential of MEMS technology as a requisite component in effectively measuring a myriad of environmental parameters across different environments.

7.6.2. Impact on Public Health and Safety

In safety and public health, MEMS is highly significant, particularly with the application of accelerometers and gyroscopes, or inertial measurement units (IMUs). Sensors have revolutionized personal health monitoring by allowing one to monitor movement and physical activity accurately. As individuals become more concerned with their health information, IMUs are able to detect anomalies in movement patterns, which may be an early indicator of potential health issues such as balance disorders or the early stages of neurological disorders. Furthermore, when coupled with wearable devices, these sensors promote preventative medical measures as they provide instant feedback. This capacity to identify early as well as improve personal health outcomes also reduces the burden on the healthcare system by enabling timely interventions, and the outstanding contribution of MEMS technology in public health interventions. Pressure sensors and flow sensors are also another fundamental component of MEMS technology that makes huge contributions to both public safety and health. These sensors are of key importance in monitoring various environmental conditions, such as air and water supply, for the purposes of community protection. A good example is pressure sensors that can detect abnormalities in blood pressure, which enables the application of early intervention in cardiovascular management. Flow sensors are also of key importance in ensuring safe drinking water through the monitoring of levels of contamination or ensuring proper water flow in urban systems. By its role in aiding detection of anomalies and facilitating timely reaction, MEMS pressure and flow sensors are not only advantageous to individual health but also to the public health system.

Figure 7.16. As individuals become increasingly aware of their health data, IMUs can help detect anomalies in movement patterns, which may indicate potential health issues such as balance disorders or early signs of neurological conditions.

Source: https://www.mdpi.com/sensors/sensors-21-06886/article_deploy/html/images/sensors-21-06886-g001.png.

Anticipatory surveillance is an important function in the prevention of public health emergencies and shows how MEMS technology enhances community safety through enhanced awareness of health-critical factors. Inertial navigation systems (INS) and environmental monitoring MEMS also contribute to public health and safety by providing an entire perspective of several considerations with respect to health and safety. INS technology, which is used in transportation solutions on a large scale, makes public transport and emergency response safe by enabling precise navigation and location tracking. This becomes especially vital during emergencies, as response at the right time is a matter of saving lives.

In addition, MEMS intended for environmental monitoring can detect dangerous toxic pollutants, allergens, and other toxic conditions, thereby offering valuable information that can guide public health initiatives. These sensors offer a three-dimensional framework for defending public health, revealing trends that could be equivalent to health advisories or prevention. The combination of all these different MEMS applications underscores their valuable contribution to public safety and health gains, and their overall potential as proactive instruments for managing health in contemporary society (Figure 7.16).

7.6.3. Future Trends in MEMS Environmental Sensors

Technologicaladvancementisdrivingthedevelopmentofmicroelectromechanical systems (MEMS) into increasingly more complex applications for environmental monitoring. As cities and businesses attempt to meet regulatory needs and enhance urban living, MEMS environmental sensors are becoming more essential in the monitoring of air quality, water quality, and soil condition. Future inertial measurement units (IMUs) comprised of accelerometers and gyroscopes will be equipped with advanced machine-learning algorithms that enable real-time processing of data and predictive modeling, leading to increased monitoring capability. Sophisticated sensor fusion technology will integrate information from multiple MEMS sensors, creating comprehensive understanding of environmental conditions. Further, urban to smart city development will stimulate integration within infrastructure, enabling not only real-time sensing, but also facilitating device-to-device communication. These advances unlock the door to optimal resource management and predictive environmental preservation, critical elements for sustainable 21st-century urban growth. Flow and pressure sensors are another key MEMS technology front, as miniaturization and sensitivity further rise and redefine environmental monitoring. As industries move into automation and real-time data capture, MEMS pressure sensors will be an indispensable tool for monitoring the effects of climate change and natural disasters. Future versions of these sensors will be characterized by smart materials that dynamically adapt to evolving environments, extending their functionality even further. Enhanced fluid dynamics modeling, such as model improvement with respect to Timoshenko beam effects, will improve the way these kinds of MEMS function in different environments, making them more reliable and accurate (Beardslee et al., 2014). Integrating these sensors into networks will provide holistic data analysis, with the capability to make real-time responses to environmental issues, ultimately helping scientific research and public safety efforts. Furthermore, use of MEMS in inertial navigation systems (INS) will greatly expand, primarily for environmental applications. For regions where global positioning systems

(GPS) fall short like in densely populated and forested areas, MEMS-based INS provides persistent navigation solutions with signal loss reliability. Sensor fusion breakthroughs will enable more precise tracking and mapping to the advantage of applications in autonomous vehicles and drones used for environmental monitoring. More universal MEMS environmental sensors launched will enable smooth interaction among various terrestrial and airborne monitoring instruments to deploy them to complex geographical territories. These sensors, coupled with new advancements in responsive materials, will enable the development of adaptive systems that will be capable of effectively responding to environmental changes and thus result in a heightened and more informed relationship with our environment (Oliver et al., 2009).

7.7. CONCLUSION

In evaluating the revolutionary impact of MEMS sensors across various domains, note needs to be taken of their gigantic contributions in numerous categories ranging from accelerometers to gyroscopes, pressure sensors, and environmental sensors. Each category has unique capabilities that are sensitive to a broad variety of applications, including inertial navigation systems (INS) that are extremely reliant on high-precision measurements to ensure accuracy and reliability. The marriage of miniaturized technology with complex algorithms enabled MEMS devices to outperform traditional sensors in being more responsive while conserving energy. The innovation has not only accelerated developments in consumer electronics and automotive industries but also underscored the significance of such sensors in safety-critical usage, especially in hostile environments, as described by Kerkhoff et al. (2009). Last but not least, continuous innovation of MEMS technologies promises to enhance the performance and reliability of measurement systems. The use of MEMS sensors transcends mere technological innovation; it answers a broader environmental imperative that focuses on real-time monitoring of physical quantities. Pressure and flow sensors, for instance, are now a necessity in industrial automation and medicine, where they deliver efficiency and safety in operations. These sensors are of primary importance in maintaining ideal conditions, reducing waste, and increasing process control. Furthermore, MEMS devices utilized in environmental monitoring enable pollutants and atmospheric changes to be sensed, being a lifeline against climate change. Academic research and conferences, as proposed in the literature and particularly conferences, extensively engage sensor design and working principal innovations as exemplified by Brandner et al. (2011). The integration of these capabilities results in a greater, much-needed knowledge of environmental conditions, and therefore MEMS sensors become part of necessity in contemporary science and engineering. Conclusion of this talk of

MEMS sensors integrated into measurement of physical quantities reveals the multiple benefits that are associated with them. With more industries turning to such innovative technologies, precision and reliability keep on increasing as a need. With the convergence of MEMS sensors and data analytics, predictive maintenance, automation, and smart systems become likely. This will be followed by more research and development, which will open up to still larger breakthroughs in sensor sensitivity and robustness. In addition, the continuous interface between industry and academia will provide solutions to the changing challenges of environmental sustainability and efficiency. The results of the research underpin the importance of MEMS technologies, validating their role as essential components in creating evolved sensor applications.

7.7.1. Summary of Key Points Discussed

MEMS sensor development has changed the face of measurement technology, especially in the case of accelerometers and gyroscopes, which form fundamental parts of inertial measurement units (IMUs). They enable accurate tracking of motion and orientation, which is important for numerous applications ranging from automotive to aerospace and consumer electronics. Miniaturization by MEMS technology not only reduces size and price but also increases capabilities for integration, with greater capability to design more complicated systems. More performance in the form of reduced power usage and greater sensitivity allows real-time data processing and enables advanced applications such as robotics and augmented reality. As these sensors keep advancing, they can potentially provide more advanced functionalities, paving the way to innovation in applications with a high reliance on motion detection and analysis. In pressure and flow sensors, MEMS technology is a cut above the rest by being able to provide high sensitivity and fast response times that are paramount for monitoring various physical quantities. MEMS sensors are utilized in medical devices to industrial processes where accurate information is important for both efficiency and safety. Their functionality is further revealed by the utilization of MEMS sensors in environmental monitoring systems. With advanced materials and manufacturing techniques, the devices can detect changes in environmental parameters such as fluid dynamics and air quality accurately. The growing need for real-time data in such applications underscores the importance of MEMS technology, as it offers the capability to continuously monitor critical parameters governing public health and industrial processes. MEMS for INS is a critical advancement in the navigation technology's flexibility and economy. INS relies on onboard sensors that detect the inertial forces experienced by a body in order to accurately position and navigate irrespective of any external references. Such capability is of extreme importance in uses such as autonomous vehicles and aircraft, where navigation has to be as precise as possible for purposes of safety and efficiency. The integration of MEMS sensors into high-end computational platforms allows data fusion, and this results in more reliability and accuracy in navigation tasks. As highlighted by recent research, hierarchical uncertainty quantification methods could significantly contribute to the simulation of complex MEMS circuits by applying stochastic techniques and further improve the system's reliability as well as push the boundary in what is achievable with MEMS-based INS technologies (Daniel et al., 2014). An improvement in MEMS technology has the potential to revolutionize navigation accuracy across various applications.

7.7.2. The Significance of MEMS Sensors in Various Fields

The multimodal character of MEMS sensors is seen in their extensive use across various technologies, particularly in the form of gyroscopes and accelerometers, which are extensively used in inertial measurement units (IMUs). MEMS sensors play a very critical role in autonomous vehicles, smartphones, and wearable tech since they enable real-time tracking of motion and orientation. With the ability to measure changes in velocity and positions, IMUs provide precise navigation and stabilization, which is essential for various applications including aerospace engineering and augmented reality. With the integration of the MEMS sensors, functionality and performance are improved since they provide the data needed for algorithms with low latency requirements and high accuracy. This technology transition in MEMS-based IMUs is a quintessential representation of the revolutionary potential of micro-electromechanical systems, which makes it pioneer components in the evolving era of smart technology and interface design in a range of industries. In addition to motion sensors, MEMS technology has also significantly enhanced pressure and flow sensing, with profound implications for industrial automation and environmental monitoring. These MEMS sensors can detect minute pressure differences and flow rates, allowing for better dynamic system control such as for HVAC, automotive engines, and medical equipment. In industry, they maximize the efficiency of operations by making conditions detectable and controllable in real time. Moreover, their reduced power requirements and minute size make them well suited for remote and portable applications, where traditional sensors are impractical. Since they are capable of delivering precise, timely data, MEMS pressure and flow sensors are transforming into critical tools in environmental monitoring for pollution tracking and assisting the transition of the world towards sustainable manufacturing and energy management technologies. The applicability of MEMS sensors also seeps into high-facility inertial navigation systems (INS) and environmental sensing, contributing to wealth in data collection and spatial orientation analysis. INS, using MEMS gyroscopes and accelerometers, are a standard of navigation for aerial and submarine vehicles and provide precise location tracking without dependence upon external sources and are extremely useful in GPS-denied environments. At the same time, in environmental monitoring, MEMS sensors are used for detecting air quality and determining chemical compounds with record sensitivity and minimal dimensions. This property addresses present needs in biomedicine and environmental science, such as the detection of chemical signals emitted by cells when exposed to any stimulus, even at trace concentrations found in circulating blood. The flexibility of MEMS technology, exemplified by its high performance across a broad spectrum of conditions, illustrates its significance as a revolutionary element in various fields of research and application.

7.7.3. Future Prospects and Advancements in MEMS Technology

Emerging trends in Micro-Electro-Mechanical Systems (MEMS) technology will transform the future of inertial measurement units (IMUs), made up of gyroscopes and accelerometers. As the complexity of our applications keeps growing, the sensing capability can be enhanced to allow these sensors not only to sense motion with high precision but also interpret high-complexity environmental inputs. Enhanced integration with machine learning algorithms enables the ability for IMUs to learn in real time to improve navigation accuracy in a variety of applications from aerospace and consumer goods. This is stimulated by the booming Internet of Things (IoT), where networked MEMS devices share data and communicate with one another to provide smart analytics. Thus, the innovation of MEMS technology in IMUs is symptomatic of a general trend towards multi-function devices that can sense cognitively their environment and, as a result, redirect their functions between industries, ranging from robotics to disease monitoring (Baaziz et al., 2015). Pressure sensors and flow sensors are also another major area anticipating gigantic leaps thanks to MEMS technology. As the global need for accuracy in applications such as healthcare and automotive increases, sensor miniaturization and integration into compact designs becomes essential. Technologies such as microfabrication enable the fabrication of highly sensitive and precise pressure sensors that can operate in harsh conditions. Furthermore, MEMS flow sensors are being developed to offer real-time sensing, which is imperative for application in environmental monitoring and industrial process control. These technologies are making new opportunities for the development of sophisticated sensing networks that can monitor air quality, leak detection, or optimization of energy efficiency. As sustainability pressure increases, these MEMS-based solutions will play a significant role in enabling better resource management and green solutions (Baaziz et al., 2015). Monitoring the environment is a bright future for MEMS technology, especially with world challenges of climate change and pollution mounting increasingly quickly. With the application of MEMS-based sensors, it becomes achievable to come up with portable and cost-effective solutions for the real-time monitoring of air and water quality. The new devices can be applied in different environments, collecting vast amounts of data that can be processed using big data techniques. The experiences gained from applying advanced capabilities, as proved through high-contrast imaging for astrosciences, highlight MEMS technology's viability to be repurposed for environmental uses. As such high-performing microsensors continue their development, they will play a vital role in our knowledge of ecological effects, allowing for faster interventions in environmental hazards and promoting a forward-looking strategy toward sustainability.

REFERENCES

1. Ahmad, A. S., Ahmar, A. S., Asmuni, H., Hishamudin, M., et al. (2018). Iris segmentation. https://core.ac.uk/download/19162806.pdf (accessed on 14th October 2025).
2. Alù, A., Chen, P. Y., Cheng, M., Cui, T. J., et al. (2018). Generalized parity-time symmetry condition for enhanced sensor telemetry. http://arxiv.org/abs/1812.08296 (accessed on 14th October 2025).
3. Alzubaidi, L., Bai, J., Al-Sabaawi, A., Santamaría, J., Albahri, A. S., Al-dabbagh, B. S. N., Fadhel, M. A., et al. (2023). A survey on deep learning tools dealing with data scarcity: Definitions, challenges, solutions, tips, and applications. *Journal of Big Data, 10*. https://doi.org/10.1186/s40537-023-00727-2 (accessed on 14th October 2025).
4. Baaziz, A., Kleiner, A., Feigelson, E. D., Waller, M. A., Boyd, S., Vandermerwe, S., & Zhou, Z. (2015). Big data and the Internet of Things. http://arxiv.org/abs/1503.07092 (accessed on 14th October 2025).
5. Barber, A. E. (2013). Estimating epipolar geometry with the use of a camera mounted orientation sensor. https://core.ac.uk/download/19567341.pdf (accessed on 14th October 2025).
6. Barrett, H., Baudoz, P., Belikov, R., Beuzit, J. L., et al. (2012). Review of small-angle coronagraphic techniques in the wake of ground-based second-generation adaptive optics systems. https://core.ac.uk/download/33120517.pdf (accessed on 14th October 2025).
7. Bartel, C., Kaharkar, A., Rodriguez, L., Sabatini, R., et al. (2012). Carrier-phase GNSS attitude determination and control system for unmanned aerial vehicle applications. https://core.ac.uk/download/pdf/18619676.pdf (accessed on 14th October 2025).
8. Bauer, P., Gáspár, P., Lantos, B., Prohászka, A., et al. (2007). Design of an embedded microcomputer-based mini quadrotor UAV. https://core.ac.uk/download/11855948.pdf (accessed on 14th October 2025).
9. Beacken, M., Bishop, D. J., Chang, J., Holyoak, T., et al. (2018). High performance, continuously tunable microwave filters using MEMS devices with very large, controlled, out-of-plane actuation. http://arxiv.org/abs/1804.08658 (accessed on 14th October 2025).
10. Beardslee, L. A., Brand, O., Dufour, I., Heinrich, S. M., et al. (2014). Lateral-mode vibration of microcantilever-based sensors in viscous fluids using Timoshenko beam theory. https://core.ac.uk/download/213078271.pdf (accessed on 14th October 2025).
11. Bertuletti, S., Caldara, M., Cereatti, A., Comotti, D., et al. (2017). Static

and dynamic accuracy of an innovative miniaturized wearable platform for short range distance measurements for human movement applications. https://core.ac.uk/download/98348275.pdf (accessed on 14th October 2025).

12. Brady, T., Crain, T., Henderson, T., Phillips, R., et al. (2012). Integrated rate isolation sensor. https://core.ac.uk/download/pdf/10571414.pdf (accessed on 14th October 2025).
13. Brandner, J. J., Dittmeyer, R., & Vittoriosi, A. (2011). Experimental design with integrated temperature sensors in MEMS: An example of application for rarefied gases. *3rd Micro and Nano Flows Conference (MNF2011).* https://core.ac.uk/download/9633296.pdf (accessed on 14th October 2025).
14. Brookhuis, R. A., de Boer, M. J., Droogendijk, H., Krijnen, G. J. M., et al. (2014). Towards a biomimetic gyroscope inspired by the fly's haltere using microelectromechanical systems technology. https://core.ac.uk/download/572159112.pdf (accessed on 14th October 2025).
15. Ciuprina, G., Isvoranu, D., & Sorohan, S. (2014). Aerodynamic behavior of the bridge of a capacitive RF MEMS switch. *4th Micro and Nano Flows Conference (MNF2014).* https://core.ac.uk/download/29139762.pdf (accessed on 14th October 2025).
16. Coste, P., Gao, Y., Gouache, T., Gourinat, Y., et al. (2010). Wood wasp inspired space and earth drill. https://core.ac.uk/download/12042352.pdf (accessed on 14th October 2025).
17. da Silva Gomes, B., Domínguez, F., Segers, L., Steenhaut, M., et al. (2014). Sound Compass: A distributed MEMS microphone array-based sensor for sound source localization. https://core.ac.uk/download/55759306.pdf (accessed on 14th October 2025).
18. Daniel, L., Karniadakis, G. E., Oseledets, I. V., Yang, X., et al. (2014). Enabling high-dimensional hierarchical uncertainty quantification by ANOVA and tensor-train decomposition. https://core.ac.uk/download/78065180.pdf (accessed on 14th October 2025).
19. de Boer, M. J., Droogendijk, H., Krijnen, G. J. M., Sanders, R., et al. (2015). Bio-inspired hair-based inertial sensors. https://core.ac.uk/download/pdf/31153370.pdf (accessed on 14th October 2025).
20. Dong, Y., Liu, H., & Song, R. (2010). Bridges structural health monitoring and deterioration detection synthesis of knowledge and technology. https://core.ac.uk/download/286607584.pdf (accessed on 14th October 2025).
21. Faragher, R., Groves, P. D., Martin, H. F. S., Newman, J., et al. (2013). A new approach to better low-cost MEMS IMU performance using sensor

arrays. https://core.ac.uk/download/16259651.pdf (accessed on 14th October 2025).

22. Fluitman, J. (1996). Microsystems technology: Objectives. https://core.ac.uk/download/pdf/11453403.pdf (accessed on 14th October 2025).
23. Gill, E., Monna, G. L. E., Scherpen, J. M. A., et al. (2007). MISAT: Designing a series of powerful small satellites based upon micro systems technology. https://core.ac.uk/download/pdf/12941826.pdf (accessed on 14th October 2025).
24. González, V., Godoy, J., Arroyo, P., Meléndez, F., Díaz, F., López, A., Suárez, J., et al. (2024). Development of a smartwatch with gas and environmental sensors for air quality monitoring. *Sensors (Basel, Switzerland), 24*. https://www.semanticscholar.org/paper/4b7ba4b4db2c6dca660574688fc300027b5cb50f (accessed on 14th October 2025).
25. Guglielmelli, E., Oddo, C. M., Saccomandi, P., Schena, E., et al. (2014). Microfabricated tactile sensors for biomedical applications: A review. https://core.ac.uk/download/pdf/54936693.pdf (accessed on 14th October 2025).
26. Hogg, T. (2006). Distributed control of microscopic robots in biomedical applications. http://arxiv.org/abs/cs/0611111 (accessed on 14th October 2025).
27. Hsieh, W. H., Miu, D. K., Tai, Y. C., Temesvary, M., et al. (1995). Design, fabrication, and testing of silicon micro gimbals for super-compact rigid disk drives. https://core.ac.uk/download/4869875.pdf (accessed on 14th October 2025).
28. Kerkhoff, H. G. (2009). Dependable reconfigurable multi-sensor poles for security. https://core.ac.uk/download/pdf/11478693.pdf (accessed on 14th October 2025).
29. Lampo, S. (2012). National MEMS technology roadmap–Markets, applications and devices. https://core.ac.uk/download/80705061.pdf (accessed on 14th October 2025).
30. Luo, Y., Abidian, M. R., Ahn, J. H., Akinwande, D., Andrews, A. M., Antonietti, M., Bao, Z., et al. (2023). Technology roadmap for flexible sensors. *ACS Nano, 17*, 5211–5295. https://doi.org/10.1021/acsnano.2c12606 (accessed on 14th October 2025).
31. Oliver, R., Tillotson, J., & Toomey, A. (2009). Bioengineered textiles and nonwovens – The convergence of bio-miniaturization and electroactive conductive polymers for assistive healthcare, portable power and design-led wearable technology. https://core.ac.uk/download/9587153.pdf (accessed on 14th October 2025).

32. Suvorkin, V., Garcia-Fernandez, M., González-Casado, G., Li, M., & Rovira-Garcia, A. (2024). Assessment of noise of MEMS IMU sensors of different grades for GNSS/IMU navigation. *Sensors (Basel, Switzerland), 24*. https://www.semanticscholar.org/paper/28130458bfdb69d19a48f230a9d824ae72efbdbd (accessed on 14th October 2025).

33. Yang, G. Z., & Zhang, Z. Q. (2014). Calibration of miniature inertial and magnetic sensor units for robust attitude estimation. https://core.ac.uk/download/42623979.pdf (accessed on 14th October 2025).

CHAPTER 8

MEMS ACTUATORS IN APPLICATIONS

Contents

INTRODUCTION

The rapid pace of technology has been generating a huge interest in the field of micro-electromechanical systems (MEMS), which is now an integral part of various applications ranging from telecommunication to biomedical devices. With the miniaturization of electronic and mechanical components, MEMS actuators are now innovation gateway enablers. These micro-machined systems use microfabrication techniques to achieve precise motion and control at the micro level, opening new avenues for high-performance applications in confined spaces. They are adaptable and can be incorporated into systems such as micromirrors and optical switches that are of tremendous importance in digital micro-mirror device (DMD) technologies and micro-opto-electromechanical systems (MOEMS). This blend of microtechnology and new applications puts the knowledge of MEMS actuators operating parameters and design principles at the forefront, thereby providing a foundation for studying their metamorphic capabilities across a wide range of industries (Fluitman et al., 1996). For communications, MEMS actuators form the basis of RF MEMS switches and tunable capacitors, transforming traditional methods of signal processing. These actuators enable improved signal integrity with low-cost, programable behavior and excellent agility of signal frequency. Selective filtering and frequency control, especially in software-defined radios (SDR), bring out the efficiency of MEMS actuators in addressing demanding engineering problems. With highly dynamic tuning, they enable one to sense desired signals at nanowatt levels without contamination by nearby transmitters of high power. As shown by recent research, the application of MEMS extreme throw actuators to the design of RF filters has demonstrated that continuous tuning over extensive frequency ranges is achievable and supporting the viability of the devices in modern communications systems (Beacken et al., 2018). Besides the telecommunications sector, the application range of MEMS actuators also extends to the biomedical sector, particularly in drug delivery systems and microfluidics. The nuances of microrobotics and microgrippers highlight MEMS technology's promise for the revolutionization of medical procedures through enabling precise manipulation on the cellular level. Through the utilization of the flexibility of MEMS actuators, innovation in the controlled release of medicines becomes viable, maximally optimizing therapeutic effects by means of targeted delivery. The integration of MEMS technology and drug-delivery mechanisms shows how far the latter goes towards not only simplifying manufacturing processes but also enhancing the efficacy of clinical interventions. Overall, MEMS actuators' diffusion across industries forms the foundation for the necessity of sustaining research and development in this area in a bid to keep it relevant and raise its applicability within contemporary technology contexts (Fluitman et al., 1996).

8.1. MICROMIRRORS AND OPTICAL SWITCHES

MEMS advancement has brought about significant advancements in optical switching technologies, particularly the use of micromirrors. The presence of these devices, which are characterized by their ability to manipulate light on a micro-scale dimension, is instrumental in establishing usable purposes in digital light processing (DLP) and micro-opto-electromechanical systems (MOEMS).

Through the assistance of advanced actuation mechanisms, micromirrors have the ability to be finely tuned in deflecting or reflecting light beams, which is significant in real-time switching functions. Their design generally involves thermally actuated bimaterials that allow out-of-plane actuation, a critical feature for performance enhancement in optical devices. As researchers continue to explore the potential of various materials for these systems, optimized configurations have been determined to significantly enhance their efficiency and reliability in operational environments, paving the way for future innovation in optical telecommunications and display technologies. In recent years, micromirrors have been instrumental in advancing the functionality of optical switches to facilitate high efficiency in signal management and routing. For instance, an array of addressable MEMS micromirrors can modulate light with high efficiency in an intracavity, with good output control of lasers. The technology allows for individual operation at the same time, which provides two distinct laser outputs of good power and quality. These configurations have been found to deliver pulse widths of as short as 28 ns at high repetition rates and are therefore ideal for applications that demand rapid response times and high-fidelity signals. Being able to generate several different output parameters with minimized crosstalk is particularly useful in communication systems, where signal integrity must be maintained (Bauer et al., 2012). As such, application of micromirrors in optical switch devices is transforming data transmission processes in industries. Continued advancement of MEMS technology is increasingly expanding the horizons for application of micromirrors in optical switching. Emerging research has introduced tunable materials and advanced fabrication techniques, which are likely to enhance the scalability and efficiency of micromirrors. Apart from this, their incorporation into miniature systems is a contributing factor to the trend of miniaturizing technology and therefore leads them to be compatible with handheld devices. The interplay between energy efficiency and flexibility of operation positions micromirrors at the foundation of next-generation optical networks and intelligent devices, where fast and accurate control of light can lead to advancements in microfluidics and drug delivery systems. With the implementation of these innovations, engineers and scientists have the potential to enhance the performance of MEMS actuators to ensure that micromirrors remain at the forefront of optical switch technologies for decades to come.

8.1.1. Functionality of Digital Micromirror Devices (DMD)

The sophisticated engineering and design of Digital Micromirror Devices (DMD) are a milestone in Micro Electromechanical Systems (MEMS) technology, which reflect their versatility in various applications. They utilize thousands of highly minuscule, accurately controlled mirrors, which are tilted to deflect light, hence forming images and patterns. The working efficiency of DMDs lies in their ability to perform high-speed switching, which is extremely crucial in products such as projection systems and digital cinema. Further, technological improvements in MEMS technology have spurred enhanced manufacturing processes, which have enabled high efficiency DMD manufacturing for various applications, such as sensing systems and communications (Sáenz et al., 2005). The rapid pace of innovation in the field continues to reveal just how important DMDs are as fundamental building blocks of modern optical systems. Scientists have increasingly realized the promise that DMDs hold in applications other than conventional imaging systems, particularly in microfabrication and rapid prototyping applications. One such standout example is the combination of DMD technology with microstereolithography machines, which allows for rapid production of intricate 3D geometries.

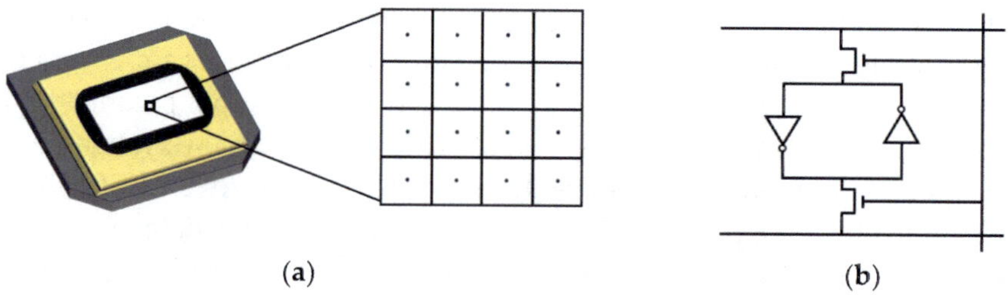

Figure 8.1. The intricate design and engineering of digital micromirror devices (DMD) represent a significant leap in the field of microelectromechanical systems (MEMS), demonstrating their versatility in various applications.

Source: https://www.mdpi.com/photonics/photonics-11-00793/article_deploy/html/images/photonics-11-00793-g001.png.

This DMD-based approach to ultraviolet (UV) microstereolithography provides the singular advantage of exposing entire layers of photopolymers simultaneously, significantly speeding up the curing process but with high resolution and low cost (Choi et al., 2003). Such a characteristic has profound implications for applications such as biomedical engineering and micro-manufacturing, as diverse materials can be handled cost-effectively to create functional micro-parts that enhance current manufacturing methods. Another perspective on DMDs shows their potential in different industries by innovative application in drug delivery and microfluidics. Being capable of precise control of light patterns, DMDs are utilized in the manipulation of micro-scale fluids with enhanced performance in microfluidic devices, which are critical in medical technology. Their ability to create customized light patterns also enables new forms of drug delivery with controlled release and targeted therapy. As there is the evolving MEMS technology, the gradual integration of DMDs into different applications shows their usability, which leads to better devices and innovative solutions that meet the constantly evolving requirements of today's technology (Sáenz et al., 2005) (Figure 8.1).

8.1.2. Applications of Micro-Electro-Mechanical Systems (MOEMS) in Telecommunications

In today's telecommunication world, the unavoidable integration of Micro-Electro-Mechanical Systems (MOEMS) is revolutionizing the data processing and transmission. Among the most captivating applications is micromirrors and optical switches that rely on regulated mechanical movement to manage light paths and speed up signal routing. The miniaturization not only facilitates component densities but also reduces the overall power consumption for optical communication systems. The utilization of Digital Micromirror Devices (DMDs) is a major manifestation of this trend and has enabled functionalities that have led to display technologies and data visualization innovations. Through the utilization of MOEMS technologies, telecommunications can obtain enhanced efficiency in bandwidth utilization and latency savings, which eventually lead to improved user experience in high-speed networking applications, particularly in data centers and cloud computing infrastructure where peak performance is critical. Aside from optical applications, RF MEMS switches and tunable capacitors play a vital role in the telecommunication sector and in the creation of adaptive networks. They capitalize on the use of mechanical actuation to switch RF signals, providing performance and reliability that surpass their traditional semiconductor counterparts. The dynamic tuning of capacitive networks allows for volumetric efficiency in antenna and signal processing unit design, allowing the rapid development of wireless communication technology such as 5G. This increased flexibility allows networks to react in real-time to changing bandwidth demands and interference signals, ultimately allowing for more robust communication systems. In turn, deployment of RF MEMS technologies can both improve personal user experiences and scalability of telecommunication infrastructure within both city and countryside. As versatile as they are, MOEMS extend their reach into microfluidics as well as drug delivery systems, illustrating the multi-dimensionality of their uses beyond traditional telecommunication. In such areas, microrobotics and microgrippers, powered by MEMS actuators, enable precision in activities such as drug delivery to targeted locations and real-time monitoring of biological samples. The revolution enhances telemedicine and remote diagnosis capabilities, with medical practitioners able to administer treatment and gather patient data more effectively. By integrating these advanced MEMS technologies with telecommunication, healthcare centers can achieve harmonious communication between devices and medical staff, therefore improving timely medical treatment and patient recovery. The ongoing research in MOEMS in various applications sheds light on the importance of multidisciplinary collaboration in the development of innovation that meets the dynamic demands of modern society and industry (Figure 8.2).

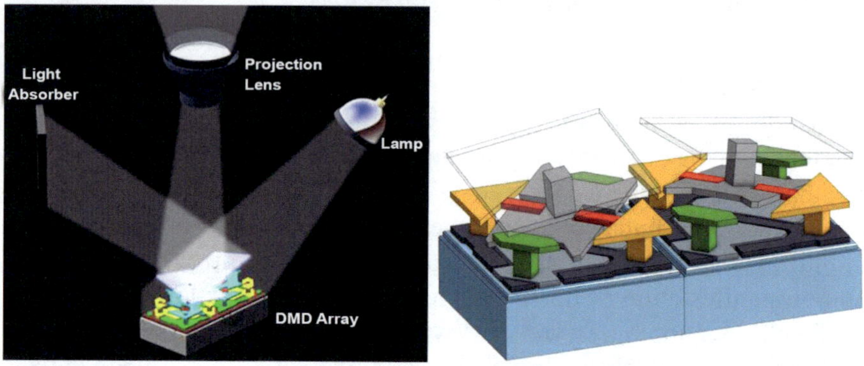

Figure 8.2. The advent of digital micromirror devices (DMDs) exemplifies this trend, offering capabilities that have led to advancements in display technologies and data visualization.

Source: https://content.cld.iop.org/journals/2515-7647/6/4/043001/revision3/jp-photonad6dc0f1_hr.jpg.

8.1.3. Advantages of Using Micromirrors in Optical Systems

Their applications for the versatility of micromirrors in optics are attested to by their use in cutting-edge technologies like Digital Micromirror Devices (DMD) and Micro-Opto-Electro-Mechanical Systems (MOEMS). The devices not only facilitate high-resolution imaging but also colossal amounts of light manipulation flexibility, which is crucial in applications ranging from projectors to adaptive telescopic optics. The MOEMS Deformable Mirrors (DM) design is very efficient as a novel due to its ability to handle extreme conditions like cryogenic and vacuum environments (Barette et al., 2017). Such adaptability makes it extremely easy for optical systems to perform optimally under various environments, making them crucial in scientific research and commercial purposes. Further, the small size of micromirrors assists in offering overall system sizes that are reduced, thereby facilitating integration into systems where spatial constraints are a top design consideration. As an added advantage to their rugged and compact nature, micromirrors offer significant energy efficiency and operational life advantages. Recent advances in MEMS technology have enabled the fabrication of angular vertical-comb-drive torsional microactuators that are strong enough to achieve large optical scan angles using low energy (Choo et al., 2006). The microactuators enhance the performance of micromirrors by allowing efficient actuation at low drive voltages, which is essential in a bid to minimize power requirements in portable devices.

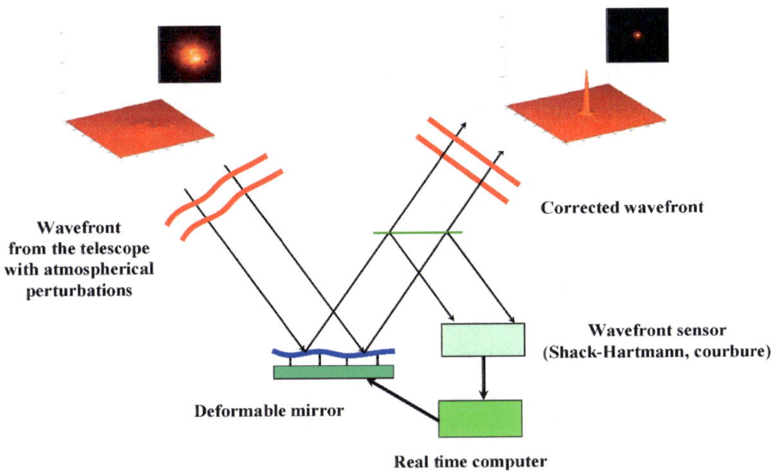

Figure 8.3. The innovative design of MOEMS deformable mirrors (DM) enables them to function effectively in harsh environments, including cryogenic and vacuum conditions.

Source: https://www.mdpi.com/micromachines/micromachines-08-00233/article_deploy/html/images/micromachines-08-00233-g001.png.

The resilience of micromirrors is attested by the fact that they can withstand high levels of operational cycles without performance degradation, thus being economically viable for extended use in a variety of applications. The longevity in this regard reassures users regarding the trustworthiness of micromirror-based systems in sensitive applications, such as telecommunications and surveillance. The precision offered by micromirrors is also vital in applications calling for high-fidelity optical control. Their capability to guide light with high accuracy enables fine functions such as beam steering, wavelength choice, and dynamic focusing. Such precision is used to the fullest in adaptive optics, where micromirrors are employed to correct wavefront aberrations in real time, significantly enhancing the quality of images. The advances in MOEMS technology have led to developments that harness such properties for use within high-performance imaging systems and laser communications to make micromirrors the blocks of next-generation optical devices (Barette et al., 2017).

The marriage of high precision, stability, and efficiency puts micromirrors at the forefront among optical system development, which ultimately influences technological advancements across the board (Figure 8.3).

8.2. RF MEMS SWITCHES

The agility of RF MEMS switches can be easily observed in a range of applications, from telecommunication to consumer electronics. The switches stand out because they can offer lower insertion loss and higher isolation compared to semiconductor switches. This is useful in applications where signal integrity and power efficiency are paramount. With increasing demand for faster and more efficient RF circuits, RF MEMS switches are extensively utilized to boost system performance while lowering overall energy consumption. They provide dynamic switching capability, allowing devices to dynamically switch between several RF signals with negligible latency, thus supporting proper management of electronic resources in consumer products and sophisticated telecommunications infrastructures. Another extremely critical domain of RF MEMS switch design is addressing issues regarding reliability and fatigue that are caused due to the mechanical nature of these devices. Optimizing the actuation pulse shapes is an extremely crucial aspect in enhancing their operational lifetime. As discussed by recent research, utilization of techniques such as Taguchi's optimization technique for actuation pulse shaping results in significant enhancement in impact velocity as well as settling time. These enhancements not only reduce mechanical stresses acquired under operating conditions but also minimize undesired effects, such as bouncing in contact, which can deteriorate performance and reliability over time. This focus on optimizing operating conditions reflects the marriage of circuit design and mechanical engineering in RF MEMS technology advancements to develop devices that possess the ability to operate under rigorous operating conditions while maintaining high performance. The innovations in RF MEMS switch technologies also involve incorporating newer materials and designs that enhance their usability even more. New device architecture methods, including the use of zipping variable capacitors, have the potential to realize higher capacitance values and dynamic tuning ranges with a small footprint. These improvements are particularly important in handheld consumer electronics, where size is of most significance. The continuous evolution of RF MEMS design mirrors not only the technological richness of these structures in responding and managing RF signals but also their sensitivity to new trends in the market. With the advancement of the technology, RF MEMS switches will be even more pivotal in the next-generation communications system to serve the fast-paced development in wireless technology and connectivity (Figure 8.4).

Figure 8.4. With a rising demand for faster and more efficient RF circuits, RF MEMS switches are increasingly employed to enhance system performance and reduce overall energy consumption.

Source: https://lumenci.com/wp-content/uploads/2023/11/RF-MEMS-cover.webp.

8.2.1. Mechanism of RF MEMS Switches

The history of RF MEMS switches has revolutionized many telecommunications applications as well as signal processing due to their compact size and low power requirements. Behind their operation is a mechanism of capacitive switching, whereby the switch element, in most cases a metal bridge, is electrostatically deflected. The bridging function allows for the device to open or close conductive paths, thereby employed as a switch. Specifically, intricate structures enable these switches to perform effectively at high frequencies, which is a critical requirement in modern wireless communications. 3D multi-physics modeling enhances the understanding of switch performance since it facilitates the analysis of fluid-structure interactions that can impact performance. The delicate interaction among mechanical deflection and electrical properties directs recent trends in RF MEMS technology in achieving higher reliability and efficiency of function in various applications (Ciuprina et al., 2014). The RF MEMS switch mechanisms also extend beyond basic actuation to sophisticated features of tunability as well as integration with other microsystems. Specifically, tunable capacitors have gained popularity in radio frequency applications by enabling dynamic variation of capacitance, hence ensuring frequency agility and optimization of signal integrity. This is facilitated by precise control over the switch's mechanical properties that govern the actuation response. Integration of RF MEMS components into larger systems, such as miniature biosensors that track variations in substance concentration, is a demonstration of the versatility of such switches. This is evidenced by the application of RF MEMS to biosensing, whereby subtle modifications in dielectric surroundings can trigger drastic shifts in resonant frequencies, allowing for sensitive mechanisms of detection (Li et al., 2009). The ongoing advancement of RF MEMS technology underscores its capability to address challenges in an array of applications, from telecommunications through biomedical devices. Optimization is typically focused on reducing actuation voltage while enhancing switching speed and reliability, thereby increasing overall efficiency. The dynamic nature of such components means that they not only have to switch effectively but also deliver consistently under evolving environmental conditions. Progress in newer research that includes sophisticated modeling techniques backed by experimental validation is essential to lead RF MEMS technology forward. These will ultimately lead to the development of devices smaller, faster, and more energy-efficient that will find broader applications and innovation in applications derived from the precise control of electromagnetic signals and interactions (Ciuprina et al., 2014) (Figure 8.5).

Figure 8.5. The implementation of 3D multi-physics modeling enhances the understanding of the switch behavior, as it allows for the analysis of fluid-structure interactions that can affect performance.

Source: https://www.ansys.com/content/dam/web/glossary/what-is-multiphysics/fluid-structure-interaction-airplane.png?wid=1200.

8.2.2. Role of RF MEMS in Wireless Communication

The growth in wireless communication has been significantly influenced by the implementation of RF MEMS technology, which has revolutionized signal processing and transmission in various aspects. One of the many applications, RF MEMS switches play a key role in enhancing efficiency and miniaturization of communication devices. The switches provide rapid signal routing and maintain low insertion loss, thereby significantly improving performance parameters such as bandwidth and power efficiency. Furthermore, the compact physical nature of RF MEMS devices is highly conducive to the movement towards lighter, thinner communication systems that is underway today. Even as wireless systems become increasingly sophisticated, there remains a persistent requirement for high-performance, high-reliability components, and RF MEMS technology is well-positioned to meet this demand on the fronts of integration and scalability.

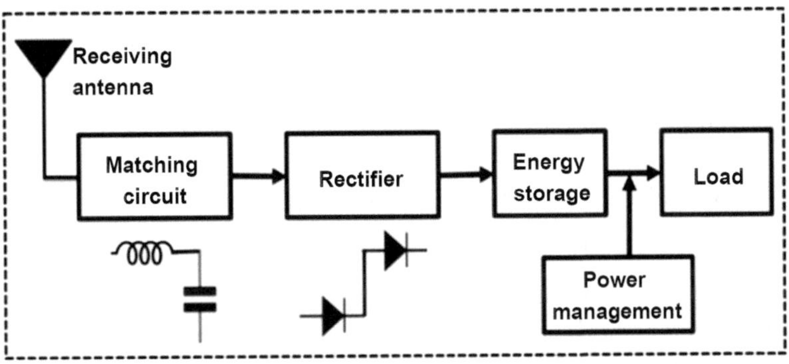

Figure 8.6. The ability to tune RF circuits leads to more versatile and efficient systems, facilitating broader bandwidth utilization and improved signal integrity.

Source: https://assets.cureusjournals.com/artifacts/upload/figure/file/9058/lightbox_bde72440e79511ef95680f7e7b2d9c56-Fig-1.png.

Yet another significant innovation attributed to RF MEMS is the development of tunable capacitors, pivotal to optimizing circuit performance in wireless devices. By assisting in dynamic adjustment of capacitance, products of this kind enhance circuit response in various frequencies, adapting to the complexity of modern communications signals. Tunability of RF circuits mean more adaptive and efficient devices, with greater bandwidth offered and improved signal integrity. This adaptability is particularly critical in systems where multiple frequency bands must be controlled dynamically. The movement toward software radio systems also enhances the significance of RF MEMS tunable devices, towards pointing to their growing importance in wireless communications of the future as devices become increasingly flexible in meeting changing and different environments. The importance of RF MEMS extends beyond the traditional communications applications, since their application in the area of mobile health (m-Health) solutions is particularly important. Of specific interest is the integration of RF MEMS into implantable medical devices, which indicates the promise of real-time monitoring of physiological processes and personalized healthcare delivery. By ensuring reliable wireless communication through such devices, RF MEMS enable data transmission in a seamless manner, which is critical to accurate medical diagnosis and treatment. Power supply issues for such systems remain of utmost concern, as illustrated in current considerations of implantable device technologies. Power problem solution and further advancement of energy efficiency are very essential to realizing the full potential of RF MEMS in medicine and to creating astounding advancements in biocompatibility and dependability in future medical devices (Amar et al., 2015). Further, RF MEMS's position in the evolution of generic technologies highlights their central contribution to addressing the twin unknowns of unKNOWN market needs and new technological opportunities, as illustrated in extant research (Kokshagina et al., 2015) (Figure 8.6).

8.2.3. Comparison of RF MEMS Switches with Traditional Electronic Switches

The development of communication devices has been focused more and more on the need for advanced technologies that offer high performance for various applications. Of these, RF MEMS switches have emerged as viable replacements for standard electronic switches. Their ability to handle RF signals efficiently at high frequencies puts them at the core of wireless communication systems today. Contrary to classical electronic switches that are primarily reliant on solid-state technology, RF MEMS switches employ mechanical motion to control signal routing, thus inducing significantly lower insertion losses and return loss improvement. Such developments provide the gateway to enhanced integration in the system, allowing product designers to support compact architecture while satisfying enhanced demands for ruggedness and reduced power consumption in communications (Casha et al., 2007). Despite the improved performance of RF MEMS switches, they are plagued by issues that face traditional electronic switches. The production of RF MEMS devices often calls for sophisticated processes that prove difficult to mass produce, particularly towards miniaturization at the cost of quality. For instance, while RF MEMS switches have better performance in speed and efficiency, the best switching reliability and long-term actuation consistency are priorities under research. By contrast, the more straightforward fabrication logistics of the conventional electronic switch often benefit from well-tested manufacturing practices guaranteeing output consistency. Moreover, the conventional switches are typically described more accurately in terms of failure modes, making it easier to troubleshoot and maintain them compared to their MEMS counterparts. When judging the larger context of MEMS applications, comparing RF MEMS switches with conventional electronic switches emphasizes their unique strengths and inherent drawbacks. With the advancement of the MEMS technology, significant advancement has been made in improving the mechanical reliability and switching speed of RF MEMS switches and bridging the performance gap between the two. Furthermore, emerging designs such as new varactors presented in recent research have demonstrated potential to improve capacitance scaling with a small footprint, beneficial in portable devices. These developments illustrate the ability of RF MEMS switches not only to substitute but even to surpass traditional electronic switching methods in some respects, particularly where increased performance and operating flexibility are needed within the domains of telecommunications and signal processing (Casha et al., 2007).

8.3. TUNABLE CAPACITORS

Greater demand for high-performance communications demands innovative components capable of responding to varying frequency signals. In the middle of this, tunable capacitors have emerged as key components, particularly in RF MEMS technology. Dynamically tunable capacitance makes it possible to control frequency response more effectively, making such components critical in software-defined radios (SDRs) where real-time frequency agility is critical. The use of tunable capacitors in these systems allows for the filtering of targeted signals, which is essential for isolation of desired frequencies from unwanted interference. For instance, application of MEMS actuators in designing tunable filters has been proven to be promising, with the devices having operation in the 4–15 GHz range without losing any significant amount and with high quality factor and hence a tremendous achievement in tunable capacitor technology (Beacken et al., 2018). It is now possible to create tunable capacitors with advanced designs for enhanced performance because of the developments in technology. Another interesting innovation is the use of split ring resonators (SRRs) and integration with microelectromechanical systems (MEMS) cantilever arrays for electrostatic frequency tuning of resonance. This is scalable as it can be fabricated monolithically and therefore diminishes the losses incurred from external components and improves efficiency as a whole. With control over the separation and geometry of the beams in these cantilevers, engineers are capable of having very high fine tuning on the resonant frequency, showing a tuning range of up to 1 GHz as a function of voltage changes across the beams. These highly advanced designs not only allow for the maximum achievable capacitance but also render tunable capacitors essential elements in state-of-the-art communication technologies (Collins et al., 2011). As MEMS develops further, tunable capacitors will find greater and greater application in fields outside RF filtering. Their applicability is very broad and extends even to areas like microfluidics, optical switching, and microgrippers used in robots, where precision control is needed. By facilitating rapid capacitance variations, these devices enhance the functioning and versatility of systems, which is another consideration towards innovations in autonomous robotics and drug delivery systems. Enhanced performance of tunable capacitors, with the escalation in MEMS technology, can revolutionize a number of industries and necessitates further research and development towards optimizing their functionality and application. Thus, the union of MEMS actuators with tunable capacitors highlights a future-oriented path of technological innovation across a wide spectrum of industries (Figure 8.7).

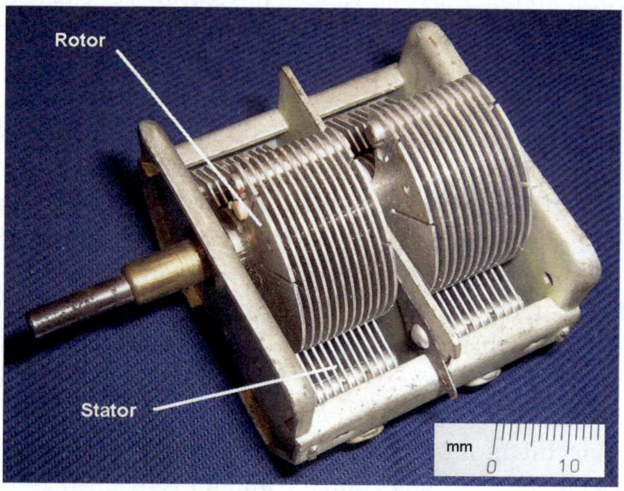

Figure 8.7. Tunable capacitors.

Source: https://upload.wikimedia.org/wikipedia/commons/6/6c/Variable_Capacitor.jpg.

8.3.1. Design and Operation of MEMS Tunable Capacitors

Advances in microelectromechanical systems (MEMS) technology have provided new applications, particularly for tunable capacitors. Such products operate on the principle of changing capacitance through mechanical displacement and thus provide control over electrical values with high precision. MEMS tunable capacitors are generally defined by intricate geometry that allows large variation in output through an infinitesimal input energy. These designs enable engineers to construct capacitors with large tuning ranges and high-quality factors, important in application such as RF communication, where the capacity to support varying signal frequencies is important. The utilization of MEMS tunable capacitors in devices such as software-defined radios (SDR) shows the potential of the devices in enhancing flexibility and efficiency in operations. One of the interesting developments in this area is the use of high permittivity dielectric materials to realize variable capacitors. Such materials lend themselves to very high capacitance ratios as evidenced by designs that are capable of realizing a capacitance ratio in excess of 40 for devices of size 100 μm × 25 μm. Using finite element simulations coupled with variational analysis, it has been proven through research that such configurations are effective in realizing continuously variable capacitance at low bias voltage levels. These technologies not only enable more compact designs but also enhance the performance of the devices in the case of high-frequency applications, i.e., RF communication systems. The quick tunability and low power consumption of these devices make them an integral part in the development of flexible electronic architectures. Incorporating MEMS tunable capacitors in larger system designs is particularly useful since they enable improved signal integrity and elimination of unwanted frequencies. For instance, as part of RF application, such as for extreme throw MEMS actuators, tunable capacitors can be the prime factor to enhance the dynamic range for software-defined radios. Through enabling fine control of narrow band filters, the capacitors are able to avoid sensitive components from saturation while also being in a position to detect weak signals when there exists significant interference. The unique capabilities of MEMS tunable capacitors not only emphasize their importance in contemporary electronic design but also point to their possibility as embedded solutions in a vast variety of applications from telecommunications to new areas like microfluidics and drug delivery (Figure 8.8).

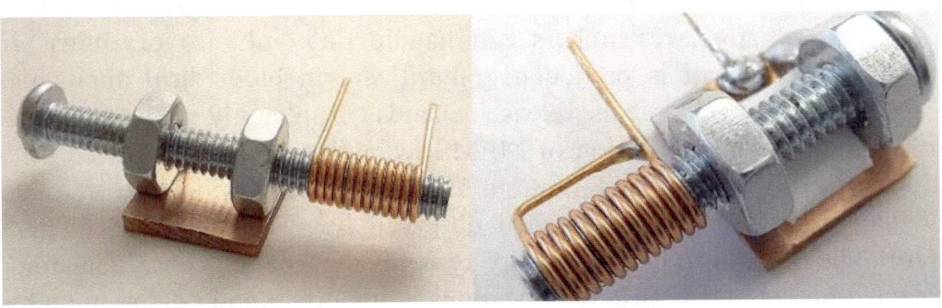

Figure 8.8. The design of MEMS tunable capacitors often involves intricate geometry that allows for significant output variation with minimal input energy.

Source: https://www.electronicproducts.com/wp-content/uploads/passive-components-capacitors-screw-variable-capacitor.jpg?fit=800%2C250.

8.3.2. Applications in Frequency Tuning and Filtering

The versatility of MEMS actuators extends to the domain of frequency filtering and tuning, where their miniature components facilitate breathtaking advances in optical and radio frequency technology. This versatility allows devices such as micromirrors and optical switches to achieve top performance within dynamic environments. By being incorporated into micro-opto-electro-mechanical systems (MOEMS) and digital micromirror devices (DMD), MEMS actuators bring enhanced functions such as wavelength selectivity and light modulation, and they are now a part of telecommunications and imaging systems. Through manipulation of light in controlled and versatile ways, these actuators support the build-up of adaptive optical networks, and they confirm the crucial role played by MEMS technologies in improving bandwidth efficiency and communication reliability in various electronic devices. Radio frequency uses show the effectiveness of MEMS switches and tunable capacitors since they suitably apply for frequency tuning and filtering processes. Electronic switching of RF MEMS switches to alter circuit paths makes them suitable for flexible reconfiguration in communication systems to achieve superior frequency selectivity. Apart from this, tunable capacitors help in dynamic impedance matching and filtering, which are extremely important in signal integrity management over large frequency bands. Above all, these MEMS devices consume very low power with high performance, which makes them an imperative in portable communication technologies. The application of such MEMS actuators not only ensures highest operation efficiency but also the total performance of devices that rely on precise frequency control. The use of MEMS in frequency tuning and filtering is not confined to communication technologies but finds immense applications in the microfluidics and drug delivery systems sectors. With the application of microrobotics and microgrippers actuated by MEMS actuators, researchers can handle tiny volumes of fluids with precision. This aspect is particularly significant in biomedical applications because therapeutic chemicals' release depends considerably on patient results. Furthermore, the incorporation of MEMS technology in such systems brings real-time feedback and compensation, thereby allowing optimal performance in dynamic situations. As has been observed by way of advancements in microfluidic devices and drug delivery systems, the role of MEMS actuators in frequency tuning and filtering applications is important and possessing a high scope for enhanced efficiency in life-critical applications based on fine-scale manipulation of fluids and frequencies.

8.3.3. Benefits of Tunable Capacitors in RF Circuits

Growing demand for more flexible and efficient components to be applied in radio frequency (RF) applications has driven extensive innovation in micromachined technologies, more significantly, in tunable capacitors. Such capacitors, forming a component of RF circuits, offer dynamic capacitance values control for enabling the improvement of signal processing and transmission. As the telecommunication system keeps evolving, the ability to guarantee signal integrity while improving operating flexibility is imperative. Tunable capacitors allow circuit performance to be tuned in real time by designers, thereby making them suitable for a range of frequencies and purposes, which plays a key role in the complicated situation of communication systems in the present day. Such flexibility, in turn, expresses improved overall performance and dependability in RF designs and meets the stringent requirements imposed by new wireless technologies (Casha et al., 2007).

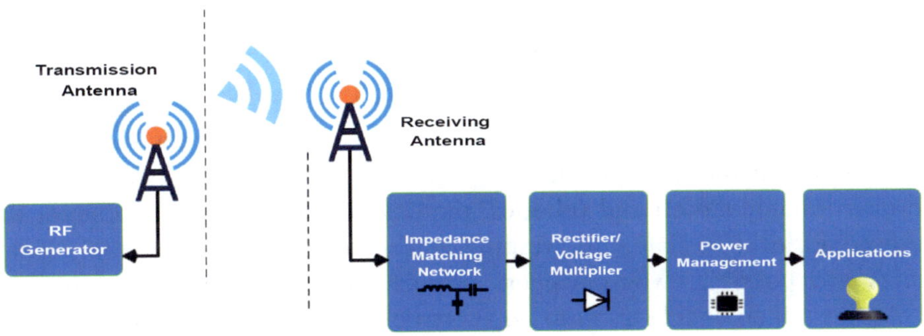

Figure 8.9. The demand for more versatile and efficient components in radio frequency (RF) applications has led to significant advancements in micromachined technologies, particularly regarding tunable capacitors.

Source: https://pub.mdpi-res.com/sensors/sensors-22-04144/article_deploy/html/images/sensors-22-04144-g001.png?1653901295.

Furthermore, tunable capacitors are very useful in miniaturization, a significant aspect to keep in mind for electronic devices nowadays. Coupled with their integration into micro-electro-mechanical systems (MEMS), engineers are able to create tiny, efficient circuits without taking a toll on performance. For instance, MEMS tunable capacitors possess high-quality factors and large tuning ranges, which are more efficient compared to traditional solid-state components of reduced size. This miniaturized idea is particularly useful to handheld consumer electronics, where limited space environments demand innovations that maximize functionality and reduce size. The ability to deliver high performance with small size renders MEMS-based tunable capacitors a necessity in the evolution of contemporary RF circuitry. In addition to their flexibility and space utilization, integrating tunable capacitors has a tendency to provide improved power efficiency in RF applications. As the significance of energy-efficient design gains popularity, tunable capacitors enhance systems' performance without affecting energy costs in general. Their unconventional actuation schemes enable them to offer excellent tight capacitance control at the cost of slight power loss, and thus they are particularly well-suited to battery-powered applications. This efficiency becomes increasingly important as consumer demand is increasing for longer battery life for handheld devices without a sacrifice to high-performance capabilities. Accordingly, with the use of tunable capacitors, not only is the performance of RF circuits improved, but electronic solutions also become more eco-friendly in a world that increasingly treasures energy efficiency (Casha et al., 2007) (Figure 8.9).

8.4. MICROGRIPPERS

The evolution of microgrippers is a significant achievement in the field of MEMS actuator technology, demonstrating the capability of microtechnology to revolutionize many applications. As accuracy is critical in applications like micro-manipulation and robotics, the devices are made to operate at the microscopic level so that safe gripping and manipulation of extremely small objects are possible. Microgrippers are usually fabricated using intricate geometries which enable them to be versatile to work on different shapes and sizes of objects being handled. This is a vital function that is greatly needed in facilities such as biological labs, where cells or small tissue samples have to be handled with exactness and gentleness—qualities that microgrippers possess alone. Because of these, applications extend outside the laboratory setting, reaching into telecommunications, pharmacy, and even biomedical engineering breakthroughs. Moreover, the integration of microgrippers into larger microsystems makes them more adaptable and versatile. The devices are likely to require integration of other technologies in order to accommodate various mechanical, optical, and electronic subsystems, resulting in an efficient and small-sized overall system. The limitation is the compatibility of processes used in the production of different components, which could complicate the assembly of these microsystems—a limitation defined in recent research activities. This is where newly emerging methods of microassembly, including self-assembly and robot assembly, come into play. As robotic technology becomes increasingly advanced, the precision and efficiency of integrating microgrippers into full MEMS systems will undoubtedly increase exponentially, providing new applications and innovations that leverage the best of microrobotics and micro-manipulation technologies. Microgrippers are transforming the art of minimally invasive procedures and drug targeting delivery systems in therapeutic applications. Their capability for controlled microscale action enables not just the manipulation of tiny medical devices but also the targeted delivery of the drug to specific areas inside the body with less side effect and greater therapeutic efficacy. Microgripper implementation in microfluidic devices complements their purpose, allowing the controlled manipulation of liquids at micro-levels, required for therapeutic and diagnostic applications. Additionally, microgripper design has been found to be adaptable to improve their functionality in handling biological materials, such that they become more applicable in medical use (Biancucci et al., 2018). As a result, the uses of microgrippers continue to expand, proving their worth in advancing both the competence and reliability of MEMS actuators in practical utilization (Figure 8.10).

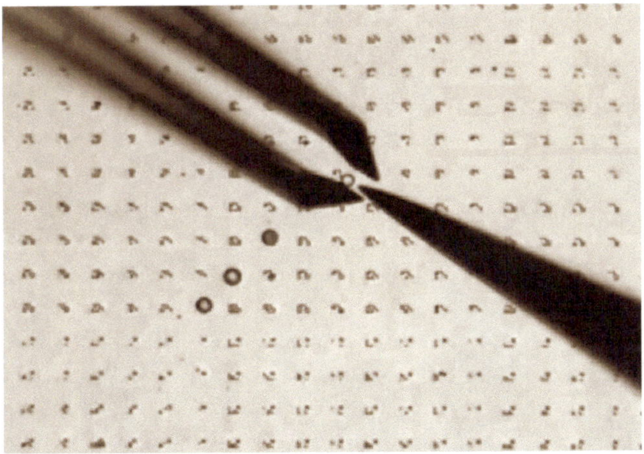

Figure 8.10. Microgrippers.

Source: https://imina.ch/storage/app/uploads/public/5f9/a92/c0f/5f9a92c0f1816938889879.jpg.

8.4.1. Design Principles of MEMS Microgrippers

MEMS advancement has played a key role in breakthrough innovations in various areas, especially where microgrippers are used. They are founded on principles of precision and flexibility that are crucial for their functionality in sensitive operations like micro-manipulation and assembly in medical equipment. One of the main design principles is the use of thermal micro-actuators, which provide linear movement with minimal handling force. By using materials like SU-8 polymer, microgrippers can be designed that are not just light in weight but are also able to perform well at low actuation voltages. The use of finite element simulations has become central in the optimization of these devices so that their electro-thermal mechanisms can be compared and improved upon. This meticulous focus underscores the interaction between actuator design and material selection in achieving intended functionality in MEMS devices (Eftime et al., 2008). One of the principal demands for MEMS microgripper design is the aspect of versatility and configurations as determined by the user. Microgrippers are expected to be designed to handle different shapes and sizes of objects, expanding their use in diverse areas, such as precision manufacturing and biotechnology. Parameters such as gripping force, range of motion, and speed are carefully set to the application at hand.

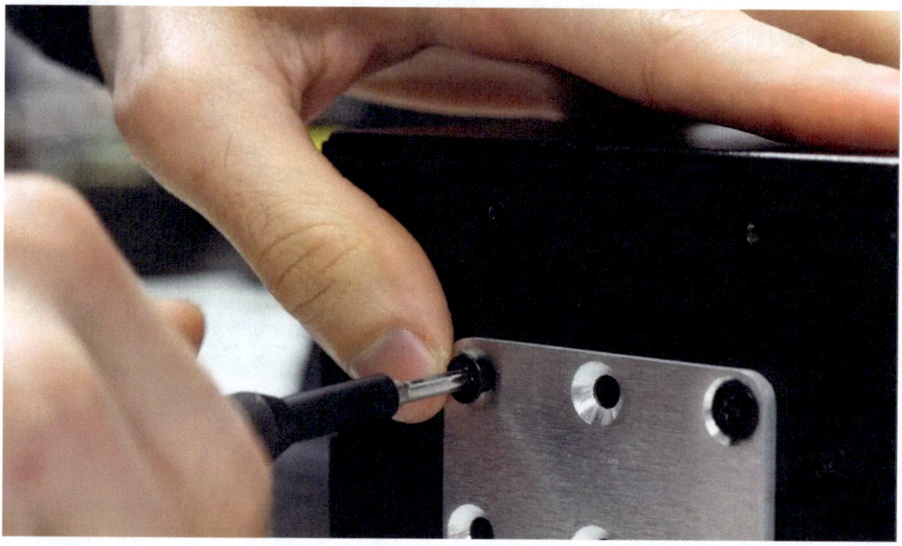

Figure 8.11. SU-8 polymer.

Source: https://www.elveflow.com/wp-content/uploads/2024/04/assembly-kit-1.webp.

Developments in electro-mechanical actuation have enabled control over these parameters to become more advanced, allowing designers to design dedicated types of actuators for specific operating needs. Furthermore, as the task becomes increasingly complex, so too does the requirement for advanced levels of feedback systems monitoring and adjusting performance in real-time. Hence, microgripper design not only reflects the accuracy needed in micro-object handling but is also a demonstration of the broader automation and sophistication of trends in MEMS technology (Eftime et al., 2008). The incorporation of user-centered design principles has also emerged as an important consideration in the development of MEMS microgrippers. Teachers and institutions increasingly pay attention to making learning experiences tailored to the different learning styles of engineering students. Instruments like Gardner's multiple intelligence inventory are employed to identify special strengths, and these can be employed to inform microgripper system design. Familiarity with the different methods can be translated to improved motivation and commitment by students when they engage with physical projects on the basis of MEMS technologies. Further, critical discourse towards the integration of practical application and theoretical models provides a complete understanding of microgripper design principles. Such convergence among application, innovation, and education assists in maximizing overall development and effectiveness of MEMS microgrippers in resolving prevailing engineering challenges (Figure 8.11).

8.4.2. Applications in Biomedical Fields and Micro-Manipulation

The use of MEMS (Micro-Electro-Mechanical Systems) actuators within bioapplications has introduced a whole new level of precision and efficiency into the healthcare sector. Microgrippers and microrobots are merely two illustrations of the way in which MEMS technology makes sophisticated action at the microlevel possible, facilitating the precise manipulation of biological cells and tissue handling. This is especially significant within surgical procedures where injury to the surrounding tissue needs to be avoided. Apart from that, R&D in the field of microfluidics has also significantly improved drug delivery precisely where needed, converting conventional drug delivery systems into highly responsive and efficient ones. By raising the accuracy and control of these devices, MEMS actuators not only enhance the efficacy of treatment but also minimize side effects, thereby exhibiting their break-through future in modern medicine, as underscored in the objectives of microsystems technology to enhance applications and solve common issues in the discipline (Fluitman et al., 1996). Furthermore, MEMS technology enhances imaging and diagnostic capabilities in biomedical applications for better patient treatment and management. MEMS optical switches like Digital Micromirror Devices (DMD) play a vital role in improving image acquisition for diagnostic imaging modalities like optical coherence tomography (OCT). Using high-speed optical switching accuracy, such devices improve the quality of eye image data needed for visual recognition operations. Experiments have indicated that advanced segmentation methods, for instance, the Circular Hough transform, optimize image processing despite environmental interference that tends to degrade image quality (Ahmad et al., 2018). The robustness of these advanced imaging tools underscores the significance of MEMS actuators to facilitate non-invasive diagnostic examination and enhance the accuracy of biomedical imaging, further positioning them at the center of modern healthcare technology. MEMS applications in drug delivery systems are particularly crucial, merging technology and therapeutic strategy for enhanced patient care. MEMS actuators are utilized by drug delivery systems to control fluid flows on micro-scales in order to permit medication targeting, which proves extremely useful in conditions requiring differentiated dosages, such as chemotherapy or alleviation of pain. In addition, MEMS-based microfluidic devices are critical for point-of-care diagnosis, facilitating rapid analysis and testing with a small sample size. It not only simplifies the diagnostic process but also facilitates timely treatment interventions. As the medical landscape evolves towards personal medicine, the role of MEMS actuators in enhancing drug delivery efficiency and enabling rapid response within a clinical setting is also set to increase manyfold, establishing new standards in patient care and treatment protocols (Figure 8.12).

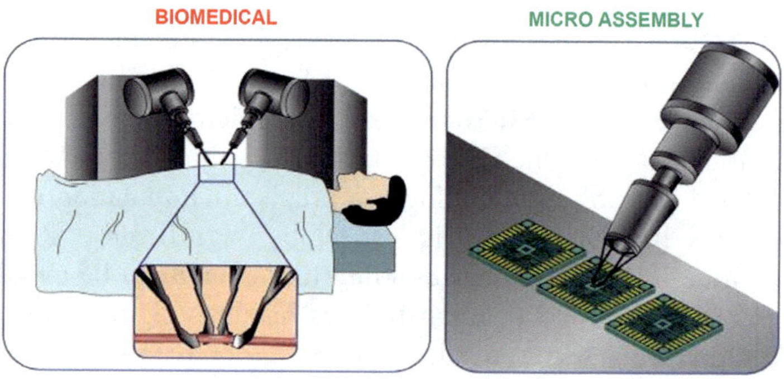

Figure 8.12. Devices such as microgrippers and microrobots exemplify how MEMS technology facilitates intricate operations at a microscale, allowing for targeted manipulation of biological cells and tissues.

Source: https://ars.els-cdn.com/content/image/1-s2.0-S0924424724007945-gr1.jpg.

8.4.3. Challenges in the Development of Microgrippers

The sophisticated technology landscape of microgrippers is a significant challenge and a key factor in the evolution of MEMS application development. With heightened miniaturization motivations, the intricacy of fabricating microgrippers capable of precise handling of tiny objects becomes vivid. Compound to these is the restraint on actuation techniques. The majority of traditional actuation techniques, such as piezoelectric and electromagnetic actuators, have limited control and integration capabilities with the micro-environment. These limitations reduce the operational efficiency needed for uses in biomedical applications, microrobotics, and microfluidics, with far-reaching consequences on the operation of these devices in actual use. New and reliable actuation methods are therefore a focus in research and development to enable microgrippers to function at maximum efficiency under controlled conditions with accuracy and reliability. Microgrippers' predictability and reliability are another setback in the development of microgrippers. According to literature, the susceptibility of microgrippers to failure mechanisms cannot be overstated, as these failures have a direct impact on their performance and working life. Advanced models of degradation are thus crucial in establishing the health state as well as predicting the remaining useful life of microgrippers so that they can operate reliably in demanding applications. Moreover, the intrinsically limited capability of MEMS technology makes a thorough understanding of the behavior and performance attributes necessary in facilitating effective fault-prognostic approaches. This need for robust health monitoring and evaluation processes emphasizes the need to integrate advanced technologies, such as Prognostic and Health Management (PHM) techniques, into microgripper development and functioning. These issues of reliability will eventually be resolved through enhanced operation fidelity in real-world applications. Lastly, integrating microgrippers into existing technologies involves many cross-disciplinary challenges, particularly in adapting them to different applications. For example, while planning their use in biomedical applications, it becomes essential to address user safety and biocompatibility. Since microgrippers are typically used for sensitive operations, e.g., tissue manipulation and drug delivery, even a slight error in their design can result in undesirable effects. On top of that, microfluidics integration requires smooth interaction between elements, necessitating increased accuracy and careful planning. The education models under which microgrippers are designed must evolve to include comprehensive preparation in all areas to prepare for such challenges. Engaging students and researchers in hands-on projects related to the needs of new technologies is at the heart of inducing innovation. This dynamic approach ensures that future professionals in the field are ready to handle the complexities of developing functional, reliable, and secure microgrippers for their intended applications (Biancucci et al., 2018).

8.5. MICROROBOTICS

Advancement in microrobotics research has made it possible to develop solutions with innovative methods to tackle micro-manipulation task complexity. One of the stunning advancements is in the development of digital microrobots that employ bistable modules to enhance the performance without having to undergo intricate assembly processes. This revolution applies microfabrication techniques to fabricate monolithic structures, avoiding the drawbacks in conventional microrobots like nonlinear control and reliance on outside energy. In doing so, the microrobots become able to maintain stable positions with minimal energy consumption, completely revolutionizing the area of microrobotics. The extent of such technology is truly limitless, with endless avenues of applications ranging from microassembly to biomedical interventions, almost ushering in avenues for the exploitation of capabilities on the microscale level (Chen et al., 2008). Alongside these advances, the microsystems technology (MST) discipline advances with a rich framework of applications and technological innovations. MST is a broad collection of topics that reflects the multidisciplinary nature of systems for controlling matter at the microscopic level. This breadth requires an orderly approach to the definition of MST components and objectives enabling researchers to develop standardized modules and provide seamless interconnection.

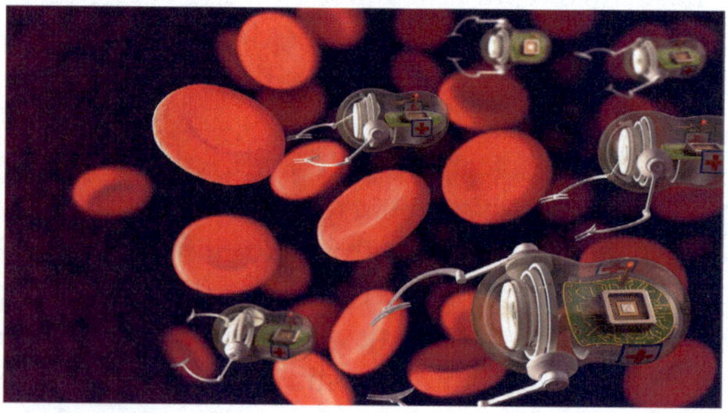

Figure 8.13. Microrobotics.

Source: https://builtin.com/sites/www.builtin.com/files/2022-08/robot-blood-cells-robotics-microbiotics_0.png.

By organizing the field upon elite applications—microgrippers in precision gripping and drug delivery systems, for instance—scientists are better placed to demarcate sub-territories for more research and standardization. In-depth knowledge of MST, as a result, fosters wider co-operation among the community overall, ultimately contributing to our general knowledge in microrobotics (Fluitman et al., 1996). Microrobotics extends its influence on other fields from optical communications to strict healthcare solutions. One of the significant uses is in microgrippers which are required to execute precision and control on a micro-scale level. These devices use MEMS actuators to move components on a micro-scale or biological samples with precision, marking significant interactions in drug delivery and diagnostics applications. In addition, MEMS technology also enhances the functionality of optical components, such as micromirrors and optical switches, to be a critical component of telecommunications and imaging systems. With the promise-full capability of microrobotics yet to be explored, the ability to design dynamic environment-sensitive systems also strengthens its importance in the realization of next-generation applications that involve complex micromanipulation capabilities. The continuous innovation and research in this sector will likely be responsible for catalyzing revolutionary impacts on various industries (Figure 8.13).

8.5.1. Overview of MEMS in Microrobotic Systems

MEMS technology has driven advancements in microrobotics systems, particularly in areas where accuracy is required at the micro and nanoscale. The application of MEMS actuators in the development of devices such as micromirrors and optical switches plays a central role to digital micro-mirror devices (DMD) and micro-opto-electro-mechanical systems (MOEMS). These devices are particularly important in most applications, including sensors and imaging, since they allow high-speed light modulation with low response time. RF MEMS switches and tunable capacitors also make it possible to achieve significant improvements in communication systems to allow efficient routing of signals and improved adaptability to changing environments. Thus, MEMS technology not only boosts performance through miniaturization but also meets the growing demands for high-speed and reliable components in microrobotic structures for different applications ranging from telecommunications to medical devices. In microrobotics, MEMS actuators play a key role in the designing and functioning of microgrippers, devices that have been custom designed for handling microscopic objects. These microgrippers take advantage of the precision of MEMS technology to clamp, push, and transfer objects as minute as cells or intricate microstructures. Therefore, they are highly useful in medical processes where micro level precision has to be controlled, like in targeted drug delivery and cellular manipulation. Despite the challenges posed by environmental factors, such as thermal fluctuations and external vibrations that interfere with performance, effective solutions have been achieved. Scientists continue to refine control schemes that employ various actuation techniques, such as piezoelectric materials, which support high-precision positioning, thereby enhancing the performance and robustness of microrobots in real-world applications. In addition, MEMS technology coupled with microfluidics is the best example of its effectiveness and application in drug delivery devices. Microfluidic devices using MEMS technology provide micron-level control over fluid volumes, which is relevant when delivering drugs in targeted therapy or conducting lab-on-a-chip analysis. This micrometer-scale control over the transport of fluids helps in more accurate dosing of drugs and improved patient results. In addition, MEMS-actuated microrobots have the capability to move within intricate biological spaces, releasing therapeutics with high specificity at targeted locations, thereby maximizing their therapeutic efficacy while minimizing side effects. The integration of MEMS technology and microrobotics not only brings medical applications to a revolutionary level but also provides the foundation for future development and research, as reflected by the ever-growing number of studies focusing on nanoscale manipulation and novel drug delivery methods, demonstrating its pivotal role in the development of medical technologies, as in.

8.5.2. Applications in Surgery and Precision Tasks

The use of MEMS actuators in surgery has vastly transformed the arena of precision medicine. The miniature devices are able to function with high precision, and hence surgeons are now able to perform complex surgeries, which were previously deemed not possible. The advancements in micromirrors and optical switches, particularly those in digital micromirror devices (DMD) and micro-opto-electromechanical systems (MOEMS), enable real-time imaging and diagnostics during surgery. Moreover, the technologies also facilitate enhanced visualization of target tissue, bettering operating room decision-making. As the demand for minimally invasive surgical procedures grows, MEMS technology allows clinicians to deliver precise interventions while inflicting less trauma to other tissues. This shift to precision-practice emphasizes the role of MEMS actuators in today's surgical models where efficiency and accuracy are paramount. The use of MEMS actuators in microgrippers and microrobotics has also improved the precision and control of surgery. The machines, utilized to grasp and manipulate biological tissues, allow accurate procedures that can be adapted to various surgical needs. For instance, the application of microgrippers in minimally invasive surgery has the capacity to reduce patients' recovery times as they allow accurate dissection and suturing within small anatomical spaces. MEMS actuators enable the development of sophisticated robotic systems employed to assist surgeons with intricate procedures. Moreover, such technology aligns with the evolving focus on automating surgery where human tremors and lack of precision compromise results. The use of microgrippers and microrobotics in modern surgery demonstrates MEMS technology's capacity to change surgical practice and improve overall patient care. A further remarkable application of MEMS actuators is in microfluidics and drug delivery. MEMS technology is used to control the fluid dynamics at the micro-level with high precision, which is critical in developing effective drug delivery systems that dispense therapeutic agents in a controlled and regulated fashion. These capabilities not only enhance the efficacy of drugs but also minimize side effects by enabling drugs to reach their targeted areas of the body. In addition, the integration of tactile sensors via microfabrication into these devices has proven very helpful, particularly in the measurement of contact properties like pressure and force, which are of high concern during drug delivery processes. By making these functions simpler, MEMS-led technologies make drug delivery revolutionary, ultimately translating into better patient-specific therapeutic regimens and more tailored healthcare solutions (Guglielmelli et al., 2014). Thus, the new applications of MEMS actuators continue to revolutionize the future of surgery and precise medical operations, moving towards a future where there is emphasis on precision and patient-centered care (Figure 8.14).

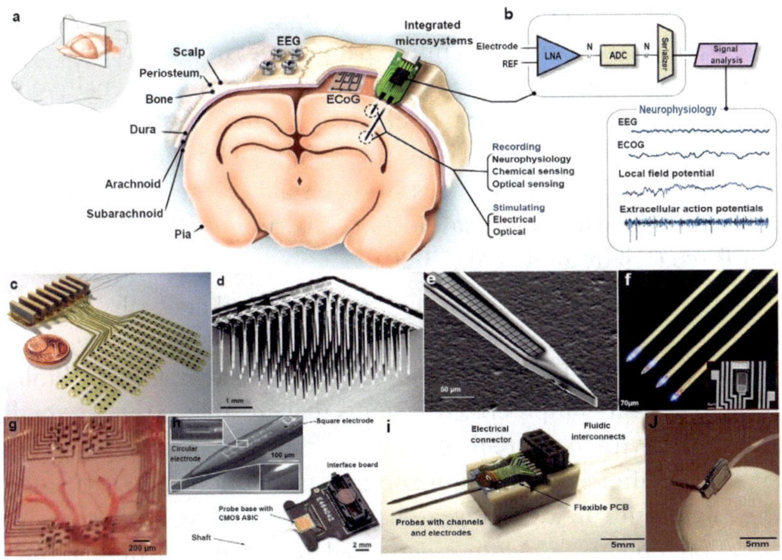

Figure 8.14. The integration of MEMS actuators in surgical applications has significantly transformed the landscape of precision medicine.

Source: https://media.springernature.com/lw1200/springer-static/image/art%3A10.1038%2Fmicronano.2016.66/MediaObjects/41378_2017_Article_BFmicronano201666_Fig1_HTML.jpg.

8.5.3. Future Prospects for Microrobotics using MEMS Technology

Miniaturization has expedited the evolution of many applications across many disciplines, particularly in microrobotics with Microelectromechanical Systems (MEMS) technology. It enables the creation of multifunctional complex systems that are able to perform multiple functions such as sensing, actuation, and processing all at microscopic levels. Optical switches and micromirrors, which play an essential role in applications for digital light processing (DLP) and micro-opto-electromechanical systems (MOEMS), exemplify the strength of MEMS. The ability to combine these components produces sophisticated optical systems with rapid modulation and control, a central feature of advancements like adaptive optics and telecommunications. Moreover, RF MEMS switch research transformed communications technology by making tunable capacitors and switches with low loss possible, increasing reliability and performance (Fluitman et al., 1996). This combination of MEMS technology and microrobotics is driving the boundaries of micro-scale application ever further forward. As technologies advance further, the incorporation of MEMS technology into microrobotics and microgrippers opens up new applications for rehabilitation, minimally invasive surgical procedures, and sensitive material handling. Precision is the essential element to these sensitive applications provided by MEMS actuators, enabling operations to be conducted with utmost care and precision. Modern advances in materials science have enabled more responsive, more agile micro-actuators that can drive small objects with varying degrees of force (Kiziroglou et al., 2020).

The combination of these technologies not only enhances their functionality but also enables sophisticated, automated systems for uses like drug delivery in microfluidic processes. These innovations are the integration of biology and engineering to deliver computerized systems that are able to release medicine in focused positions with high precision, minimizing side effects while improving overall treatment efficiency. The future prospects for microrobotics using MEMS technology are promising, with researchers always finding innovative materials and fabrication processes. Motion-activated material and higher integration methods are developing highly effective actuators that significantly magnify motion in the microscale range. Flexures and other mechanical amplification concepts are breakthroughs that enable actuators to provide remarkable stroke capabilities relative to their size (Kiziroglou et al., 2020). Potential uses of these innovations are varied, ranging from micro-vehicles to energy harvesting platforms, potentially transforming medical and environmental monitoring sectors. Furthermore, the ongoing development of miniaturization methods signals that the next generation of MEMS-based microrobots shall be capable of operating independently in various environments, providing valuable insights and interventions wherein human intervention may be unfeasible or infeasible. This research and development trend accentuates the revolutionary potential of MEMS technology in determining the future of microrobotics.

8.6. CONCLUSION

The exploration of MEMS actuators through disparate applications reveals a multifaceted encounter of function and technology. As the advances of microsystems technology (MST) keep changing, the field has been amazing regarding diversification across industries. For example, micromirrors and optical switches, such as digital micromirror devices (DMD) and micro-opto-electro-mechanical systems (MOEMS), have proven pivotal in the enhancement of imaging systems and telecommunication. This growth in applications of MEMS is not just a reflection of technological possibilities but also of the broader ramifications across sectors ranging from medicine to computing. Furthermore, the research and development efforts in progress point towards a hopeful future for MST such that standardization and modularization can lead to even more innovations and high-order system integrations, as noted within the vast corpus of literature in the field (Fluitman et al., 1996).

As viewed in the light of present developments, the specific case of closed-loop control systems for XY micro positioning stages represents the precision and reactivity common to MEMS actuators. Such systems, created with parallel-kinematic mechanisms both in concept and realization, not only illustrate the intricacies of the engineering that goes into MEMS devices but also highlight their prospect to be used in techniques such as microfluidics as well as drug delivery. The performance characteristics of such devices, whose uncertainty of position is given by a standard deviation of 5.24 nm and closed-loop bandwidth that accommodates high rates of response, highlight the effectiveness of MEMS in executing high-resolution actuation. These characteristics are crucial in applications requiring exact control yet again highlighting the importance of MEMS actuators in technology advancements (Koo et al., 2014). In summary, based on the analysis of MEMS actuators, one reaches the same conclusion that these technologies are promising future technologies. As applications expand and diversify, the integration of MEMS technology in numerous applications suggests a future with more functionality and efficiency.

The general theme is palpable: the versatility and efficiency of MEMS actuators not only invite innovation in current applications but also propose new horizons waiting to be explored. Collaborative efforts, standardization, and investment in research will enable participants in the MST ecosystem to realize the full potential of MEMS technology, and with it, revolutionary innovations that will transform industries in the coming years. This vast understanding of MEMS actuators once again emphasizes their role in current applications as well as future innovations, making them one of the pillars of 21st-century future technologies.

8.6.1. Summary of Key Applications of MEMS Actuators

The application of MEMS actuators has revolutionized numerous fields by enabling miniaturization and precision in high-risk applications. One of the most significant fields that have been impacted by using MEMS actuators is in optical technology. Micro-Opto-Electromechanical Systems (MOEMS) and Digital Micromirror Devices (DMD) are some of the devices utilizing micromirrors to manipulate light—offers enhanced capabilities for display technology and projectors. Through the utilization of such MEMS actuators, optical switches can respond at fast rates and with excellent fidelity, leading to improved performance of telecommunications and imaging systems. Manipulation of light at the microscale allows for complex optical routing and signal processing, defining the revolutionary role of MEMS technology in advancing optical communication systems and lighting applications. This integration of accurate parts has been a great step forward in terms of performance and efficiency in these technology fields. Besides optical applications, MEMS actuators have found widespread use in radio frequency (RF) systems. RF MEMS switches and reconfigurable capacitors have been especially remarkable for their ability to control signal integrity and routing in wireless communication. These chips offer low insertion loss, high isolation, and reduced voltage levels compared to traditional semiconductor technology. The swift and predictable switching offered by MEMS technology enables dynamic reconfiguration of RF circuits and thus more efficient use of the spectrum and improved overall performance of mobile device and base stations. This is an incredibly significant application because faster and more efficient communication systems are required now more than ever before. By providing higher performance attributes in the form of higher-quality signal processing, MEMS actuators play a crucial part in the modern wireless technology innovation. Furthermore, MEMS actuators are at the heart of microfluidic and drug delivery system applications. Microgrippers and microrobotics have provided new opportunities for microscale manipulation necessary for biomedical applications. Employing MEMS actuators in microfluidic systems enables accurate control over the movement of fluids, enabling one to perform different biochemical reactions in a controlled manner. This function is highly useful in drug delivery systems where focused therapy and dosage control are essential. Automatic optimization of fluid handling processes enhances the progress of personalized medicine and diagnostics. The ongoing R&D in such fields is a testament to the significant role played by MEMS actuators in developing state-of-the-art medical solutions that are efficient and effective in treating modern medical issues (Cherta et al., 2020; Kopylov et al., 2015).

8.6.2. Future Trends in MEMS Technology

Micro-Electro-Mechanical Systems (MEMS) technology continues to remain at the core of numerous applications ranging from telecommunications to medical equipment. With the growing demand for miniaturization and high-precision applications, future indicators are towards integrating MEMS actuators with advanced materials to make them rugged and functional. Progress in materials science, particularly the synthesis of polymers and ceramics with excellent piezoelectric characteristics, will push MEMS actuators to areas heretofore unknown. This material science will not only improve the energy efficiency of devices but also increase their ruggedness and performance under various environmental conditions. Furthermore, standardization efforts in line with the extensive applications in various industries need to be highlighted, as Fluitman et al. (1996) observed, in order to create opportunities for more integrated and standardized MEMS actuators to promote widespread utilization and technology incorporation in all industries. As the applications of MEMS actuators are on the rise, new evolving electrotechnology trends need to transform their mode of operation. The combination of machine learning and artificial intelligence in MEMS systems will make it possible for autonomous operation that can be tuned in response to immediate changes in the environment. Microgrippers and microrobotics systems, for instance, which are increasingly utilized in minimally invasive surgeries, will be able to dynamically adapt to the state they are in within biological environments.

This adaptation may improve accuracy and reduce procedural risk significantly. In addition, advances in drug delivery systems and microfluidics, on the basis of MEMS technology developments, will continue to enable targeted therapy, hence maximizing patient results. This sort of synergy of MEMS actuators with new computational technologies is the next development step for them and the platform for even more integrated health care technologies. Of greatest expansion potential is the field of MEMS optical technologies, such as micromirrors and optical switches.

All these factors are being increasingly applied in telecommunications for applications like Digital Micromirror Devices (DMD) and Micro-Opto-Electro-Mechanical Systems (MOEMS), which allow high-resolution display and sophisticated imaging systems. The trend towards faster, more efficient MEMS-based RF switches and tunable capacitors will also alter the communication system dynamics by enabling multi-frequency operations without making them large in size. Furthermore, the capacity of MEMS technology to seek alternatives for precision agriculture, environmental observation, and wearables places it at the forefront as an enabler in sustainable practice. This diversity is evidence of the disruptive nature of MEMS actuators that transcends many domains, indicating a stimulating trajectory that will redefine our methodology towards technology in the next couple of decades.

8.6.3. Final Thoughts on the Impact of MEMS Actuators on Various Industries

The application of MEMS actuators is much wider than current applications, revolutionizing industries with innovation in the telecommunications and medicine industries. With the creation of micromirrors and optical switches, MEMS technology improves data communications speed and efficiency, which is essential in the more connected world today. RF MEMS switches and tunable capacitors also reshaped the nature of RF systems as more flexible and power-efficient. Such versatility is particularly beneficial in applications with rapid frequency changes, such as modern telecommunications networks. Not only do such innovations lead to improved performance but also pave the way for innovative technologies that can effectively meet the demands of current and future technology environments. Consequently, the use of MEMS actuators for telecommunications is a savvy pointer to their extensive impact on numerous other areas. In healthcare, the application of MEMS actuators into devices has significantly enhanced patient care and treatment. Microrobotics and microgrippers, for instance, make minimally invasive surgery possible, which reduces recovery time and improves patient outcomes. With continued development of such advancements, their application in drug delivery systems is becoming more nuanced, providing targeted medication that can be delivered with accuracy in controlled dosages. This precision not only enhances the efficacy of treatments but also minimizes any potential side effects, proving MEMS actuators' contribution to customized medicine. In addition, having the capacity to monitor patients' conditions in real-time via these actuator-based technologies is setting new standards for preventive care, creating a new era of more interactive and dynamic medicine. The revolutionary function of MEMS in medical devices is hence transforming the healthcare industry. Besides consumer products and smart environments, MEMS actuator applications extend further, pointing to the versatility of design and operation. As highlighted in the latest studies, the combination of MEMS technology and smart materials and interactive devices is paving the way for new applications, which enhance users' interaction with everyday products. Smart fabric and smart material development improve customer experience across many industries but also create opportunities in assistive healthcare and renewable energy. Interactive consumer products like interactive textiles demonstrate that MEMS actuators have the potential to build environments responsive to specific needs, thereby allowing a sustainable society in the future. The application of technologies such as organic thin-film electronic devices (OFEDs) demonstrates the ongoing experimentation with MEMS potential, signaling the age of technology being organically integrated into daily life. Thus, the impact of MEMS actuators on these sectors illustrates their role as part and parcel drivers of innovation and functionality.

REFERENCES

1. Abdulla, S. M. C., Berenschot, E., de Boer, M. J., Dijkstra, M., et al. (2011). Tuning a racetrack ring resonator by an integrated dielectric MEMS cantilever. https://core.ac.uk/download/pdf/11481181.pdf (accessed on 14th October 2025).
2. Ahmad, A. S., Ahmar, A. S., Asmuni, H., Hishamudin, M., et al. (2018). Iris segmentation. https://core.ac.uk/download/19162806.pdf (accessed on 14th October 2025).
3. Barette, R., Grassi, E., Helmbrecht, M., Lanzoni, L., et al. (2017). MOEMS deformable mirror testing in cryo for future optical instrumentation. http://arxiv.org/abs/1704.03836 (accessed on 14th October 2025).
4. Bauer, B., Couderc, C., Crunteanu, A., Uttamchandani, D., Fabert, M., et al. (2012). Dual Q-switched laser outputs from a single lasing medium using an intracavity MEMS micromirror array. https://core.ac.uk/download/9323606.pdf (accessed on 14th October 2025).
5. Beacken, M. J., Bishop, D. J., Chang, J., Holyoak, T., et al. (2017). MEMS device with large out-of-plane actuation and low-resistance interconnect and methods of use. https://open.bu.edu/bitstream/2144/31164/1/US20170141755A1.pdf (accessed on 14th October 2025).
6. Ben Amar, A., Cao, H., & Kouki, A. B. (2015). Power approaches for implantable medical devices. https://core.ac.uk/download/237711608.pdf (accessed on 14th October 2025).
7. Berenschot, J. W., de Boer, J. H., Elwenspoek, M. C., Fazal, A., et al. (2005). Bond strength tests between silicon wafers and Duran tubes (fusion bonded fluidic interconnects). https://core.ac.uk/download/pdf/11459582.pdf (accessed on 14th October 2025).
8. Biancucci, G., Binni, A., Bonciani, G., Fioravanti, A., et al. (2018). Learning micromanipulation, part 1: An approach based on multidimensional ability inventories and text mining. https://core.ac.uk/download/188823063.pdf (accessed on 14th October 2025).
9. Bogaerts, W., Edinger, P., Errando-Herranz, C., Gylfason, K. B., et al. (2020). MEMS-enabled silicon photonic integrated devices and circuits. https://core.ac.uk/download/287939692.pdf (accessed on 14th October 2025).
10. Brand, O., Caillard, B., Debeda, H., Dufour, I., et al. (2013). Effect of hydrodynamic force on microcantilever vibrations: Applications to liquid-phase chemical sensing. https://core.ac.uk/download/213077385.pdf (accessed on 14th October 2025).

11. Brand, O., Dufour, I., Heinrich, S. M., Josse, F., et al. (2013). Resonant characteristics of rectangular hammerhead microcantilevers vibrating laterally in viscous liquid media. https://core.ac.uk/download/213075122.pdf (accessed on 14th October 2025).

12. Bright, V. M., Syms, R. R. A., Whitesides, G. M., Yeatman, E. M., et al. (2003). Surface tension-powered self-assembly of micro structures–The state-of-the-art. https://core.ac.uk (accessed on 14th October 2025).

13. Casha, O., Gatt, E., Grech, I., Micallef, A., et al. (2007). Utilization of microsystems technology in radio frequency and microwave applications. https://core.ac.uk/download/83022043.pdf (accessed on 14th October 2025).

14. Chaillet, N., Gauthier, M., Lutz, P., Régnier, S., et al. (2009). Recent advances in the study of micro/nano robotics in France.

15. Chen, D. W., El-Fatatry, F., Wardle, F., Gabler, H., Huang, W., Ikawa, M., et al. (2005). Design of ultraprecision machine tools with application to manufacturing of miniature and micro components. https://core.ac.uk/download/334084.pdf (accessed on 14th October 2025).

16. Chen, Q., Haddab, Y., & Lutz, P. (2008). Digital micro robotics based on bistable modules: Design of compliant bistable structures.

17. Cherta, A., Gomez Vargas, D., Pedraza, R. C., Pozo Montero, C., et al. (2020). Damage identification in structural health monitoring: A brief review from its implementation to the use of data-driven applications. https://core.ac.uk/download/288625955.pdf (accessed on 14th October 2025).

18. Choi, H., Hadipoespito, G. W., Li, X., Ning, Y., et al. (2003). Digital micromirror device based micro stereolithography for micro structures of transparent photopolymer and nanocomposites. https://core.ac.uk/download/238846504.pdf (accessed on 14th October 2025).

19. Choo, H., Kim, J., Lin, L., Muller, R., et al. (2006). Microfabricated torsional actuators using self-aligned plastic deformation of silicon. https://core.ac.uk/download/33121701.pdf (accessed on 14th October 2025).

20. Ciuprina, G., Isvoranu, D., & Sorohan, S. (2014). Aerodynamic behavior of the bridge of a capacitive RF MEMS switch. *4th Micro and Nano Flows Conference (MNF2014)*. https://core.ac.uk/download/29139762.pdf (accessed on 14th October 2025).

21. Collins, P. J., Coutu, R. A., Jussaume, M. E., et al. (2011). Electrostatically tunable meta-atoms integrated with *in situ* fabricated MEMS cantilever beam arrays. https://core.ac.uk/download/213087973.pdf (accessed on 14th October 2025).

22. Czaplewski, D., Hyman, D., Karatzidis, D., Rebeiz, G., Newman, H., Sumali, H., McKillop, J., et al. (2011). RF-MEMS switch actuation pulse optimization using Taguchi's method. https://core.ac.uk/download/337977.pdf (accessed on 14th October 2025).

23. Diaz Lantada, A., & Muslija, A. (2014). Deep reactive ion etching of auxetic structures: Present capabilities and challenges. http://oa.upm.es/35846/1/INVE_MEM_2014_191226.pdf (accessed on 14th October 2025).

24. Ducrée, J., Koltay, P., & Zengerle, R. (2004). Microfluidics: An enabling technology for the life sciences. https://core.ac.uk/download/11309631.pdf (accessed on 14th October 2025).

25. Eftime, L., Muller, R., & Voicu, R. (2008). Design optimization for an electro-thermally actuated polymeric microgripper. https://core.ac.uk/download/15497668.pdf (accessed on 14th October 2025).

26. Fluitman, J. (1996). Microsystems technology: Objectives. https://core.ac.uk/download/pdf/11453403.pdf (accessed on 14th October 2025).

27. Gauthier-Manuel, B., & Pichonat, T. (2006). Recent developments in MEMS-based micro fuel cells. https://core.ac.uk/download/15488679.pdf (accessed on 14th October 2025).

28. Guglielmelli, E., Oddo, C. M., Saccomandi, P., Schena, E., et al. (2014). Microfabricated tactile sensors for biomedical applications: A review. https://core.ac.uk/download/pdf/54936693.pdf (accessed on 14th October 2025).

29. Holmes, A. S., Pu, S. H., & Yeatman, E. M. (2007). Design and simulation of zipping variable capacitors.

30. Kirchberger, H., Lindler, P., & Wimpliger, M. (2007). Novel bonding technologies for wafer-level transparent packaging of MOEMS. https://core.ac.uk/download/15495620.pdf (accessed on 14th October 2025).

31. Kiziroglou, M., Temelkuran, B., Yang, G. Z., Yeatman, E. M., et al. (2020). Micro motion amplification–A review. https://core.ac.uk/download/305117006.pdf (accessed on 14th October 2025).

32. Kokshagina, O., Le Masson, P., & Weil, B. (2015). Portfolio management in double unknown situations: Technological platforms and the role of cross-application managers.

33. Koo, B. J. (2014). A 2 degree-of-freedom SOI-MEMS translation stage with closed loop positioning. https://core.ac.uk/download/29175019.pdf (accessed on 14th October 2025).

34. Kopylov, G. (2015). Overview of sensors suitable for active flow control

methods. https://core.ac.uk/download/30294554.pdf (accessed on 14th October 2025).

35. Li, L., & Uttamchandani, D. (2009). A microwave dielectric biosensor based on suspended distributed MEMS transmission lines. https://core.ac.uk/download/9023884.pdf (accessed on 14th October 2025).
36. Oliver, R., Tillotson, J., & Toomey, A. (2009). Bioengineered textiles and nonwovens – The convergence of bio-miniaturization and electroactive conductive polymers for assistive healthcare, portable power and design-led wearable technology. https://core.ac.uk/download/9587153.pdf (accessed on 14th October 2025).
37. Sáenz, J. P. (2005). An introduction to micro electro mechanical systems (MEMS). https://core.ac.uk/download/39120350.pdf (accessed on 14th October 2025).

CHAPTER 9

PACKAGING AND RELIABILITY

Contents

INTRODUCTION

The increasing applications of Micro-Electro-Mechanical Systems (MEMS) in various fields — from consumer electronics to the field of medicine — have resulted in heated discussions regarding packaging and reliability. As MEMS devices continue to become more and more complex, effective packaging solutions are needed that can effectively protect sensitive elements while ensuring the best performance. Conventional packaging methods cannot fulfill the demands because they do not provide sufficient integration between electrical integrity and device mechanics. Packaging issues for MEMS extend beyond mere physical enclosure and, rather, involve the need to address electrical parasitics, thermal issues, and environmental conditions that can cause functional degradation. The unique nature of MEMS necessitates a specialized packaging strategy that not only addresses risk but enhances global reliability under diverse operating conditions. Reliability testing is the pillar for confirming MEMS devices will endure the stresses of everyday utilization and environmental conditions. Various failure mechanisms, such as stiction, fatigue, and corrosion, demand critical examination by advanced testing methods. The relevance of accelerated life testing has grown increasingly evident, enabling manufacturers to predict the lifetime of a MEMS device under simulated environmental stress. Furthermore, packaging forms—like wafer-level packaging and hermetic sealing—are evaluated by playing an integral role in establishing the manner in which parameters like humidity, temperature, and mechanical stress may impact device performance and lifetime. As a result of the growth of advanced computational platforms and simulation techniques, researchers can now model these parameters more effectively, which means improved designs able to withstand destructive environmental conditions. Selecting the right packaging form has a major impact on MEMS device reliability and functionality. Each packaging method possesses its own advantages and drawbacks, which are usually determined by the application demands. Wafer-level packaging, for instance, has received considerable interest due to its capabilities of minimizing electrical parasitic phenomena while allowing more integration of components, ultimately leading to the miniaturization of the overall device (Bartek et al., 2006). Alternatively, traditional packaging technologies may be more suitable in applications where environmental protection is of critical importance because they are stronger against harsh conditions. Enhancements in materials science and engineering processes have opened up new avenues for the development of new packaging concepts that enhance the reliability of MEMS devices. By possessing a complete understanding of the capabilities of various packaging technologies, MEMS solutions that not only perform at their best but are also durable for a long period in actual applications can be developed (Achkar et al., 2006).

Importance of Packaging in MEMS Technology

The intricate interdependency between Micro-Electro-Mechanical Systems (MEMS) and packaging is emphasized by the profound technical challenges in maintaining functionality and reliability. MEMS device miniaturization has led to highly complex packaging requirements that must be capable of withstanding mechanical, thermal, and chemical stresses. For instance, environmental conditions like temperature fluctuation and humidity can lead to degradation of MEMS performance in the long term. Overcoming these difficulties requires an appreciation of numerous failure modes, such as fatigue, corrosion, and mechanical wear that degrade the sensitive MEMS components. Furthermore, the packaging type plays a critical role in determining the shelf life as well as operation lifetime of MEMS devices since inadequate packaging would result in metal diffusion or water penetration, ultimately resulting in device failure and variable performance in applications ranging from automotive sensors to consumer devices. Reliability testing is a critical method used for assessing the performance and service life of MEMS devices in real operating conditions. With techniques such as thermal cycling tests and highly accelerated stress tests (HAST), scientists can now test packaged MEMS under extreme conditions for long periods.

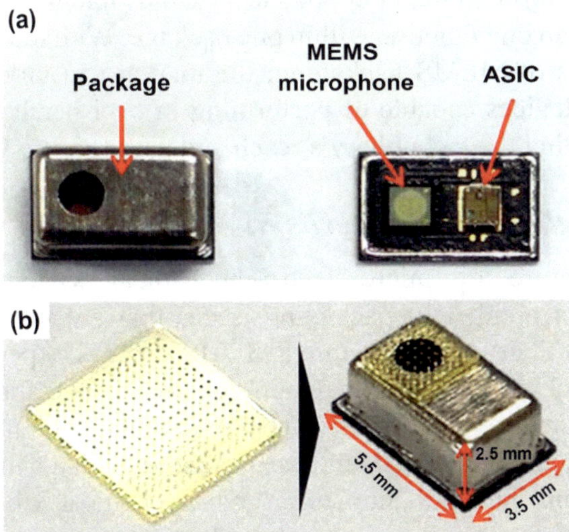

Figure 9.1. The intricate relationship between micro-electro-mechanical systems (MEMS) and their packaging is underscored by the significant challenges inherent in maintaining functionality and reliability.

Source: https://encrypted-tbn0.gstatic.com/images?q=tbn:ANd9GcSevJBQHZQ0khN32nYZk81XDqnnqhoL7p9K-JqfCEJ_5_ctAuUXPckDU8DMXYuOTeTiJjU&usqp=CAU.

As in the case of microwave curing processes, lifetime expectancy improvements have been seen at high levels together with lower residual stress in packaged MEMS, an indication of the significance of curing processes towards performance improvement (Adamietz et al., 2018). Moreover, they can offer insight into the long-term effect of different packaging materials on mechanical stability, illustrating the inherent trade-off between factors like cost, weight, and functionality. Ultimately, these tests are essential in informing design decisions and ensuring that MEMS technologies have the potential to perform reliably across a broad range of applications. The selection of a suitable packaging form is not a question of engineering preference but a key consideration that can make or break the success of MEMS technology in any sector. The packaging forms may be hermetic seals, overmolds, or micro-machined supports, each of which has unique advantages. Hermetic packaging, for example, offers better blocking of environmental impurities and hence greater device reliability, whereas overmolding techniques may offer cost-effective mass production. Yet the trade-offs among competing packaging schemes must be compared seriously against anticipated use environments and lifecycle requirements. MEMS packaging remains to evolve through novel approaches such as integrated packaging that combines more than one function within one package. With the solution of these fundamental areas of MEMS packaging, the market can unlock more robust and dependable devices capable of performing in ever-harsher environments, ultimately expanding the technology's reach and uses (Figure 9.1).

Overview of Reliability Concerns in MEMS Applications

The intricate nature of micro-electro-mechanical systems (MEMS) is characterized by a number of packaging issues that are important to ensure reliability. Due to their compact nature, MEMS devices experience a number of problems in packaging, that is, maintaining performance and shielding them from external environmental factors. Such equipment must be functional under varied conditions, which demands the use of materials that can endure stresses such as temperature changes, moisture, and mechanical vibrations. Besides this, several packaging types exist, such as laminates, ceramics, and metals, designed for various applications; they must be compatible with the MEMS design to ensure maximum robustness and functionality and minimum potential harm during fabrication. The combination of these problems necessitates the inclusion of appropriate packaging as a component of a broader reliability strategy for MEMS applications. Testing for reliability helps define potential modes of failure inherent to MEMS devices. Because these systems are intended to undergo severe operating conditions, understanding their fatigue, wear, and environmental degradation resistance becomes increasingly critical. Common failure mechanisms are mechanical fatigue, where repeated use may lead the structure to fail, and stiction, where two surfaces stick together unknowingly, preventing operation. More advanced test methods, including accelerated life tests and environmental tests, are often employed to assess long-term performance. Of particular interest is the optimization of actuation pulses as a technique for minimizing contact impact forces that, according to Czaplewski et al. (2011), have the potential to significantly lower wear and extend the service life of MEMS switches. This testing not only directs design improvement but also establishes reliability targets crucial to real-world use. Environmental effects on MEMS reliability cannot be overstated. Because such systems are meant for a range of applications from automobile sensors to medical devices, they must be operational in different environments. Humidity can lead to corrosion of metal components, and temperature fluctuations can affect material properties and performance stability. Adding to reliability issues are contamination threats from dust or chemicals that can seep into protective packaging and compromise device performance. As briefly outlined by Charalampidis et al. (2010), the MEMS switch configuration has evolved to include features that enhance its ability to withstand such environmental conditions. For instance, some design alterations for ensuring robust actuation reduce the risk of material failure from stress, which is crucial when implemented in real-world applications. Mitigation of these ecological problems is important not only for the operation reliability but also for the MEMS device life in most applications.

9.1. MEMS PACKAGING CHALLENGES

The intricacy of MEMS (Micro-Electro-Mechanical Systems) is especially prominent, particularly in consideration of the packaging challenges they present. The process from microfabrication to packaging is not only extremely precise in requirement but also requires relief from numerous reliability issues that have the potential to be introduced during and following manufacture. Packaging materials are not only necessary to suspend fragile MEMS devices from the environment but also to allow them to operate. For example, the choice of biocompatible materials employed in biomedical devices is crucial, exemplified by the work that reported fabrication of a parylene-based pressure sensor for intraocular application as an example of the necessity to employ suitable materials to ensure device longevity and minimization of invasiveness (Chen et al., 2008). Furthermore, maintaining structural integrity of MEMS structures during packaging is critical to preventing mechanical failure, which typically becomes more accentuated under adverse environmental conditions. Hence, understanding these complications is vital to MEMS technology progress. Testing for reliability plays a significant role in the successful deployment of MEMS devices, with the significance being the guarantee that packaging options safeguard and preserve device performance in an appropriate manner. There are several failure mechanisms that contribute to MEMS performance degradation, including mechanical stress, thermal cycling, and moisture ingress. They can result in sensitivity and device performance declines overall. One such example is in applications where MEMS technologies need to discover operation in adverse environments, e.g., in software-defined radios. They require selective RF filters so that they can protect sensitive parts from overwhelming external signals (Beacken et al., 2018). Failure to address such constraints jeopardizes realizing devastating device failures, which ultimately impact their reliability in actual applications. Test procedures therefore need to be carefully designed to simulate field conditions and assess long-term reliability, providing hints on failure modes. The diversity of packaging technologies applied to MEMS processing makes these even more challenging. From the traditional ceramic and plastic packaging to new ones such as flip-chip bonding and wafer-level packaging, each has its own advantages and disadvantages in reliability and performance.

Figure 9.2. Packaging materials must not only protect sensitive MEMS devices from environmental factors but also facilitate their performance.

Source: https://media.springernature.com/lw685/springer-static/image/prt%3A978-90-481-9751-4%2F13/MediaObjects/978-90-481-9751-4_13_Part_Fig1-319_HTML.jpg.

The preferred package type generally relies on the needs of a specific application, i.e., size constraints, environmental exposure, and mechanical performance demands. These are particularly crucial in compact integration-focused applications such as wireless communication systems where MEMS filters must be operational under various conditions (Beacken et al., 2018). Overcoming packaging challenges demands an interdisciplinary approach, integrating material science, mechanical engineering, and electrical engineering knowledge to achieve reliable MEMS devices in all applications. Packaging innovation in the future will be a key factor in ensuring MEMS reliability, paving the way for further application in emerging technologies (Figure 9.2).

9.1.1. Size and Complexity of MEMS Devices

Microelectromechanical systems (MEMS) devices occupy the intersection of electronic and mechanical engineering, with extremely intricate structures usually in the range of micrometers or less. Such miniaturization makes new functions possible across a broad variety of applications, including sensors and actuators. However, the extremely high complexity in their design and fabrication poses extremely high packaging challenges. Effective MEMS packaging must offer protection from the environment, address thermal performance, and offer mechanical integrity with device functionality retention. Since MEMS devices are more advanced, with features such as onboard actuators and sensors, there is an immediate need for advanced packaging methods. A good example of this is the integration of MEMS-capable components within standard silicon photonics processes, where there is the need for specialized packaging methods to accommodate evolving functionality and prevent degradation in performance under varying operation conditions (Bogaerts et al., 2020). Reliability of MEMS devices depends on rigorous test procedures to uncover likely failure modes. Because of their minute size, defects in performance and functionality can have major impacts, and therefore a careful analysis of reliability is needed. Environmental factors such as temperature fluctuations, humidity, and mechanical vibrations cause failure modes to get magnified. For instance, the mechanical motion of wave-guiding structures, which form a crucial component of the majority of MEMS applications, must withstand repetitive operational stresses without failure (Bogaerts et al., 2020). In addition, the packaging methods themselves must seal these delicate structures away from ambient influences that would compromise device reliability. Therefore, an explicit understanding of potential failure mechanisms together with innovative packaging methods is essential to achieving long-term performance and reliability for MEMS devices. Furthermore, the different types of packaging can significantly impact both the performance and reliability of MEMS devices. These include hermetic packaging, providing a very strong shield against moisture and contaminants, and adhesive bonding, which can be utilized to integrate with other systems. However, all types of packaging come with their own set of issues; for example, hermetic packages are costly and difficult to mass-produce, perhaps limiting their accessibility in various applications. The use of advanced materials, such as organic thin-film electronic devices and bio-responsive polymers, offers innovative opportunities while creating further challenges in MEMS packaging and design. By the development of these interdisciplinary avenues, engineers can design solutions not only that enhance device functionality but also contribute to efficiency and sustainability in MEMS deployment and fabrication (Figure 9.3).

Figure 9.3. Effective MEMS packaging must ensure environmental protection, manage thermal performance, and maintain mechanical integrity while preserving device functionality.

Source: https://media.springernature.com/lw685/springer-static/image/prt%3A978-90-481-9751-4%2F13/MediaObjects/978-90-481-9751-4_13_Part_Fig3-319_HTML.jpg.

9.1.2. Integration of Multiple Components in a Single Package

The integration of multiple constituents in a single package causes significant issues in the arena of Micro-Electro-Mechanical Systems (MEMS). As various functionalities are merged in compact forms, issues such as thermal dispersion, mechanical stress, and electromagnetic interference become critical. For instance, the presence of multiple sensors and actuators within a package necessitates careful consideration for their placement to achieve maximum performance at the cost of no reliability. Also, the sophisticated manufacturing processes required for such integration must satisfy the demands of accurate alignment and bonding, which are critical to the integrity of the system. Satisfying these demands is crucial not only to enhance the effectiveness of MEMS devices but also to make them durable and dependable in numerous applications. Generally speaking, the success of integrating multiple components relies on overcoming such gargantuan engineering challenges. MEMS package reliability testing is critical in establishing their actual performance, particularly when subjected to extreme environmental conditions. Whole-chip testing methodology must involve a variety of failure modes, including mechanical fatigue, thermal cycling, and moisture ingress, which may significantly impair device operation. With advancing MEMS technology, simulation models based on failure rate prediction under simultaneous stresses become increasingly relevant. These models are useful and can guide design improvements as well as shape packaging to result in more robust systems. Not only are the packaged forms' reliability dictated by what is contained within them but also by the external world into which they are put. Furthermore, test methodologies must keep pace with technology advancement to support emerging MEMS applications in a way that chip-scale packages provide reliable performance for their design lifecycle. Packaging forms significantly affect the reliability and performance of MEMS devices, particularly when integrating more than one component. Various packaging solutions such as hermetic, organic, and hybrid packages possess different strengths and weaknesses that must be evaluated against the intended application. As an instance, though hermetic packages offer better protection against environment conditions, at the same time they would raise the issues of thermal management and manufacturing complexity. In wearable computing, miniaturization and power efficiency are given top priority; hence, the choice of packaging should be such that it is a balance among functionality, size, and cost. With progress in research, the use of advanced materials and new designs will be a game-changer in overcoming such challenges, facilitating the production of highly reliable MEMS packages for numerous applications—from the latest developments in cardiovascular medical implants (Bussooa et al., 2018) to uses in wearable technology (Bharatula et al., 2018).

9.1.3. Cost Implications of Advanced Packaging Techniques

MEMS packaging is a complex field responsible for causing a whole gamut of issues that contribute meaningfully towards the cost aspects. New-generation packaging technologies, while guaranteeing better device reliability and performance, are capital-intensive in terms of material as well as process costs. Hermetic sealing for protection of the delicate MEMS components against the outside environmental parameters such as humidity and particulate matter increases the costs. In addition, the technological sophistication of packaging technology calls for sophisticated equipment and expert labor, further propelling costs. Inadequate accounting for such expenses can result in cost overruns and inefficient use of resources, potentially rendering competitive edge MEMS applications less competitive. Therefore, there must be a balance between the short-term cost drivers of advanced packaging and the long-term benefits of increased reliability that lies at the center of the application of MEMS technologies across industries. Reliability testing becomes a key area of interest in understanding cost drivers associated with advanced packaging techniques used for MEMS devices. Thorough reliability studies are important in a bid to identify potential failure mechanisms that may arise because of the packaging process itself. There are differing levels of sensitivity, for instance, to environmental conditions such as thermal cycling and mechanical stress, amongst differing packaging technologies, such as chip-on-board, flip-chip, and wafer-level packaging. All of these packaging technologies come with an associated cost structure dependent on the necessary testing protocols required to ensure long-term function. In addition, complete reliability testing not only serves to defend against product failure, but can also support market confidence, allowing widespread acceptance of MEMS technologies.

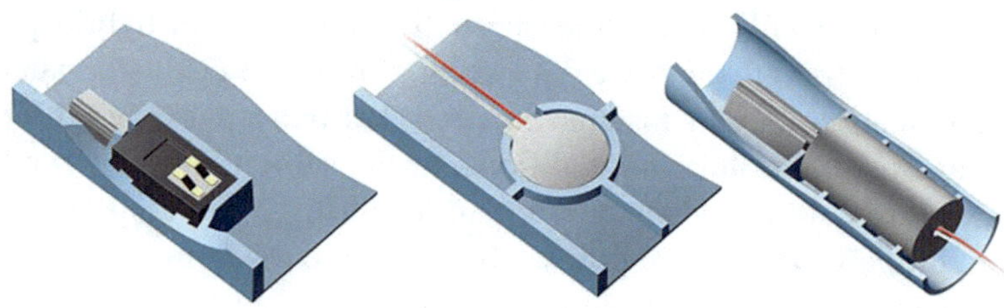

Figure 9.4. The need for hermetic sealing to protect sensitive MEMS components from environmental influences such as humidity and particulate matter compounds these costs.

Source: https://www.electronicadhesive.com/wp-content/uploads/2023/05/1200x350-051a.jpg.

Consequently, while the first test phase may seem to be cost-intensive, the likelihood of preventing costly recalls and warranty claims in the long term improves a product's profitability and reputation for a brand in competitive markets. The impact of the environment on MEMS reliability also underscores the significance of advanced packaging techniques in mitigating outside forces and offering durability. Severe environments, such as extreme temperature and chemical exposure, have the ability to compromise the integrity of MEMS devices, establishing better requirements for robust packaging solutions. Application of new materials such as organic thin film electronic devices and bio-responsive polymers can instill greater resistance to these environmental stress factors, along with incorporating a sense of adaptability within the packaging. However, application of such advanced materials in practice is sometimes coupled with higher production and material costs, hence making cost-benefit calculation more challenging for producers. As this field continues to move forward toward the use of even more advanced materials, identifying trade-offs between initial investment and potential durability gains will be paramount. This interaction will ultimately decide the industry's approach to developing reliable MEMS systems with the capability to survive within harsh environments, affecting both the longevity and economic feasibility within the industry (Figure 9.4).

9.2. TYPES OF MEMS PACKAGING

Among the several challenges that encumber MEMS packaging, device reliability takes the lead. The MEMS community is typically faced with environmental factors that can induce performance degradations in the long term. Failure mechanisms like fatigue, stress, and contamination are commonly encountered by virtue of materials utilized in packaging and manufacturing. For instance, integration into systems at a higher level subjects the MEMS devices to stress beyond mechanical tolerances, most often exacerbated by temperature fluctuations or exposure to humidity. Rugged packaging must withstand these environmental conditions while still providing thermal dissipation and protection required. Without proper thought of such challenges, even the latest MEMS technologies can be marred with compromised performance and longevity, which makes it imperative to seek potent packaging solutions to ensure the reliability of MEMS devices in real-world implementations. Reliability testing is a core activity in stimulating MEMS packaging solutions and ensuring devices operate well under different conditions. Various techniques are being used to assess the performance and durability of MEMS packages, including thermal cycling, moisture sensitivity testing, and stress testing. The latter are intended to simulate potential real-world applications and therefore can display how the packaging material responds to mechanical stress, environmental stress, and aging effects. Furthermore, the use of health and usage monitoring systems (HUMS) for MEMS technology introduces a new aspect of reliability testing. The reliability architecture under consideration in HUMS can enable on-the-fly diagnosis so that degradation can be quickly identified before failure is catastrophic. For example, embedding a dual-mode MEMS sensor with in-board self-test capabilities illustrates how advanced test mechanisms improve performance as well as toughness under hostile conditions, emphasizing the requirement for reliability testing of MEMS packaging (Begbie et al., 2010). MEMS packaging environment is controlled by diverse methodology, each of which is characterized to fulfill specific functional, and reliability needs. Hermetic packaging providing robust protection against environmental contamination and non-hermetic packaging allowing less expensive solutions but possibly with compromise in long-term reliability are some of the typical forms of MEMS packaging. MEMS devices, especially those for wireless applications, tend to use flip-chip packaging methods to save space while providing better electrical performance. Moreover, developments in 3D packaging technologies are opening up pathways to integrate MEMS devices with their related electronics more compactly and efficiently. Above all, the most important challenge remains the production of packaging configurations capable of maintaining the microstructures free from physical and chemical stress, while still keeping their intended functionality under operating conditions. On the focus of especially tailored packaging solutions, performance and reliability of MEMS devices for applications across a broad spectrum can be significantly enhanced, as exemplified by RF MEMS switches for antenna systems (Charalampidis et al., 2008).

9.2.1. Hermetic Packaging Solutions

The packaging challenges of MEMS are multi-faceted, primarily due to the unique operating environments under which these devices function. MEMS occasionally require reliable hermetic packaging structures to protect sensitive components from ambient contaminants such as moisture and particulate matter that cause degradation and could eventually fail the device. The protection is critical as even trace exposure can affect the performance and lifetime of MEMS devices significantly. Also, the inherent mechanical stresses caused by temperature cycling operation can further increase failure modes, and therefore robust packaging solutions are necessary. Use of sophisticated technologies such as Through Silicon Via (TSV) provides new avenues for minimizing device area while maximizing protection, thus facilitating the solution of these packaging challenges effectively (Bauer et al., 2008). Reliability testing is an important step towards ensuring that MEMS devices maintain their stated performance in varying environments. Hermetic package characterization, particularly in RF-MEMS, involves not just optimizing mechanical strength but also electrical performance. Parasitic effects need to be minimized to a bare minimum with careful design and material selection to guarantee that hermetic sealing does not interfere with the devices' operational capability. Advanced finite-element electromagnetic simulations are typically employed to make intelligent design choices, especially regarding through-substrate electrical interconnect via size and location. These methods provide critical information for improving MEMS reliability in the form of improved hermeticity without device functionality degradation under real-world conditions (Bartek et al., 2006). There are different packaging types available for MEMS devices, each with specific advantages based on application needs and exposure scenarios. Hermetic packages to a large extent involve utilization of materials offering impermeability to gases and particles, typically achieved by utilizing glass or metal enclosures. These packaging forms are especially important in applications where environmental exposure is considerable and can influence sensitivities, e.g., automotive and aerospace technologies. Further, implementation of wafer-level packaging processes allows for simpler manufacturing, reducing cost of production without any compromise in overall reliability. By tackling specific environmental concerns by way of innovative hermetic packaging methods, the MEMS industry continues to thrive, consistently enhancing the performance and resilience of these devices under the threat of competition (Figure 9.5).

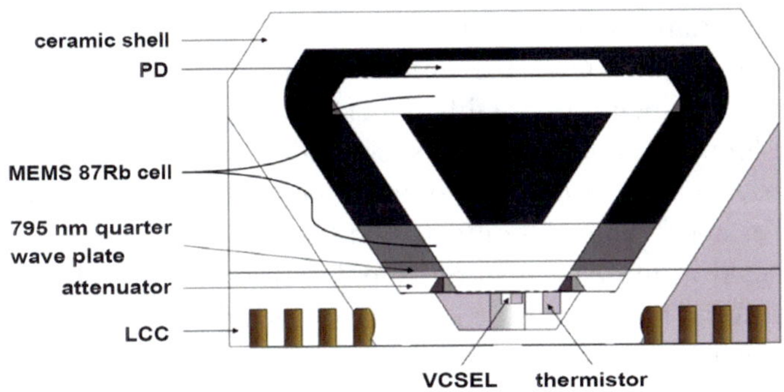

Figure 9.5. MEMS often require reliable hermetic packaging solutions to protect sensitive components from external contaminants, such as moisture and particulate matter, which can cause degradation and ultimately lead to device failure.

Source: https://www.mdpi.com/applsci/applsci-11-08536/article_deploy/html/images/applsci-11-08536-g001.png.

9.2.2. Non-Hermetic Packaging Options

MEMS development has brought to focus severe packaging issues, particularly with non-hermetic packages. These kinds of solutions, while economical, involve special reliability concerns due to their exposure to the environment. For instance, the entry of moisture will lead to corrosion and mechanical failure, affecting device performance as well as lifetime. Thus, understanding implications for such packaging is extremely significant to MEMS designers and manufacturers. Packaging failure is achieved through numerous mechanisms, among them thermal cycling and vibration as the most important, which are also exaggerated by non-hermetic packages. The challenge, thus, is to have cost efficiency balanced with reliability required to sustain performance over time, thereby ensuring overall device reliability for MEMS. An in-depth exploration of specific failure modes for non-hermetic solutions, thus, is required to propel this technology sector forward. Reliability testing is a critical component in assessing the performance of MEMS under different environmental conditions, especially when encapsulated with non-hermetic packaging. While traditional hermetic packages provide very good shielding of the interior from the outside environment, non-hermetic packages demand a different paradigm for testing depending on long exposure to contamination. Advanced accelerated life testing methods can simulate long-term environmental stressors such as humidity, temperature cycling, and particulates. These tests are essential to defining failure conditions and guiding design modifications that will mitigate risks. With a system where MEMS devices are increasingly relied upon for accuracy-based applications, the information acquired through reliability testing helps manufacturers establish performance standards. Furthermore, the feedback loop generated through such testing enhances the current design as well as provide critical data for future innovation in non-hermetic MEMS packaging. Yet another good point of non-hermetic packaging for MEMS is that it promises revolutionary material and design innovations that can enhance device capability at the same time as it can encapsulate reliability risk. Recent advances in materials science have provided biocompatible polymers and advanced coatings that can provide adequate protection from environmental stressors while enabling the miniaturization of parts. As reported in literature, the convergence of nanotechnology, biotechnology, and high-end materials can facilitate the creation of multifunctional MEMS devices with enhanced performance requirements under non-hermeticity conditions. For instance, the integration of biosensors with non-hermetic packaging has the potential to develop new MEMS applications in medical diagnostics, including real-time monitoring and improved patient care. Hence, research into such non-hermetic opportunities not only addresses existing packaging challenges but also opens up the avenues for new MEMS technologies, which are critical in an environment where rising expectations for reliability and functionality in many applications (Colomer et al., 2015).

9.2.3. Emerging Packaging Technologies

Emerging packaging technologies provide key solutions for resolving the intricacies involved in Microelectromechanical Systems (MEMS) packaging. With MEMS devices tending to become more miniature and integrated into varied applications, corresponding packaging challenges tend to grow stronger. Conventional packaging mechanisms tend to be incapable of coping with the situation, leading to enhanced vulnerability to environmental stresses and mechanical failures. Under these circumstances, innovative packaging mechanisms featuring cutting-edge materials and designs are required. Not only are these new technologies focused on physical protection of MEMS devices but also reliability through reducing failure mechanisms like mechanical fatigue and thermal cycling. Some of the new strategies include the use of encapsulation materials with high resistance to temperature cycling and humidity, which are crucial in guaranteeing the performance and lifespan of MEMS devices under varying operating conditions. MEMS package's reliability testing has much to do with the viable implementation of next-generation packaging technologies. With the increasing demand for MEMS applications, particularly in mission-critical fields such as healthcare and aerospace, rigorous test protocols are needed to ensure the dependability and functionality of these systems. More advanced methods of reliability assessment are employed to try the effects of environmental aspects, such as moisture absorption and heat expansion, which tend to exacerbate failure modes. Accelerated life testing and stress screening are a few of the methods employed to simulate actual environmental conditions so that researchers can predict potential points of failure. With the use of these advanced test methods, manufacturers can enhance their package designs and thereby achieve higher reliability.

The use of health and usage monitoring systems (HUMS) is another example, as they provide real-time data on the condition of the MEMS packages, allowing anticipatory maintenance, and ensuring system integrity (Begbie et al., 2010). Package type development, led by technical innovation in materials science and engineering, has a direct impact on the usability and reliability of MEMS technologies. One of the new packaging technologies is the development of smart materials such as electroactive polymers and bio-responsive hydrogels that provide new opportunities for integration with MEMS devices. These materials not only provide minimum protective functions of traditional packaging but also have adaptive properties that can dynamically adapt to changing conditions. These adaptive packaging solutions are able to enable the functionality of MEMS in assistive health technology through to interactive consumer products, extending the range of potential applications for MEMS systems. Moreover, as these kinds of technologies become increasingly advanced, they drive the development of ambient living platforms designed to meet human-centric needs, exemplifying the real influence of evolving packaging technologies on overall reliability and usability of MEMS (Oliver et al., 2009) (Figure 9.6).

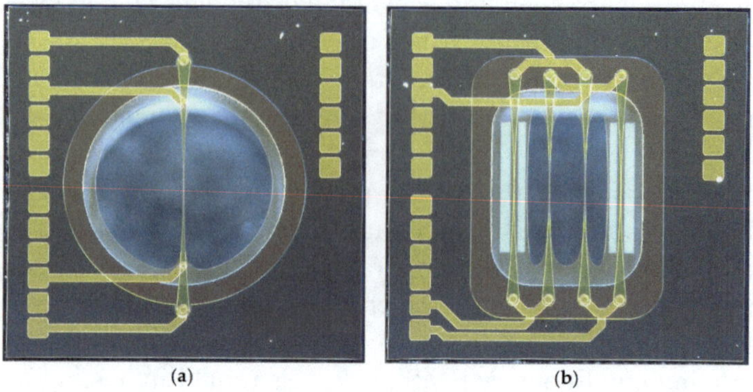

Figure 9.6. As MEMS devices become increasingly miniaturized and integrated within diverse applications, the associated packaging challenges intensify.

Source: https://www.mdpi.com/micromachines/micromachines-15-01404/article_deploy/html/images/micromachines-15-01404-g008.png.

9.3. RELIABILITY TESTING OF MEMS

Micro-Electro-Mechanical Systems (MEMS) development has increased the significance of reliable packaging solutions that safeguard complex devices against potential failure. Possibly one of the prime challenges for packaging MEMS is to overcome discrepancies between the physical properties of MEMS devices and protective covers. Such mismatches may result in mechanical stresses that compromise device integrity over time. Additionally, varied package types, such as ceramic and polymer-based packages, are used, with each possessing specific strengths and weaknesses in the thermal stability, moisture resistance, and material compatibility aspects. These challenges in packaging need to be overcome as they have direct effects on the performance and longevity of MEMS devices and its reliability across many applications, for example, in healthcare monitoring systems and infrastructure monitoring systems (Buss et al., 2013). As MEMS demand increases, the necessity for developing and optimizing packaging techniques that improve reliability brings forth a crucial research and innovation domain. Reliability testing, a bedrock of MEMS design, offers a comprehensive examination of MEMS robustness under dynamic environmental conditions and operational stress. Accelerated life testing and fatigue testing are major techniques that simulate long-term operational conditions to uncover failure mechanisms. These modes of failure, such as device wear, material degradation, and packaging stress, need to be understood carefully to enhance reliability. Environmental stresses, such as temperature cycling, humidity, and corrosive exposure, possess a wide potential for their effects to impact MEMS performance and life. The test regimens need to incorporate these items to ensure proper assessments. With the evolution of MEMS technology, the consideration of reliability testing becomes very critical in confirming the ruggedness of the device and, consequently, guaranteeing manufacturers and end-users alike their reliability in operating in mission-critical applications. Furthermore, the integration of advanced monitoring techniques in the MEMS reliability testing process is growing to be the center of attention as scientists seek innovative solutions for combating potential failures. Through the application of real-time feedback technologies, such as wireless sensor networks, MEMS-based systems can be monitored in real-time throughout their lifespan of operation to provide valuable performance and integrity information. The process aligns with present trends in sustainable infrastructure solutions that have MEMS at their core, especially in structural health monitoring. The addition of environmental sensors in MEMS devices offers feedback on long-term reliability over numerous different operating conditions (Buss et al., 2013). Furthermore, the interdisciplinary strategy towards microassembly techniques, such as robotic microassembly, promises to break the limitations of traditional packaging, allowing precise device integration that improves reliability and performance. These advancements highlight the pivotal relationship between packaging technologies and the successful implementation of reliable MEMS solutions in the current technological landscape.

9.3.1. Standard Testing Methods for MEMS Reliability

The very intricate nature of MEMS technology necessitates the application of aggressive packaging strategies to combat built-in challenges that would otherwise debase reliability. MEMS devices are continually subjected to environmental stresses like temperature variations, humidity, and mechanical shock, which can lead to degradation as well as failure in function. Traditional test methods become critical to ensure the efficacy of encapsulation in protecting such micro-scale products. Various packages, such as hermetic and non-hermetic, provide varying degrees of protection but must be inspected to ensure their resistance to the provided operational environment. Thorough examinations of these packaging techniques can illustrate inherent weaknesses and assist with improving design, enhancing performance, and adding life to MEMS. The procedure for reliability testing should, therefore, encompass not just the physical toughness of the package but also the endurance of the devices' functioning under simulated environmental conditions. Reliability testing of MEMS is inherently challenging due to the range of possible failure mechanisms that arise in operation. These mechanisms would include mechanical fatigue, thermal stress, and chemical degradation, impacting performance in a manner unpredictable over time. Traditional test protocols commonly entail a series of accelerated life tests designed to replicate years of operation within a condensed period. In this manner, engineers can observe failure mechanisms and quantify reliability metrics over a variety of stress levels. Advanced techniques such as Weibull analysis are typically employed to process failure data, providing vital information regarding the probabilistic nature of MEMS reliability. Furthermore, methods like Prognostic and Health Management (PHM) have been devised to detect real-time monitoring and predictive maintenance to even lower risks to a greater extent. The combination of traditional reliability testing and predictive analytical algorithms is an end-to-end approach to characterizing and enhancing MEMS reliability during the lifespan of their operation.

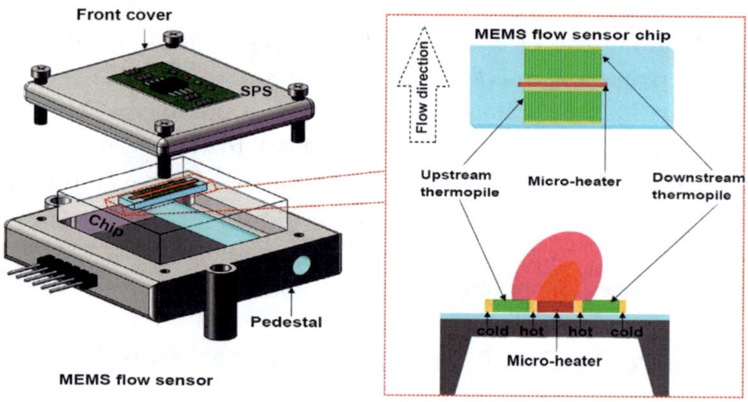

Figure 9.7. MEMS devices are often subjected to various environmental stressors, including temperature extremes, humidity, and mechanical vibrations, which can lead to performance degradation and failure.

Source: https://www.mdpi.com/sensors/sensors-23-08733/article_deploy/html/images/sensors-23-08733-g001.png.

Evaluating the overall reliability of MEMS also necessitates a thorough knowledge of their package and interconnect materials. The interaction between soldering technologies, such as Au–20 wt.% Sn, and contact metallizations like Ni, Cu, and Pt is a predominant factor that controls device reliability, particularly under thermal aging conditions. According to recent studies, it is evident that the choice of metallization has a significant bearing on the stability of interconnection, with Au–20 wt.% Sn|Pt exhibiting high thermal stability compared to others under prolonged exposure to elevated temperatures. Presence of intermetallic compounds (IMCs) at solder interfaces can influence mechanical properties and, by extension, MEMS reliability. In this regard, phase diagram knowledge and experimental analysis of interfacial reactions are essential in determining the selection of proper materials. This rigorous testing of solder joint reliability, combined with established reliability testing methodologies, enables the production of long-lived MEMS devices to endure operation in a broad range of applications, from automotive sensors to advanced biomedical systems (Figure 9.7).

9.3.2. Accelerated Life Testing Procedures

Reliability in microelectromechanical systems (MEMS) involves intricate considerations towards their packaging and then subsequent testing. Accelerated life testing (ALT) is a significant methodology applied to testing the performance of MEMS under conditions simulating prolonged usage. ALT focuses on identifying failure mechanisms that may arise due to environmental factors such as temperature fluctuations, humidity, and mechanical stress. By subjecting packaged MEMS devices to adverse environments, researchers are in a position to identify vulnerabilities in the materials and designs used. Besides exposing vulnerabilities inherent in the devices themselves, such exposure also reveals packaging integrity, an important aspect in safeguarding sensitive components. As MEMS technologies evolve and expand into many areas, the growing need to further develop ALT processes is necessary to ensure that the MEMS devices, as well as their packaging, are robust enough to withstand the rigors of real-world usage. In MEMS technology, many forms of packaging have been developed to ensure greater reliability and performance. These packages range from generic ceramic and plastic enclosures to more unusual implementations aimed at minimizing size and maximizing functionality. The packaging choice directly affects the performance of MEMS devices, particularly when considering the impact of thermal cycling and mechanical loading. Accelerated life test procedures therefore must be created for each package type in order to accurately quantify their ruggedness. For example, stress concentration phenomena at dissimilar material boundaries may be exacerbated by thermal expansion mismatches. Quantifying these interactions via rigid testing methodologies supports the generation of improved packaging solutions that are capable of mitigating against common failure mechanisms such as fatigue and fracture. The evolving nature of MEMS applications requires ongoing investigation into packaging technologies prioritizing reliability over miniaturization. Investigation of reliability through accelerated life testing helps to illuminate the intricate nature of failure mechanisms in MEMS devices and their packaging. This is a critical awareness, particularly as environmental factors, such as moisture entry and temperature extremes, can have significant impacts on device life. For instance, the behavior of MEMS switches, which are primarily micro-contact interfaces in nature, may be greatly affected by contact resistance properties and surface roughness under stress conditions (Coutu et al., 2009). Further, it has been established that residual stresses of the solder materials and variations in coefficient of thermal expansion (CTE) can induce mechanical failures, primarily under the reflow packaging process (Sharon et al., 2011). With focused study of these failure modes on ALT, industry participants are able to optimize MEMS design and packaging solutions, ultimately enhancing reliability and performance in real-world end-use environments where the systems are deployed (Figure 9.8).

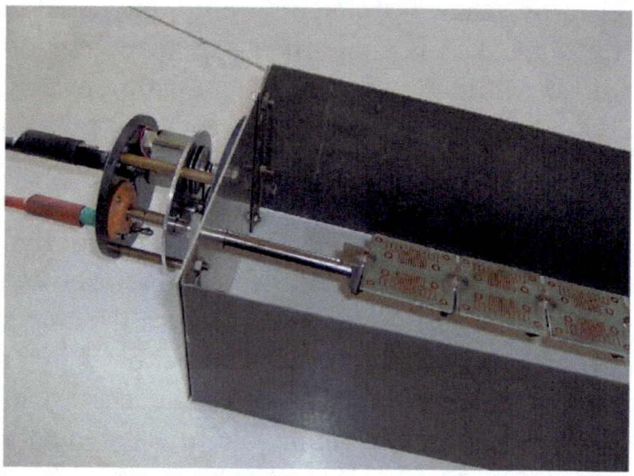

Figure 9.8. Accelerated life testing (ALT) serves as a crucial methodology in evaluating MEMS performance under conditions that simulate long-term usage.

Source: https://www.mdpi.com/sensors/sensors-07-02846/article_deploy/html/images/sensors-07-02846f2.png.

9.3.3. Evaluation of Long-Term Performance

With MEMS (micro-electro-mechanical systems), the finesse of packaging is paramount to long-term performance. Since devices get smaller and more functional, the challenge of achieving robust packaging becomes increasingly complex. As a vital element of the functionality of MEMS devices are package forms that not only will protect the delicate components from mechanical abrasion but also enable the required environmental interaction. For instance, thermal and humidity variations can have a profound influence on performance, necessitating extreme reliability testing to simulate and offset different failure mechanisms. A complete analysis of such packaging solutions emphasizes the need to match material and configuration to the operational requirements presented by these systems and to ensure that any measurement of long-term performance includes every variable that goes into MEMS packaging. Reliability testing is among the cornerstones of quantifying the long-term performance of MEMS devices since it allows scientists to measure the way the systems perform in terms of withstanding operation stressors over time. This would involve subjecting the devices to extreme conditions like thermal cycling and humidity testing, which replicate real-world operating conditions. One of the biggest challenges is identifying the precise failure mechanisms that can happen under these conditions.

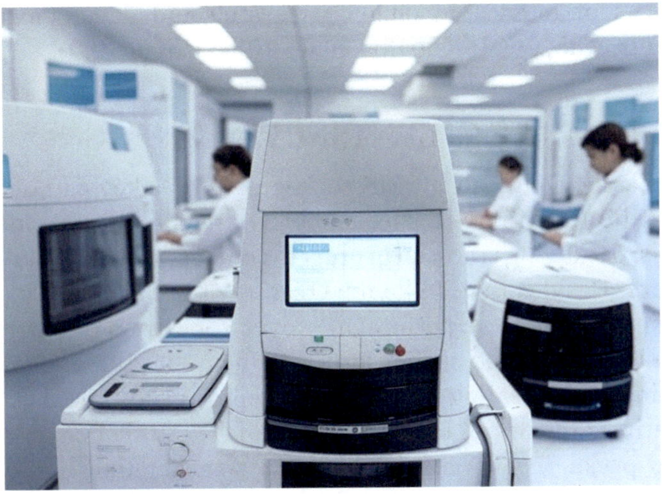

Figure 9.9. Reliability testing is a cornerstone of assessing the long-term performance of MEMS devices, as it allows researchers to determine how well these systems withstand operational stressors over time.

Source: https://www.caplinq.com/images/editor/67e6d75d65bc7-Highly%20Accelerated%20Stress%20Test%20(HAST).jpg.

These mechanisms may include thermal shock fatigue or environmental exposure corrosion, and each of these can impair device functionality and integrity. Of crucial importance is the fact that the results of reliability testing are fed back into the design processes, prompting redesigning of packaging strategies. An assessment/test redesign cycle is therefore required for the development of MEMS systems that not only function optimally but also sustain performance over extended durations of operation. Proper types of packaging are a crucial selection in enhancing the reliability and longevity of MEMS devices. Various packaging options, such as ceramic, plastic, and metal enclosures, each have advantages and disadvantages. For instance, while ceramic packages offer enhanced environmental protection, they are expensive and may complicate manufacturing processes. Conversely, plastic packaging is lighter and usually less costly but may not be able to provide adequate protection in adverse conditions. The evaluation of long-term performance needs to be integrative in character towards handling the impact of package type on the functionality and reliability of the device. In addition, some applications, such as in reconfigurable microstrip array antennas with RF MEMS switches, require explicit design parameters to determine reliability under varying operation requirements. The relationship between package choices and MEMS performance thus remains a critical focus of interest in ongoing research and development activities (Charalampidis et al., 2008; Kopylov et al., 2015) (Figure 9.9).

9.4. FAILURE MECHANISMS IN MEMS

Complexities associated with MEMS packaging form a key part of their overall reliability and life. MEMS devices are extremely susceptible to outside influences, and due to this, packaging is crucial in order to combat environmental phenomena causing failure. Besides issues like the creation of hermetic seals, which form a key part of ensuring device performance over a spectrum of temperatures and pressures. For instance, the sealing process often encounters challenges such as poor wetting of the sealing material to the silicon substrate that may lead to voids and intrinsic fractures in the hermetic seal. These challenges significantly contribute to packaging failure risks, with a consideration for robust design strategies that enhance strength and functionality in different environmental conditions (Harrison et al., 2004). Consequently, awareness of these packaging problems is part of improving MEMS reliability. Reliability testing is a key area in scrutinizing the performance of MEMS devices, particularly when establishing the efficiency of their packaging solution. Accelerated life testing and stress testing are some of the techniques employed to mimic operational conditions and predict potential failure mechanisms. These tests are also crucial in identifying weaknesses in packaging systems to enable manufacturers to remediate issues before releasing devices into real applications. Failure modes can often be due to mechanical fatigue and thermal cycling, leading to material degradation. For example, with increasing temperature, the deformation from brittle fracture to material yielding becomes a reality, which can fundamentally alter the device's as well as packaging's mechanical stability (Harrison et al., 2004). It is essential to gain a thorough understanding of such failure mechanisms to enhance the reliability of MEMS and ensure them to be in compliance with industry-high standards. Environmental factors also make MEMS reliability more complicated than mere packaging-related issues. MEMS devices are usually exposed to varying temperatures, humidity, and mechanical stresses, all of which have an influence on their lifetime and functioning. Coupling these environmental factors with material properties could lead to corrosion, material fatigue, and even failure. All packaging type, whether plastic, ceramic, or metal, possesses inherent advantages and disadvantages with regards to environmental ruggedness. Ceramic packages, for instance, offer superior mechanical reliability but may not possess the flexibility to operate within varying temperature conditions. Therefore, packaging choice is to be finalized in accordance with an understanding of environmental conditions to provide optimal device reliability with minimal failure risk. Lastly, an integrated approach to tackle these multi-disciplinary issues will be essential in the successful integration of MEMS in various applications.

9.4.1. Mechanical Failure Modes

The intricate world of Micro-Electro-Mechanical Systems (MEMS) carries with it a plethora of issues, primarily in the realm of packaging. Packaging is a critical barrier that protects MEMS devices from outside influences, but at the same time is also a source of mechanical failure. Different packaging types, such as hermetic and non-hermetic, have respective advantages along with limitations. For example, whereas hermetic packaging provides superior environmental protection, it may induce stress on the MEMS structure due to thermal expansion coefficients' mismatch. This is mechanical stress that may lead to delamination, cracking, or device failure. MEMS issues must be comprehended in order to optimize the design and manufacturing process for MEMS, furnishing engineers with the tools necessary to face mechanical failure modes and deliver greater overall reliability for practical applications. Reliability testing is not only a requirement, but also a methodical way to determine the numerous failure mechanisms present in MEMS devices. Very highly standardized tests, like shock and vibration tests, are used to simulate the environmental effects these systems would be subjected to during their operational lifetime. Such testing identifies the ways in which mechanical failure can occur, either as material failure under fatigue or surprise stress under external vibration. Further, the inclusion of emerging technologies, such as fault-tolerant sensor architectures discussed by Begbie et al. (2010), increases the need for ongoing reliability assessment. By encouraging a climate of vigorous testing and ongoing optimization, MEMS devices can be designed by engineers who are not only resilient to environmental stresses but also highly reliable over their life cycle. Other environmental factors, like humidity, temperature fluctuations, and exposure to contaminants, pose further complexity to MEMS system reliability. The properties of the material can significantly degrade under corrosive environments, and such degradation can lead to hard-to-detect failures unless they happen to take place. For instance, MEMS switches used in reconfigurable microstrip antennas should ensure robust performance under a defined set of operating conditions, as demonstrated by Charalampidis et al. (2008). Unless environmental stresses are controlled properly at packaging time, the device not only may be unable to function as desired but can also pose new hazards in applications that rely on MEMS technology. Thus, engineers must not only design against mechanical failure modes but also anticipate environmental interactions to ensure the lifetime and reliability of MEMS devices, thus creating safer and more efficient mechanical systems (Figure 9.10).

Figure 9.10. Packaging serves as a crucial barrier protecting MEMS devices from external influences, yet it can also become a point of mechanical failure.

Source: https://media.schott.com/api/public/content/c312df5f20984808bccd-93300d51c1d4.

9.4.2. Electrical Failure Mechanisms

The convergence of microelectromechanical systems (MEMS) and packaging has become significant since their applications cut across various industries. The most significant aspect of this convergence lies in the inherent nature of MEMS packaging challenges that lead to several electrical failure modes. These may be induced by material fatigue, mechanical stress, or process variation, which all degrade system performance. The use of correctly designed packaging can alleviate some of the aforementioned issues by providing environmental protection and physical stability. Nevertheless, with mounting requirements for miniaturization and functionality, the designers must contend with a host of substrate materials and structures in a bid to ensure the long-term reliability of MEMS devices and highlighting the importance of robust packaging solutions. Reliability testing becomes a valuable tool of understanding and predicting the failure mechanisms in MEMS. Through accelerated life testing under conditions that mimic operating environments, scientists can ascertain how environmental conditions can lead to electrical failure in the long term. Such environmental phenomena as humidity, temperature cycling, and mechanical shock exposure can significantly influence device function and lifespan. For instance, moisture entry can lead to corrosion or interfacial delamination, while abusive temperatures can induce thermal stress leading to electrical short circuits. Comprehensive reliability testing can reveal MEMS design weaknesses so that packaging techniques and the device's overall durability can be enhanced by designers (Lo et al., 2008). Proper packaging format selection is also vital for strengthening MEMS reliability against electrical failure mechanisms. There are different kinds of packaging that offer exclusive size-reduction and performance-efficiency advantages, such as cavity packages or wafer-level packages. Each of these forms, however, presents its own specific challenge, e.g., the availability of adequate thermal dissipation while having electrical insulation. Material advances, such as the application of parylene HT® in electret generators, have indicated promising progress in stability and reliability under conditions of changing operation. By resolving such packaging challenges and giving high priority to an aggressive reliability testing strategy, the long-term performance of MEMS can be significantly enhanced, ultimately allowing for their integration into high-risk applications across sectors (Lo et al., 2008).

9.4.3. Environmental-Induced Failures

The MEMS packaging and design naturally accrue a multitude of problems, most notably relating to the environment to which these devices are subjected over their service life. MEMS packaging dictates critically excluding the environmental stressors of temperature change, humidity, and particulate contamination from influencing the sensitive elements. With these devices shrinking in size—in keeping with Moore's Law principles—the vulnerability to environmental failures increases. The call for innovative packaging technologies, not only to address these threats but to advance the agenda of miniaturization and higher functionality, has become front and center in the industry. Besides, with the shift to More than Moore, more emphasis is being placed on the integration of advanced packaging technologies that are able to withstand hostile environmental conditions while maintaining MEMS device reliability (Evans et al., 2025). Reliability testing is a very important component in addressing potential environment-related failures for MEMS devices. More reliability testing is employed to replicate stressors occurring over the long term and provide information on long-term behavior under conditions that are essentially real. The tests must consider thermal, mechanical, and chemical stressors that mimic the environmental threats MEMS would experience when put into actual applications. Most importantly, the data obtained from these tests guide manufacturers to make packaging schemes more robust against failure mechanisms such as moisture intrusion and thermal cycling. Therefore, knowledge of the interaction between environmental conditions and packaging materials becomes critical.

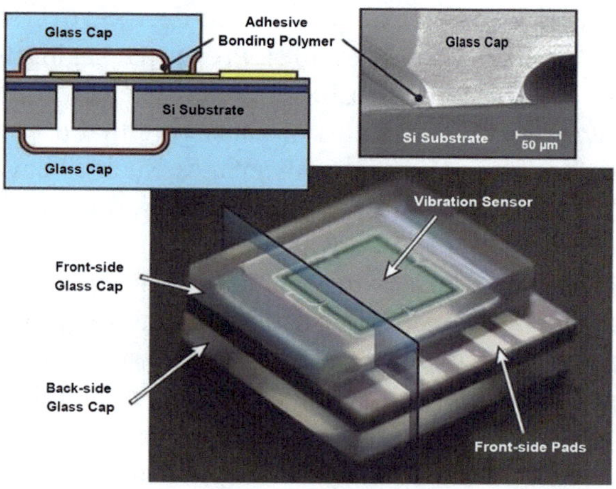

Figure 9.11. A critical aspect of MEMS packaging is ensuring that the encapsulation methods effectively shield the sensitive components from environmental stressors such as humidity, temperature variations, and particulate contamination.

Source: https://pub.mdpi-res.com/micromachines/micromachines-07-00192/article_deploy/html/images/micromachines-07-00192-ag-550.jpg?1569115104.

In an effective testing procedure and analysis, as well demonstrated in industry standards, engineers have the capability to develop packaging solutions that not only meet IPC/JEDEC/Mil requirements but also take proactive steps to mitigate risks due to possible environmental failure (Evans et al., 2025).

The environmental failure mechanisms can substantially erode the reliability of MEMS devices. Environmental factors such as corrosion, delamination, and stress fractures may occur due to direct causes of improper packaging that do not also treat the special environmental concerns. The importance of using materials with low surface roughness during packaging is realized, particularly for devices such as MEMS switches where resistance at the micro-contact directly affects performance. Progress in package designs, such as the application of moisture-repellent coatings and temperature-stable materials, can serve to mitigate these hazards without compromising efficient operation and durability. Furthermore, ongoing research on the topic, i.e., more reliable models for micro-contact resistance prediction, signifies a straightforward path for the progress of MEMS reliability under the influence of the environment. By taking these decisive factors into account, designers can ensure the overall life span and performance of MEMS devices stronger in varied environments (Figure 9.11).

9.5. ENVIRONMENTAL EFFECTS ON MEMS RELIABILITY

The intricate interaction between package and environmental conditions is of significant implication for Micro-electromechanical Systems (MEMS) reliability. Temperature fluctuation, humidity, and particulate contamination are significant factors in establishing the life and performance of MEMS devices. Degradation results in various failure mechanisms, e.g., stiction, corrosion, and bond degradation, which can compromise the operational functions of MEMS. In packaging designs, the requirements for high-strength configurations that minimize exposures to adverse environmental conditions have been at the forefront. These configurations typically involve the inclusion of protective mechanisms such as hermetic seals and desiccants to avoid moisture intrusion and particulate intrusion, thus ensuring the integrity of the MEMS devices over their lifetime. Reliability testing is the foundation of MEMS packaging when the impact of environmental factors needs to be assessed. Thermal cycling and accelerated life testing are common techniques used in attempts to predict actual performance under various environments. These tests allow engineers to identify failure points and implement design enhancements particular to specific environmental conditions. Because the products are typically exposed to high thermal and mechanical stresses, understanding how the tests equate to long-term reliability is important. The challenge lies in simulating conditions that accurately reflect what MEMS would encounter in their use across consumer goods to space systems. Addressing these areas of reliability testing, MEMS manufacturers can design packages that withstand diverse environmental stresses with impunity without sacrificing performance integrity. MEMS technology advancement has resulted in various packaging types with their advantages and appropriateness depending on the application and environment needs. On-wafer packaging strategies, for instance, have been shown to reduce loss and improve reliability through integrated designs that eliminate the need for extrinsic interconnects such as solder bumps or bond wires, as in the case of RF MEMS switches (Herrick et al., 2001). Additionally, the need for small, versatile filters in software-defined radios emphasizes the demand for MEMS technology that ensures high performance under tough conditions but provides higher performance. The development of high-Q, low-loss MEMS filters with wide tuning capabilities indicates an adaptive strategy in addressing challenging operating conditions, reflecting the demand for developing novel packaging techniques that can withstand both environment and operating challenges (Beacken et al., 2018).

9.5.1. Impact of Temperature Variations

MEMS packaging challenges are brought out by the temperature variation sensitivity of these devices. Temperature changes may lead to thermal expansion mismatches between the materials of the MEMS device and package materials and consequently result in long-term mechanical stress and performance degradation. These stresses may speed up failure mechanisms to such factors as material fatigue and actuation capability degradation. For instance, in the case of MEMS switches, performance can deteriorate drastically with longer exposure time under high-temperature conditions, affecting actuation reliability. It is crucial to design the package material and structure optimally so as to offset these temperature stresses in order to enhance the life and reliability of MEMS devices so that they function as desired under varying operating conditions (Czaplewski et al., 2011). The reliability test of particular importance in the context of MEMS is impossible to overestimate when temperature variations are accounted for. Comprehensive test processes must simulate ambient conditions where MEMS device's function, including drastic thermal fluctuations. This testing seeks to identify potential failure mechanisms caused by thermal cycling, such as material embrittlement and material interface delamination. Such rigorous testing enables engineers to discover flaws in the design and make the requisite changes to the packaging. By utilizing high-end materials that have lower thermal coefficients of expansion, or thermal management techniques, MEMS device reliability can be significantly increased. Hence, addressing temperature-related issues by controlled testing and design optimization of packaging solutions is of utmost importance in enhancing operational reliability (Beacken et al., 2018). Environmental impact on MEMS performance is further aggravated by temperature variations, which can create both short- and long-term reliability issues. Devices targeted for mission-critical applications need to have a robust package to withstand the influence of different thermal environments. For example, with RF-MEMS switches in applications, temperature variations can lead to shifting of actuation profiles, which results in instability of switching characteristics. An optimized MEMS implementation may involve tailored actuation strategies which account for temperature effects, e.g., pulse shaping to minimize the impact of thermal expansion on control and switching times (Czaplewski et al., 2011). Thus, advancements in MEMS packaging technology in terms of integrating adaptive components that can compensate for temperature fluctuations would enhance the reliability of devices, such that devices can keep pace with dynamic operation demands while ensuring performance levels.

9.5.2. Humidity and Moisture Effects

The application of microelectromechanical systems (MEMS) in various applications has led to heightened focus on the problems presented by moisture and humidity to packaging. As most of these devices operate in humid environments, it is imperative to get a complete understanding of the consequences of moisture penetration. Inadequate packaging will have adverse implications such as internal corrosion, which compromises the functional integrity of MEMS. This corrosion can change operating conditions, resulting in unstable performance results. Therefore, choosing a durable packaging material and design is essential in order to maintain the endurance and consistency of MEMS devices under humid environments. This need has been underlined by ongoing research into cheap, reliable integration methods that allow for the integration of fluidic and electronic devices, as in research on using Printed Circuit Board material for microfluidic channel fabrication (Akkerman et al., 2012). Testing for reliability of MEMS devices has also advanced with the need to understand how durable these devices are against moisture intrusion. Since most applications necessitate MEMS to function in humid environments, the likelihood of failure mechanisms due to humidity must be assessed systematically. The existing industry practices put lot of stress on hermetically sealed packages with a view to preventing moisture exposure, since it is appreciated that the reliability of the devices contained within these packages relies directly on the effectiveness of the seals. Introduction of specifically configured packages is done to address such severe environmental conditions in a way that emerging MEMS devices are prepared to face rugged testing. Following industry reports, it is important that testing of these packages considers not only exposure to external moisture but also internal ones that could enhance potential failure modes (Rencz et al., 2008). Packaging form choice is an important consideration in determining MEMS device susceptibility to environmental stress such as humidity. Traditional packaging materials may be unable to provide suitable barriers to moisture, leading to a search for emerging technologies that will further reliability. As the field continues to evolve, advancements in packaging technology, including the utilization of advanced polymer composites or coatings, are being pursued in order to offer greater protection against humidity. These enhancements are designed to provide a stronger interface between the MEMS and the external world, therefore extending their usage lifespan. Understanding how different packaging designs and moisture behavior interact will be important in enabling future MEMS to work and perform under fluctuating humidity levels (Akkerman et al., 2012) (Figure 9.12).

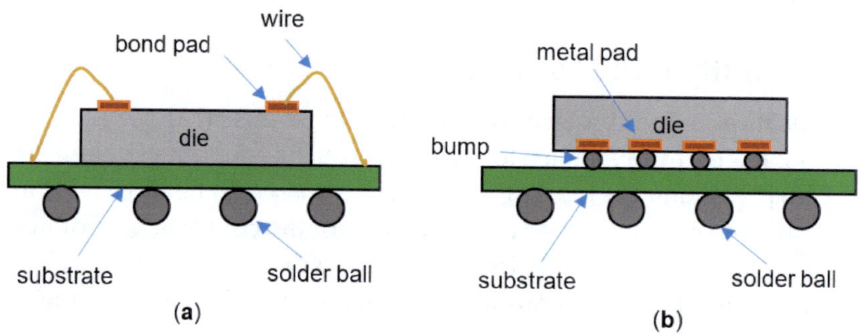

Figure 9.12. Inadequate packaging can result in detrimental effects such as internal corrosion, which compromises the functional integrity of MEMS.

Source: https://www.mdpi.com/sustainability/sustainability-14-15730/article_deploy/html/images/sustainability-14-15730-g001.png.

9.5.3. Chemical Exposure and Contamination Risks

The dependability of microelectromechanical systems (MEMS) hinges mostly on the utilization of correct packaging techniques, which serve as a shield against chemical exposure and contamination threats. The danger posed by these hazards is heightened by the miniature nature of MEMS, whereby any failure in a package can lead to system failures of disastrous magnitude. For instance, exposure of MEMS devices to strong chemicals can alter their mechanical properties and lead to corrosion or delamination. Not only does the packaging protect from environmental threats, but it is also required to provide a stable and reliable interface for long-term operation of such devices. The packaging thus becomes critical in preventing the chemical interaction, thus ensuring the reliability of MEMS, particularly in applications where the environment is hostile, such as automotive or medical applications (Coutu et al., 2009). MEMS reliability testing plays an important role in understanding failure mechanisms after exposure to chemicals. All of these test methods, including thermal cycling, humidity testing, and chemical resistance testing, are utilized to understand packaging's resistance to environmental stress. The results of these reliability tests inform the designers about weaknesses in the MEMS structure, which allows further improvement in packaging material and design. For example, chemical degradable materials can be quantified prior to commercial deployment, which is extremely crucial in preventing unexpected failure in the field. Quantification of chemical exposure hazards and their long-term consequences also allow estimation of device lifetimes and performance parameters, allowing MEMS to run within specifications designed for its entire lifetime (Fernandes et al., 2017). Environmental effects, the interaction between MEMS and environmental chemical substances, demand a sophisticated solution to packaging selection. Different types of packaging, e.g., hermetic and non-hermetic, provide varying levels of protection against contaminants. Hermetic packaging is usually superior in barrier properties, and hence it is most appropriate for applications where moisture or reactive gas exposure poses a concern. Whereas non-hermetic packaging will introduce some cost reduction, it also potentially introduces lower reliability of a greater likelihood to contain contaminants. Knowledge of the actual application environment aids in making intelligent choices in packaging types and thus improves overall MEMS reliability. As demand for MEMS technology rises in multiple industries, the key to chemical exposure and contamination risk levels during packaging will be central to moving and maintaining MEMS applications (Coutu et al., 2009) (Figure 9.13).

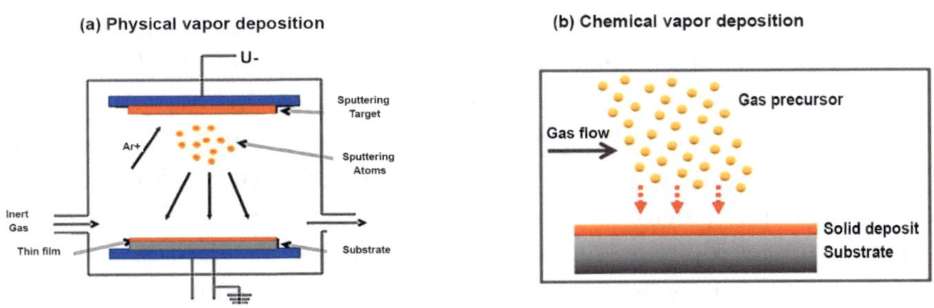

Figure 9.13. The exposure of MEMS components to aggressive chemicals can alter their mechanical properties and lead to corrosion or delamination.

Source: https://pub.mdpi-res.com/coatings/coatings-15-00878/article_deploy/html/images/coatings-15-00878-g005.png?1753522404.

9.6. STRATEGIES FOR ENHANCING MEMS RELIABILITY

The intricacy of Micro-Electro-Mechanical Systems (MEMS) places stringent packaging demands on them since many challenges arise due to the need to protect sensitive components without compromising performance. The reduction of the impact of environmental conditions, i.e., humidity and particulate contamination, is one of the main issues since these have the potential to drastically degrade the device's functionality. Packaging must address the likely failure mechanisms resulting from the environmental stresses of corrosion and mechanical fatigue. A well-designed package is also a protective barrier from the external forces and helps to maintain microstructures in a stress-free condition during operation stresses. To achieve this, it is essential to integrate novel materials and fabrication techniques into packaging solutions. Novel material-based designing with improved barrier qualities and mechanical robustness can enhance the reliability of MEMS substantially, such that devices respond reliably throughout their lifespan, even under harsh conditions (Oliver et al., 2009). Severe reliability testing is needed to understand the life and performance of MEMS devices under various operating conditions. Such tests encompass several techniques to simulate real application environments, i.e., thermal cycling, humidity profile, and mechanical stress application. By employing rigorous testing procedures, designers are able to identify likely failure modes at an initial phase of development. Temperature changes, for example, can further accelerate the deterioration of materials in MEMS packaging towards earlier failure if not properly addressed. Thus, utilizing accelerated life testing methods can give a good insight into the durability and lifespan of MEMS under different stressors. Analysis of data from such tests enables engineers to take informed decisions on design modifications and material selection, which ultimately leads to more reliable and long-lasting MEMS devices (Brand et al., 2013). The various challenges of MEMS reliability have necessitated the development of a range of packaging forms each addressing unique application needs. For example, hermetic packaging is a suitable remedy for reducing the impact of environmental considerations on delicate devices, while non-hermetic packaging may be cost-effective and easy to integrate. Packaging type choice exerts an enormous influence on general performance and reliability for MEMS devices, considering the importance of tailored approaches. Additionally, newer packaging trends, such as chip-scale packaging and hybrid integration, enable better thermal management and miniaturization, essential for modern applications. With the use of technology developments in packaging and flexibility in application-specific requirements, engineers will be in a position to make MEMS more trustworthy to guarantee successful integration in numerous industries. Further research on novel techniques will be crucial in overcoming the existing restrictions and achieving the complete potential of MEMS technology (Oliver et al., 2009).

9.6.1. Design Improvements for Robustness

Design resilience in Micro-Electro-Mechanical Systems (MEMS) is paramount to the system's functionality and reliability. MEMS are exposed to numerous packaging issues, some of which are sensitivity to environmental stresses such as humidity, temperature variations, and mechanical stress. A key factor in enhancing reliability is the selection of appropriate packaging materials that can withstand the stressors and provide adequate protection to delicate MEMS devices. For instance, parylene coatings are discovered to be of great benefit since they can coat devices and shield them from corrosive environments and also physical abrasion. Through the application of new materials and design methods like those exhibited in embedded chip integration, MEMS packaging can extend the life and performance of critical components when operating under harsh conditions. This pre-emptive approach addresses microscopic vulnerabilities that are the source of failure modes and provides assurance to the reliability of MEMS devices in their intended application. Reliability testing is a significant phase of developing robust MEMS packaging solutions as it allows engineers to simulate real-world operating conditions and identify probable failure modes. Techniques such as accelerated-lifetime testing, including thermal cycling and vibration testing, can detect how various environmental stresses influence function and integrity. To illustrate, certain works have demonstrated that certain RF MEMS devices, under constant usage and thermal shock, show significant performance variation, pointing to the need for rigorous testing in design processes. More broadly, integrating RF MEMS components into Low Temperature Co-Fired Ceramics (LTCC) technology has the possibility of making them tougher while simultaneously maintaining operational performance under diverse conditions. Such multi-dimensional test strategy not only demonstrates the ruggedness of different packaging shapes but also enlightens designers on best practices for MEMS design so that end products remain viable under harsh operating conditions without complete deterioration. Mitigation of various failure modes encountered in MEMS also requires a proper understanding of environmental factors that can compromise device life. Problems such as moisture penetration, particulate contamination, and mechanical stress are typical issues to be addressed at the design phase. For instance, inclusion of silicon encasements combined with protective layers minimizes some of the adverse effects with environmental exposure, such as successful implementations in RF identification chip applications (Li et al., 2008). Moreover, exploring different topologies and configurations, including cantilever and clamped–clamped beam geometries, can optimize best performance under conditions, delivering withstood reliability performance (Bartolucci et al., 2012). By meticulous planning of these parameters at the design level, engineers can develop MEMS packaging that not only achieves but exceeds levels of reliability, thereby making the overall feasibility of MEMS applications in industries more viable (Figure 9.14).

Figure 9.14. Coatings like parylene have been shown to offer substantial benefits, as they encapsulate devices, safeguarding them from corrosive environments and physical damage.

Source: https://4615657.fs1.hubspotusercontent-na1.net/hubfs/4615657/Imported_Blog_Media/BL0111_Parylene_Right_Featured.jpg.

9.6.2. Advanced Materials for Packaging

In the case of microelectromechanical systems (MEMS), the packaging problems are crucial to their performance and reliability. Miniaturization in MEMS components requires sophisticated materials that not only serve to shield sensitive structures but also support the complex designs typical of contemporary devices. Conventional packaging materials lack sufficient thermal stability and mechanical strength in response to changing environmental conditions. This deficiency can increase failure modes, such as delamination, water ingress, and mechanical stress, which significantly influence the device performance. To counter this, scientists are looking for novel materials—such as ceramics, polymers, and metals—that will provide better barrier capabilities and thermal management. Application of such materials in packaging designs resolves the reliability issues of MEMS directly and ensures their successful implementation in a variety of applications, for example, medical devices and automotive sensors, where environmental factors are most dominant. Testing for MEMS packaging reliability has also come to encompass a variety of methods for prediction under real conditions. Life testing at acceleration, thermal cycling, and mechanical shock are significant in seeing how materials respond to stressors under the test of time. One of the most crucial issues might be the influence of environmental conditions on packaged MEMS; variations in humidity, temperature, and mechanical loading can lead to increased material degradation. By simulating these conditions in a laboratory environment, engineers can identify where failure mechanisms could occur early in the design cycle. Moreover, ultra-high-performance materials suitable for the harsh environment are being tested under stringent test protocols, providing a route to the optimization of packaging strategies. With increasing complexity of MEMS devices, emphasis on understanding deeply the role of packaging in the reliability becomes increasingly relevant, and the need to apply novel materials in future designs is imperative. Various packaging styles like wafer-level packaging (WLP), chip-scale packaging (CSP), and hermetic sealing offer varying degrees of protection and reliability for MEMS devices. Each of them possesses unique strengths and weaknesses that dictate the selection of the materials for package building. For instance, WLP is in favor of miniaturization but may not perform exceptionally on moisture barrier capability compared to hermetically sealed packaging, which may be heavier and stronger. As more MEMS applications are made in fields such as high-performance environments in photonics and telecommunications, there will be more need for package solutions that span size, weight, and reliability, fueling innovation in material science. The use of MEMS-capable components, particularly silicon photonics—a technology that has only recently seen the introduction of novel materials in packaging to offer more functionality—illustrates the value added by innovative solutions in this regard, as represented by efforts employing IMEC's standard iSiPP50G silicon photonics platform (Bogaerts et al., 2020).

9.6.3. Implementation of Redundancy and Fault Tolerance

In Micro-Electro-Mechanical Systems (MEMS), increasing device complexity and shrinking pose enormous packaging challenges that directly affect reliability. The encapsulation methods employed must safeguard MEMS from environmental stresses in the form of humidity, temperature fluctuation, and particulate contamination, all of which worsens device performance. The package must also meet stringent mechanical requirements to avoid fragility during device operation and manufacturing process. Effective redundancy practices in packaging help enhance operational integrity by providing redundant pathways or components to take over duties upon failure. Resolution of the packaging problems through innovative material selection and design plays an essential role in minimizing failure mechanisms such as fatigue and fracture, which are exacerbated by environmental stress (Chen et al., 2025). Strategic use of redundancy in packaging is an important component in the development of MEMS reliability in a range of applications. Reliability testing is a critical component of MEMS analysis because it informs designers about possible modes of failure that can arise under operation. This involves subjecting devices to a variety of mechanical, thermal, and electrical stress tests to ensure performance in less-than-ideal operating conditions. By identifying and prioritizing possible failure modes, engineers can implement targeted redundancy controls to minimize risk. As an illustration, the implementation of dual sensor architectures in MEMS can offer critical data redundancy, such that operation remains possible even upon failure in one sensor. Furthermore, investigation of environmental influences, such as radiation or extreme temperatures, is at the heart of making fault-tolerant MEMS packaging possible. Knowledge of how different packaging structures impact overall reliability justifies selection of material and structure that enhances device ruggedness, as part of the overall objectives of improved aerospace programs like NASA's Vision for Space Exploration (Adams et al., 2008). Redundancy and fault tolerance are not only technical necessities but strategic necessities in advancing MEMS technology in critical applications. As MEMS devices find broader application in high-reliability environments, the failure ramifications become more severe. Fault tolerance methods can significantly improve operational reliability, and these can include hardware as well as software approaches designed to detect and recover faults autonomously. As an example, employing self-diagnostics algorithms and redundant channels ensures that MEMS systems remain operational successfully despite any anomalies. This synergy between packaging approaches and fault tolerance techniques is vital to the process of innovation in the MEMS field, particularly in the context of strict reliability environments, such as those found in aerospace and medical fields. If engineers focus on these, they can create reliable MEMS technologies that meet existing reliability standards of emerging applications while laying the foundation for future innovation in this rapidly changing sector.

9.7. CONCLUSION

While designing the Micro-electro-mechanical systems (MEMS) technology, special attention needs to be paid to the packaging issues. Since MEMS devices are being integrated with newer and complex functions, the packaging aspect plays a very important role in determining their reliability and performance. The packaging must ensure that these small, sophisticated devices are safe from environmental stresses such as humidity and temperature fluctuations, which have adverse effects on their functioning. Further, various forms of packaging—ranging from hermetic to non-hermetic—boast certain advantages and disadvantages that influence the overall reliability of MEMS. Lacking in solving packaging issues adequately, the development of MEMS use in common industries might be precluded, with the issue of continued research and technological innovation in the field of packaging technologies to enhance device life and operation. Reliability testing emerges as an important component in the determination of MEMS device performance over their intended life span. Very stringent test methods aim to uncover prospective failure mechanisms that compromise device integrity when put under operational conditions. These could include mechanical wear and fatigue, along with environmental factors contributing to degradation. By analyzing these modes of failure, scientists are able to design more dependable packaging solutions aimed at lessening risk from MEMS deployment. Further, the establishment of standard testing protocols is important in offering consistency and equivalence of reliability information between different MEMS devices. This not only encourages manufacturers' accountability but also helps establish confidence in MEMS technology among consumers and stakeholders, thus encouraging increased usage of MEMS applications. Environmental impact knowledge is important in ensuring confidence that MEMS devices will remain reliable even after being exposed to actual environmental conditions. As highlighted in the case of next-generation technologies, sensor integration and wireless communication will bring MEMS to more dynamic environments, which demand strong packaging solutions. Future-generation technologies such as flexible sensors are being created to address the vulnerability of conventional rigid sensors, but their marketplace uptake remains low due to several bottlenecks achieved in current research. The development of smart flexible sensors, together with a comprehensive roadmap for MEMS packaging, can lend guidance to collaborative efforts toward meeting these challenges. By addressing collective work across disciplines, advancements in MEMS technology can be expedited, helping to keep pace with the overall vision of an interconnected world driven by progress in wireless communication and IoT technologies, as emphasized by Luo et al. (2023) and Ian et al. (2020).

9.7.1. Summary of Key Points Discussed

The dynamic interaction between the operating conditions and structural stability of micro-electromechanical systems (MEMS) necessitates an intimate knowledge of packaging issues. Specifically, MEMS devices are prone to extreme pressure and temperature, risking their functionality. For instance, in fluidic packaging, the reliability of hermetic seals is critical. Critical thermal and pressure limits need to be ensured, as exemplified by research in which seals with temperatures of 20–600°C and pressures between 100 and 300 atm were examined (Harrison et al., 2004). Fracture failures commonly result from pre-existing flaws or cavities within the seal material, and this highlights the need for careful processing and proper choice of material to provide reliability. Thus, overcoming these issues is important not just for device performance but also to extend the operating life of MEMS technology in applications. Reliability testing is a significant activity that is employed to calculate the effective performance of MEMS packaging. It is the technique of simulating operational conditions to predict potential failure modes during the device lifetime. A number of factors like mechanical stress, thermal cycling, and environmental exposure play a strong impact on reliability results. Applying Weibull statistical analysis allows researchers to model the probability of failure, a useful measurement of packaging integrity under reference operating pressures, as indicated by results with a determination of 0.99 survival at 125 atm for individual tubes (Harrison et al., 2004). The residual stress analysis by finite element methods further assists in ascertaining the critical locations of failure. As the MEMS market continues to mature, reliability testing will need to take precedence in order to develop robust packaging solutions that have withstood the stresses of real-world operation. Packaging form development for MEMS devices is obviously moving ahead, and low-cost integration processes with high reliability are now the focus. Historically, glass and silicon have been the dominant conventional materials in MEMS manufacturing; however, improvement in flexibility and versatility in deployments is now called for. New breakthroughs have indicated the use of Printed Circuit Board (PCB) materials as substrates when creating microfluidic channels, which will provide cost-effective integration methods that bring together electronic and fluidic components in harmony. This approach enhances the serious inadequacies of existing packaging schemes by ensuring better scalability and reliability. With the progress of technology, material scientists and MEMS engineers in collaboration will be essential to address the issues of environmental concerns as well as the lifespan of such miniaturized systems across diverse application areas.

9.7.2. Future Trends in MEMS Packaging and Reliability

In the field of microelectromechanical systems (MEMS), the packaging issues present significant challenges that become increasingly important to address. The issues arise due to the need to maintain performance while designing the package to become durable for use across different environments. As MEMS devices are finding applications in increasing fields, such as consumer electronics, automotive systems, and healthcare, their packaging must evolve to thrive in dynamic environments. Standard failure modes such as mechanical overstress, fatigue, and the effect of moisture ingress necessitate state-of-the-art packaging techniques. Also, as miniaturization in MEMS devices continues to advance, standard packaging would result in deteriorated reliability due to increased thermal and mechanical stress. Future technologies should focus on the integration of intelligent packaging technology that can provide real-time detection of device integrity in order to guarantee long-term performance and rule out the risk of premature failure (Charalampidis et al., 2010). Testing for reliability is now the cornerstone in the design of MEMS devices, which must meet rigid industry standards to meet the requirements of the marketplace. Historical approaches to reliability testing have had a past reliance on established standards based on larger semiconductor devices. However, because of the unique nature of MEMS, such as their micro-scale dimensions and sensitivity to environmental conditions, unique test methodologies are necessary. Accelerated life tests and environmental test protocols of particular interest are as they simulate actual operating conditions under which MEMS devices are utilized in various applications. Reliability testing for the future will likely see more sophisticated, data-elicited approaches such as machine learning models that are able to predict failure mechanisms based on historical performance data. These technologies will not only improve the reliability of MEMS devices but also learn how different packaging configurations influence overall durability and operational life (Oliver et al., 2009). The environmental effect on MEMS reliability cannot be overemphasized, as the systems are exposed to their environment and subject to thermal, humidity, and mechanical shock variations. As MEMS applications find more applications in wearable technology and smart sensors, effects of ambient conditions will be even more important. This necessitates the development of highly robust packaging solutions able to efficiently mitigate environmental stressors. New forms of packaging, such as those based on bio-responsive materials, might offer dual functionality—protection by acting as barriers and improving device performance through adaptability to changes in the environment. Such advancements will be critical in providing reliability, particularly in applications designed for extended use in harsh environments. The integration of such advanced high materials in MEMS packaging is set to revolutionize reliability measures and longevity standards, paving the way for more durable microdevices (Charalampidis et al., 2010).

9.7.3. Final Thoughts on the Importance of Addressing These Challenges

MEMS packaging complexity needs timely attention and innovative solutions to enhance performance and reliability. Given that MEMS devices are increasingly integrated into a variety of applications, from medical equipment to automotive systems, packaging challenges need to be looked into. Mechanical stress, thermal management, and encapsulation processes are merely some of the concerns that affect device performance considerably. The unique character of MEMS, including their miniaturized sizes and sensitive structures, increases the importance of efficient packaging solutions that are capable of shielding devices from outside factors while facilitating durability and efficiency. Without addressing these concerns, the mass adoption and commercialization of MEMS technology may be halted and, in the long run, impair likely advances in industrial areas and the quality of life as seen in recent controversies surrounding sensors' applications (Luo et al., 2023). In addition, information on MEMS devices' reliability cannot be ignored. Reliability testing is key to identifying failure mechanisms that may appear during operation, particularly under harsh environmental conditions. Factors such as humidity, temperature gradient, and mechanical vibrations play high impact on the reliability and lifespan of MEMS devices. Moreover, the complexities in combining different materials call for high-level inspection as the interaction between them could result in unforeseen effects on the failure modes. Solutions to these problems through effective testing strategies enable the manufacturers to achieve the highest possible performance and shelf life of MEMS devices. Thus, proactive reliability strategy not only increases the level of confidence in consumers but also makes it easier to transfer from lab-scale research to real application in field conditions, thus fostering innovation and progress in various sectors (Nikoli et al., 2020). Lastly, the different packaging forms of MEMS devices must be researched to adequately deal with new sensor technology and reliability challenges. Other packaging options, including hermetic, hybrid, and integrated packaging, possess particular strengths and limitations. The proper choice of packaging depends upon the application's specific needs, for example, sensitivity, environmental responsiveness, and cost. Moreover, the establishment of knowledge for new materials and methods optimizes the capability for creating more effective packaging solutions. It is crucial that industries and researchers collaborate in this effort, as comprehensive strategies will have a significant impact on the trajectory of MEMS reliability and adoption. Ultimately, resolving the interconnected problems of MEMS packaging, reliability, and environmental conditions will realize the complete potential of such systems to address the demands of an increasingly digitalizing and networked world (Luo et al., 2023).

REFERENCES

1. Achkar, H., Aubert, H., Coccetti, F., et al. (2006). BCB based packaging for low actuation voltage RF MEMS devices. http://arxiv.org/abs/0709.3022 (accessed on 14th October 2025).
2. Adamietz, R., Bailey, C., Desmulliez, M. P. Y., et al. (2018). Reliability testing and stress measurement of QFN packages encapsulated by an open-ended microwave curing system. https://core.ac.uk/download/160605962.pdf (accessed on 14th October 2025).
3. Adams, J. H., Cressler, J. D., Darty, R. C., et al. (2008). High-performance, radiation-hardened electronics for space and lunar environments. https://core.ac.uk/download/pdf/10543416.pdf (accessed on 14th October 2025).
4. Akkerman, R., Bastemeijer, J., Bossche, A., et al. (2012). Low-cost technology for the integration of micro- and nanochips into fluidic systems on printed circuit board: Fabrication challenges. https://core.ac.uk/download/pdf/11483959.pdf (accessed on 14th October 2025).
5. Akyildiz, I. F., Kak, A. C., & Nie, S. (2020). 6G and beyond: The future of wireless communications systems. *IEEE Access, 8*, 133995–134030. https://doi.org/10.1109/access.2020.3010896 (accessed on 14th October 2025).
6. Bartek, M., Gaddi, R., Gnudi, A., et al. (2006). Parasitic effects reduction for wafer-level packaging of RF-MEMS. https://core.ac.uk/download/15488624.pdf (accessed on 14th October 2025).
7. Bartolucci, G., Casini, F., De Angelis, G., et al. (2012). RF MEMS ohmic switches for matrix configurations. https://core.ac.uk/download/pdf/53861235.pdf (accessed on 14th October 2025).
8. Bauer, T., & Rimskog, M. (2008). High density through silicon via (TSV). https://core.ac.uk/download/15497651.pdf (accessed on 14th October 2025).
9. Beacken, M., Bishop, D. J., Chang, J., et al. (2018). High performance, continuously tunable microwave filters using MEMS devices with very large, controlled, out-of-plane actuation. http://arxiv.org/abs/1804.08658 (accessed on 14th October 2025).
10. Begbie, M., Kotsov, D., Li, L., et al. (2010). Design and simulation of a multi-function MEMS sensor for health and usage monitoring. https://core.ac.uk/download/68020.pdf (accessed on 14th October 2025).
11. Bharatula, N., Lukowicz, P., & Tröster, G. (2018). Functionality-power-packaging considerations in context aware wearable systems. https://core.ac.uk/download/159152136.pdf (accessed on 14th October 2025).

12. Bogaerts, W., Edinger, P., Errando-Herranz, C., et al. (2020). MEMS-enabled silicon photonic integrated devices and circuits. https://core.ac.uk/download/287939692.pdf (accessed on 14th October 2025).
13. Brand, O., Caillard, B., Debeda, H., et al. (2013). Effect of hydrodynamic force on microcantilever vibrations: Applications to liquid-phase chemical sensing. https://core.ac.uk/download/213077385.pdf (accessed on 14th October 2025).
14. Buss, A., Ceylan, H., Ceylan, H., et al. (2013). Highway infrastructure health monitoring using micro-electromechanical sensors and systems (MEMS). https://core.ac.uk/download/38934593.pdf (accessed on 14th October 2025).
15. Bussooa, A., Mercer, J. R., & Neale, S. (2018). Future of smart cardiovascular implants. https://core.ac.uk/download/161513601.pdf (accessed on 14th October 2025).
16. Charalampidis, N., Kampitaki, D., Mallios, N., et al. (2008). An ohmic RF MEMS switch for reconfigurable microstrip array antennas built on PCB. https://core.ac.uk/download/336424.pdf (accessed on 14th October 2025).
17. Charalampidis, N., Nilavalan, R., Spasos, M., et al. (2010). Analysis and design of an all metal in line series ohmic RF MEMS switch for microwave applications. https://core.ac.uk/download/336972.pdf (accessed on 14th October 2025).
18. Chen, P. J., Humayun, M. S., Rodger, D. C., et al. (2008). Microfabricated implantable parylene-based wireless passive intraocular pressure sensors. https://core.ac.uk/download/4880373.pdf (accessed on 14th October 2025).
19. Chen, Y., Hodson, R., Hughes, M., et al. (2025). Flight avionics hardware roadmap. https://core.ac.uk/download/pdf/10574787.pdf (accessed on 14th October 2025).
20. Colomer i Farrarons, J., González-Piñero, M., Juanola, E., et al. (2015). Design of a customized multipurpose nano-enabled implantable system for *in-vivo* theranostics. https://core.ac.uk/download/43551928.pdf (accessed on 14th October 2025).
21. Coutu, R. A., McBride, J. W., & Starman, L. A. (2009). Improved micro-contact resistance model that considers material deformation, electron transport and thin film characteristics.
22. Czaplewski, D., Hyman, D., Karatzidis, D., Rebeiz, G., Newman, H., Sumali, H., McKillop, J., et al. (2011). RF-MEMS switch actuation pulse optimization using Taguchi's method. https://core.ac.uk/download/337977.

pdf (accessed on 14th October 2025).

23. Dennehy, N. (2025). Vision for micro technology space missions. https://core.ac.uk/download/pdf/10513661.pdf (accessed on 14th October 2025).
24. Dong, H. (2016). Design of the contact metallization's for gold-tin eutectic solder–A thermodynamic-kinetic analysis. https://core.ac.uk/download/80720756.pdf (accessed on 14th October 2025).
25. Dorado Castaño, A. D., Buguña, D. G., Buguña, G., et al. (2015). Biofilm dynamics characterization using a novel DO-MEA sensor: Mass transport and biokinetics. https://core.ac.uk/download/41777652.pdf (accessed on 14th October 2025).
26. Evans, J. W., & Ghaffarian, R. (2025). Enabling more than Moore: Accelerated reliability testing and risk analysis for advanced electronics packaging. https://core.ac.uk/download/pdf/42725204.pdf (accessed on 14th October 2025).
27. Fernandes, L. A. L. (2017). The development of a miniaturized hydration sensor based on the principle of osmotic pressure. https://core.ac.uk/download/650685548.pdf (accessed on 14th October 2025).
28. Gao, S., & Holmes, A. S. (2006). Thermosonic flip chip interconnection using electroplated copper column arrays.
29. Harrison, T. S., Mracek, A., Peles, Y., & Spearing, et al. (2004). Fluidic packaging of microengine and micro rocket devices for high pressure and high temperature operation.
30. Herrick, K. J., Katehi, L. P. B., Margomenos, A., et al. (2001). Silicon micromachined packages for RF MEMS switches. https://core.ac.uk/download/11067921.pdf (accessed on 14th October 2025).
31. Kopylov, G. (2015). Overview of sensors suitable for active flow control methods. https://core.ac.uk/download/30294554.pdf (accessed on 14th October 2025).
32. LaBel, K. A., & Sampson, M. J. (2025). The NASA Electronic Parts and Packaging (NEPP) Program: Roadmap for FY15 and beyond. https://core.ac.uk/download/pdf/42707403.pdf (accessed on 14th October 2025).
33. Li, W., Rodger, D. C., & Tai, Y. C. (2008). Implantable RF-coiled chip packaging. https://core.ac.uk/download/4885310.pdf (accessed on 14th October 2025).
34. Lo, H., & Tai, Y. C. (2008). Parylene-based electret power generators. https://core.ac.uk/download/4879299.pdf (accessed on 14th October 2025).
35. Luo, Y., Abidian, M. R., Ahn, J. H., Akinwande, D., Andrews, A.

M., Antonietti, M., Bao, Z., et al. (2023). Technology roadmap for flexible sensors. *ACS Nano, 17*, 5211–5295. https://doi.org/10.1021/acsnano.2c12606 (accessed on 14th October 2025).

36. Nikolić, M. V., Milovanović, V., Vasiljević, Z. Ž., & Stamenković, Z. (2020). Semiconductor gas sensors: Materials, technology, design, and application. *Sensors, 20*, 6694–6694. https://doi.org/10.3390/s20226694 (accessed on 14th October 2025).
37. Oliver, R., Tillotson, J., & Toomey, A. (2009). Bioengineered textiles and nonwovens – The convergence of bio-miniaturization and electroactive conductive polymers for assistive healthcare, portable power and design-led wearable technology. https://core.ac.uk/download/9587153.pdf (accessed on 14th October 2025).
38. Rencz, M., & Vass-Varnai, A. (2008). Package hermeticity testing with thermal transient measurements. https://core.ac.uk/download/15497634.pdf (accessed on 14th October 2025).
39. Sharon, G. (2011). Modeling the physics of failure for electronic packaging components subjected to thermal and mechanical loading. https://core.ac.uk/download/56108471.pdf (accessed on 14th October 2025).

CHAPTER 10

SIGNAL CONDITIONING AND INTERFACE ELECTRONICS

Contents

INTRODUCTION

The intricate field of signal conditioning and interface electronics is an essential component of modern technology, significantly contributing to a wide range of applications across various fields. With sensor systems becoming a vital part of industrial as well as consumer goods, signal processing methodologies have become most vital. Good signal conditioning facilitates data conversion and transmission from sensors, achieving optimum performance from microelectromechanical systems (MEMS) with higher functionalities. This encompasses crucial processes like amplification and filtering, which preserve the transmitted signals unaltered and without noise, critical to accurate interpretation of data. Further, advancements in microelectronics ever-dynamically reproduce the signal conditioning landscape through innovative methods to address the complexities that come with MEMS signal processing. The transition between analog and digital formats demands intensive A/D and D/A conversion operations, indicating the need for interface electronics. Such conversions are needed since they facilitate the seamless integrating of various sensor data types to digital systems applied in analysis as well as control. Therefore, the exploration of noise considerations becomes paramount because it directly relates to the reliability of data acquisition. These problems concern the operation of MEMS devices with regard to external electromagnetic interference and temperature changes. By solving these issues, engineers and scientists will be able to design more robust systems withstanding exacting performance specifications in real-world applications, in line with the forces of technology driving existing electronics. Power management then also takes on a prominent role within the discipline of signal conditioning and interface electronics, specifically in the context of the durability and functionality of MEMS devices. As these systems are often employed in environments where energy resources are not readily available, effective power management techniques need to be used to optimize performance yet sustain function. Finding various circuit configurations and mechanisms of energy optimization, such as energy harvesting principles, is most important for enhancing the overall efficiency of MEMS applications. Moreover, innovations in device structure, as seen in RF and digital hierarchical designs exemplified in recent studies, are a turning point that solves rising demands for power efficiency and computing power in sophisticated electronic systems (Adami et al., 2010). By integrating such cutting-edge technologies, the field of signal conditioning is poised to make gigantic leaps that not only enhance existing systems but also pave the way for innovation in the future (Chan et al., 2025).

Definition of Signal Conditioning

Signal conditioning in signal processing has a central role to play as early as possible, in that it guarantees sensor signals are correctly processed before they can be utilized by control systems or data acquisition hardware. The process involves different tweaks—powering down small signals, filtering noise, and converting analog signals into digital format via A/D (analog-to-digital) converters. Especially in Micro-Electro-Mechanical Systems (MEMS), signal amplification is required due to the typically low levels of signals produced by these tiny sensors. Proper signal management allows for more stable and reliable readings, enabling better system performance across applications ranging from automotive electronics to industrial control. The importance of such conditioning processes cannot be overstated as they are the foundation for proper interpretation of data and robust electronic interfacing. Furthermore, the interface electronics being employed along with signal conditioning are also crucial in addressing the intricate problems being posed by embedding sensors.

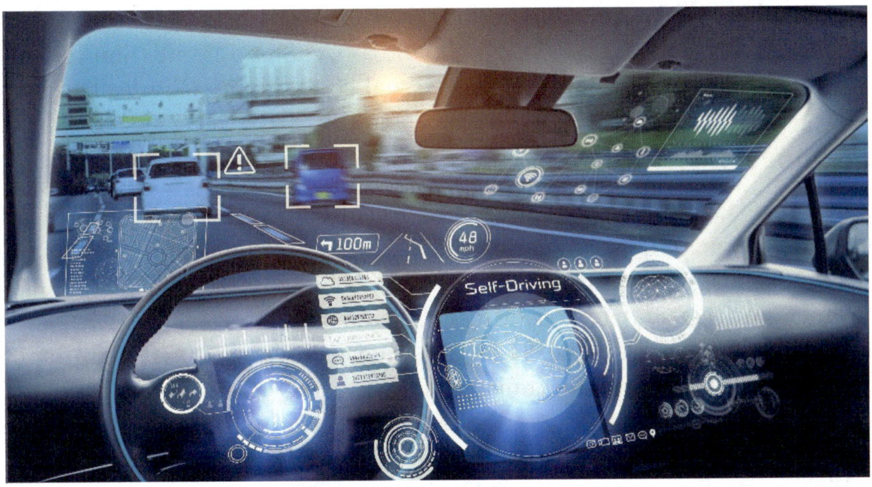

Figure 10.1. The role of signal conditioning becomes crucial as it ensures that sensor outputs are adequately processed before being utilized by control systems or data acquisition devices.

Source: https://www.volersystems.com/hubfs/Blog/Sensor%20Signal%20Conditioning-1.jpg.

For instance, noise cancellation techniques are essential in improving data quality transmitted, particularly in electromagnetic interference-heavy environments. Signal conditioning employs various filtering techniques to eliminate undesirable frequency components such that the system can receive only the desired signal. In addition, power management within MEMS devices is crucial. Power conditioning solutions such as those outlined within research on electronic control units indicate the ability of effective power management in decreasing hurdles for heat dissipation and system reliability. These advancements promote the functional capabilities of signal conditioning and affirm its place within the overall framework of interface electronics. Lastly, the combination of different technologies necessitates a globalized model towards modelling signal conditioning systems, particularly as they are applied to different applications. A shift towards generic modelling frameworks, as exemplified in some electro-optic sensor systems, provides one with the big picture in terms of how these systems operate under different conditions and specifications. By imagining a shared model, it is possible to look into design trends, ultimately enhancing both the effectiveness and efficiency of signal conditioning practice. The convergence of multiple kinds of models, from parametric to image-flow descriptions, allows engineers to tailor their systems for specific functions within the infrared waveband or some other range. This adaptability is not only crucial to optimize performance but also to look forward to changes throughout the product life cycle, providing a foundation for future innovations in interface electronics and signal conditioning (Figure 10.1).

Importance of Interface Electronics

The interface electronics are crucial in the use of MEMS (Micro-Electro-Mechanical Systems), particularly in processes of signal amplification and filtering. Sophisticated MEMS devices with integrated onboard interface electronics are essential for ensuring precision in sensor readings as well as minimizing noise interference. By employing techniques such as differential signaling and algorithms for noise filtering, such systems can effectively eliminate weak signals to processing levels. An example of improvement can be seen in the novel POSFET tactile sensing chip that integrates multiplexers and high compliance current sinks into interface electronics to reduce performance loss and interconnections. This. This. This technology not only provides increased sensitivity and linearity of response within specified dynamic ranges but also illustrates how better interface electronics can reduce the complexity of MEMS operation to make it more applicable in real-world applications (Adami et al., 2014). Power management is another key area of interface electronics, particularly with regards to maintaining MEMS device performance across various operating environments. Power consumption plays a central role in overall system performance, and for battery-powered systems, it is especially important. Effective interface electronics allow sophisticated power conditioning techniques to achieve maximum utilization of energy and minimize heat dissipation, as discussed in research on engine control systems. The systems are assisted by next-generation power supply concepts and actuation, giving long-term reliability and reduced lifecycle cost to electronic control units (ECUs). With the application of switching shunt power conditioners, thermal generation within the ECU is largely prevented, an important component of overall reliability in such control systems. Proper power control in interface electronics thus is not an ancillary function but an elementary component enabling efficiency of operation (Beattie et al., 2025). Lastly, the integration of analog-to-digital (A/D) and digital-to-analog (D/A) conversion circuits into interface electronics is essential in today's sensor applications. Analog signal conversion of the MEMS sensors into the digital realm allows further advanced processing and analysis by microcontrollers or digital signal processors for enhanced data interpretation and decision-making. Moreover, D/A conversion facilitates the rebuilding of processed data into actuation analog signals, thus creating a continuous input and output cycle that facilitates responsive systems. The sophisticated interface electronics thus play a central role not only in making accurate conversions but also in managing the resultant noise factors involved within such procedures. An integrated strategy towards signal conditioning and interface capability is thus needed for the development of robust MEMS applications in a variety of industries, from medical devices to automotive engineering (Figure 10.2).

Figure 10.2. Power management emerges as another critical aspect of interface electronics, particularly for sustaining the functionality of MEMS devices under varying operational conditions.

Source: https://www.azosensors.com/images/Article_Images/ImageForArticle_3138_17337809726788234.jpg.

Overview of Applications in Modern Technology

The use of Micro-Electro-Mechanical Systems (MEMS) in modern technology has revolutionized the operation in regard to signal conditioning and interface electronics. Due to the creation of MEMS signal amplifiers and filtering, engineers are now able to achieve high-quality data acquisition with minimal noise interference. This is particularly important in medical equipment and environmental monitoring where precision in measurement is essential. To complement this, A/D and D/A conversion technologies have been instrumental in converting analog signals into digital form so that any noise which may have been introduced by the sensors, or the environment is properly managed. Accuracy focus makes systems which rely on accurately interpreting data more dependable, hence providing wider applications for the use of signal conditioning technology in aerospace, robotics, and consumer electronics industries. Noise issues have to be tackled if sensor interfacing is to be effective, especially with the increased complexity of modern electronic systems. Signal integrity can be compromised by several external and internal noise sources, and so new techniques need to be evolved to counteract them. Techniques such as differential signaling, filtering, and the employment of special materials are employed to reduce the noise floor so that the transmitted signals remain in their original state. As the complexity of sensor designs increases, the ability to implement efficient solutions for noise mitigation directly affects their applicability in demanding applications, such as high-speed data acquisition systems utilized in astronomical observations, as observed in the successful astronomical observations of the Crab Nebula Pulsar utilizing high-technology tools like the Peripheral Component Interconnect bus and Field Programmable Gate Arrays (Eikenberry et al., 2001). MEMS device power management also indicates the growing importance that signal conditioning has within today's technology. As devices become smaller and power use is a controlling factor, efficient power management methods must be used in order to extend functionality without sacrificing performance. Low-power signal conditioning circuitry has been improved in an effort to allow MEMS devices to operate efficiently under limiting energy conditions. This not only enhances battery life in mobile applications but also minimizes thermal issues that would be generating noise and signal loss. As the aviation industry progresses toward Integrated Modular Avionics (IMA) for commercial airframes, implications of power management of multi-functional systems are even more significant. The aviation sector's ongoing migration toward high-end avionics structures necessitates robust and reliable signal conditioning compatible with the very high reliability and safety standards inherent in this application (Ananda et al., 2009).

10.1. FUNDAMENTALS OF SIGNAL CONDITIONING

In signal conditioning, managing noise considerations within sensor interfacing is of utmost importance. Since sensors are normally designed to measure infinitesimal changes in their surroundings, any unwanted noise can be catastrophic, making their output unreliable. To overcome this problem, engineers use numerous approaches, one of which is filtering techniques, which are instrumental in improving signal integrity. More advanced filtering techniques not only strike the noise frequencies of interest but also preserve the desired signal undisturbed so that the sensed data can be approximated more accurately. The application of signal amplification is also a sine qua non to amplify these low-level signals in order to improve the signal-to-noise ratio before the analog-to-digital (A/D) conversion occurs. With the incorporation of comprehensive noise management solutions, sensor interfaces are optimized for performance with precise data capture for further inspection. Signal integrity also finds application in the field of MEMS devices, where robust signal amplification and filtering means are required. MEMS sensor characteristics are often varied, for example, lower voltage outputs, and hence it is crucial to have robust signal conditioning mechanisms.

The techniques applied—on a scale from operational amplifiers to digital filtering programs—play a crucial role in extracting useful data out of these devices. Additionally, the use of specialized A/D and D/A converters plays a fundamental role in transferring analog information into and from digital forms. Such converters facilitate effortless data interfacing between MEMS devices and digital circuitry, enabling real-time signal processing. Besides signal amplification and correct signal conversion, power management must be paid attention to by designers in such a way that MEMS devices are operated efficiently and sustainably for a very long time without compromising performance. Besides signal conversion and amplification, the very spirit of signal conditioning highlights the importance of power management in the context of MEMS devices. Power management methods that are power efficient not only add extended working duration to such devices but also enhance performance characteristics in applications. This is a serious concern in wearable and portable systems where power constraint is the primary topic of concern. Keeping sensors in power-optimal condition while generating high-fidelity signals needs proper design considerations, including low-power circuits and energy harvesting techniques. Also, as the necessity for real-time data processing is strengthened with the progress in bio-electronics interfaces, as described in recent literature, there is a necessity to incorporate smart power management systems. With effective balance between power consumption, signal quality, and device lifetime, engineers can unlock the full potential of MEMS-based signal condition technologies in many fields, including healthcare and environmental monitoring, identified by current research and development directions (Muniandi et al., 2024; Zhao et al., 2023) (Figure 10.3).

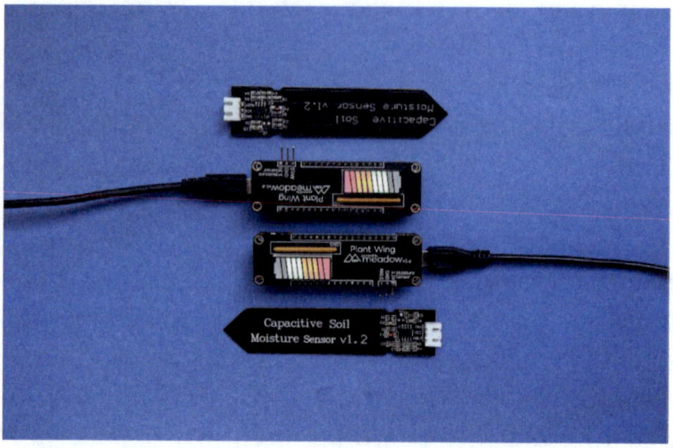

Figure 10.3. MEMS sensors often exhibit unique characteristics such as low output voltage levels, making it imperative to implement robust signal conditioning techniques.

Source: https://wevolver-project-images.s3.us-west-1.amazonaws.com/0.lleww-2p5qvjorge-ramirez-VNfiAyRoKv4-unsplash.jpg.

10.1.1. Purpose and Functionality

In signal conditioning and interface electronics, the complex interaction of differing components has a direct bearing on the performance of electronic systems. A key part of this process is MEMS (Micro-Electro-Mechanical Systems) signal amplification and filtering, ensuring raw sensor data is converted to useful forms. The amplification process enhances weak signals produced by MEMS devices and therefore helps them overcome noise to produce higher fidelity outputs. Secondly, filtering methods are an important feature in eliminating unwanted frequency contents and therefore cleansing the signal prior to further processing. This is required since residual noise distorts data interpretation results into erroneous system responses. Hence, the delivery of a high-quality stream of signals not only enhances the reliability of measurement but also extends the functional life of various applications in robotics, consumer electronics, and many others. With the advancement of technology, effective A/D (Analog-to-Digital) and D/A (Digital-to-Analog) conversion is a critical element in the operation of signal conditioning systems. These conversions facilitate the bridging of the analog realm in which physical signals are experienced and the digital realm in which processing and analysis occur. The level of precision in these conversions has a direct impact on the quality of the processed output that is then utilized in applications ranging from medical devices to car systems. Moreover, the use of strong algorithms in these conversions is also capable of enhancing overall performance, making processing of complex signals more efficient with reduced latency. All of this technological cooperation between A/D and D/A conversions effectively facilitates the ideal usage of sensor technologies such that not only are the signals captured but also idealized in a form compatible with the application of the system. Noise in sensor interfacing is of utmost concern, since it greatly affects the reliability and performance of electronic devices. The correct signal conditioning needs a complete understanding of potential sources of noise, including electromagnetic interference, thermal noise, and environmental-caused fluctuations. With the inclusion of robust filtering and amplification techniques, designers can minimize these sources of noise to a fraction, hence achieving the highest integrity of sensor outputs. This focus on noise reduction is especially critical in those applications wherein accuracy is a concern, such as with medical instrument or industrial monitoring system uses. Furthermore, sound power management for MEMS devices also remains a key consideration, as inadequate levels of power can further amplify noise issues and decrease overall performance. With the integration of effective noise management methods with sound power management, engineers are able to optimize the functional efficiency and life of sensor-based systems, with new fields of creative applications in various domains.

10.1.2. Types of Signal Conditioning Techniques

In signal conditioning, several methods are significant in enhancing the performance and reliability of electronic systems, particularly with Micro-Electro-Mechanical Systems (MEMS). MEMS signal filtering and amplification is one significant method where the objective is to clean the outputs from sensors by strengthening the signals and reducing unwanted noises. Various amplifier arrangements can be employed, like operational amplifiers and differential amplifiers, which can amplify signals with high precision. In addition, filtering methods—both digital and analog—are a critical component of eliminating noise that can hide the relevant signal. These methods of filtering further increase the fidelity and accuracy of measurements by MEMS devices in a way that signals propagated to subsequent stages of processing are not damaged and remain valuable in practical applications, an aspect that is paramount in instances of biomedical sensing and environmental monitoring. Another significant area that falls under signal conditioning is the Analog-to-Digital (A/D) and Digital-to-Analog (D/A) conversion processes. Such operations form the foundation for enabling the seamless interfacing of analog sensor signals to digital processing platforms. Analog output from MEMS devices is converted to digital by A/D converters so that digital signal processors can manipulate and analyze data. D/A converters, conversely, translate digital signal into analog form, often required for applications requiring controlled actuation or signal output. The effectiveness of these converters depends not only on resolution but also on sampling rates and linearity, which have significant implications for the signal quality and for the overall system performance. With the pervasiveness of digital systems, it is important that these conversion processes be maximized to facilitate true analog-digital interfacing. In addition, power management is a challenge and opportunity in the situation of MEMS signal conditioning techniques. Energy efficiency for MEMS devices is essential, particularly when used in battery-driven systems or environments where power must be saved. Power management schemes can use a number of mechanisms to scale down energy usage, including dynamic voltage scaling and duty cycling, which enable MEMS devices to operate effectively without performance compromise. Additionally, integrating power management techniques with filtering and amplification processes assists the overall system in attaining optimal performance while maximizing sensors' operational life. As seen in contemporary literature, efficient handling of power not only increases duration but also enhances signal processing reliability and quality, particularly for complex applications such as those applied in the Internet of Things (IoT) and smart wearable technology, which necessitate accuracy and effectiveness in signal maintenance and utilization (Bojoi et al., 2009) (Figure 10.4).

Figure 10.4. Various amplifier configurations can be employed, including operational amplifiers and differential amplifiers, which can enhance signals with high precision.

Source: https://upload.wikimedia.org/wikipedia/commons/thumb/4/43/Ua741_opamp.jpg/1200px-Ua741_opamp.jpg.

10.1.3. Key Components in Signal Conditioning

Amplification and filtering in signal conditioning are baseline operations that increase sensor output quality. Amplifiers are critical in magnifying the signal from a MEMS device in such a way that downstream processing stages can be supplied with a robust signal with little noise in abundance. This is particularly the case when considering integrating MEMS devices into air system structures, as described in assessing digital electric systems. The improvement in fuel consumption and cost of operation means that conditioned signals can largely influence performance as well as system efficiency overall. Filters are vital in addressing unwanted high-frequency interference that can hide valuable information. By effective matching of amplification and filtering, the engineers are in a position to come up with a more efficient and stable signal path, which is vital towards the effective functioning of many applications, most significantly in advanced technology such as parts of aircraft (Bailey et al., 2025). Analog to digital signal conversion plays a significant role in signal conditioning and renders it simple for sensors to exchange information with digital processing equipment. Analog-to-Digital (A/D) and Digital-to-Analog (D/A) converters serve to overcome the gap between digital devices that are dominant in modern electronics and analog signals from MEMS sensors. Careful design of converters is important as they directly impact the systems overall resolution and accuracy. For instance, local processing circuits for tactile sensors can be provided via several platforms such as microcontrollers, PSoCs, and FPGAs, each having a unique contribution to systems performance. The application of FPGAs, for instance, offers high processing speed and flexibility and closely approximates specific signal conditioning requirements and helps to demonstrate the key interaction between signal conditioning components and system architecture (Ramos et al., 2001). Noise problems must not be taken lightly in the case of sensor interfacing since they have the potential to greatly impair the performance and reliability of signal conditioning systems. As MEMS devices are sensitive devices, they will by themselves cause environmental noise and electrical interference issues. Noise reduction countermeasures include the implementation of shielding, grounding procedures, and differential signaling. Moreover, power management is essential to the well-being of MEMS devices because proper power distribution can do a great deal to eradicate noise interference. The trade-off between power management and signal integrity is paramount where high reliability is required, for example in modern aviation and robotics. Understanding these dynamics is how it becomes possible to design robust interface electronics capable of effectively enabling involved signal processing operations without compromising system stability to disturbances (Bailey et al., 2025; Ramos et al., 2001) (Figure 10.5).

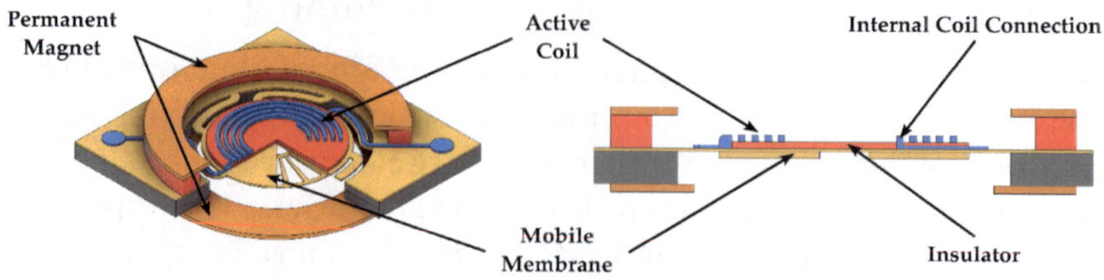

Figure 10.5. Amplifiers are crucial for increasing the signal strength from a MEMS (micro-electro-mechanical systems) device, ensuring that subsequent processing stages receive a robust signal devoid of excessive noise.

Source: https://www.mdpi.com/micromachines/micromachines-14-00847/article_deploy/html/images/micromachines-14-00847-g003.png.

10.2. MEMS SIGNAL AMPLIFICATION

The incorporation of Micro-Electro-Mechanical Systems (MEMS) signal amplification into signal conditioning geometries is a pivotal moment within the subject of sensor technology. The effectiveness of MEMS devices is dependent upon their ability to process and relay signals from various sensors effectively. Likewise, MEMS signal amplifiers play an important role in not just amplifying the amplitude of output signals but also suppressing noise, which is vital for effective interpretation of signals. With increasingly sophisticated sensors being developed, amplification technology that can carry a wide frequency range and range of signal types is needed. Enhanced performance is achievable with more advanced design methods so that better filtering and amplification can be provided to offer sensor output that is reliable and repeatable and thus create a safe interface for electronics to integrate freely into existing digital infrastructures. From the role of amplification, the analog-to-digital (A/D) and digital-to-analog (D/A) conversion process takes front-stage billing in facilitating MEMS output consistency with digital signal processing devices. Sophisticated A/D converters translate amplified signals into digital representations, where they can be treated by a variety of algorithms for analysis. At the same time, MEMS-based D/A converters facilitate the inverse process, allowing controlled analog output from digital input. This dualism is particularly critical in uses like biomedical sensing, where accurate conversions of signals can lead to enhanced sensitivity in monitoring applications. These conversion steps must then be integrated with stringent noise control methods, especially because the low amplitude signals that are most typical outputs from MEMS sensors. These features lie at the core of developing optimized MEMS devices that can address the multi-dimensional issues of modern signal conditioning. Power management is yet another significant consideration in the deployment of MEMS signal amplification technology. Effective power management systems are critical to maintaining the operational integrity of MEMS devices, especially in instances where low energy consumption needs to be weighed against performance. Many systems in such instances feature various strategies such as dynamic voltage scaling and low-power operation modes that help address changing power demands. Improved energy efficiency does not only extend the life of MEMS devices but also their applicability to various environments, from consumer devices to monitoring critical infrastructure. In all such usages, integrating MEMS signal amplification with power management strategies encourages more sustainable practices in electronics design. Also, through the intricacies of MEMS signal amplification, continuous research brings to the forefront the ability to create even more flexible and effective power management technologies that have the ability to further enhance their application scope in today's digital age (Figure 10.6).

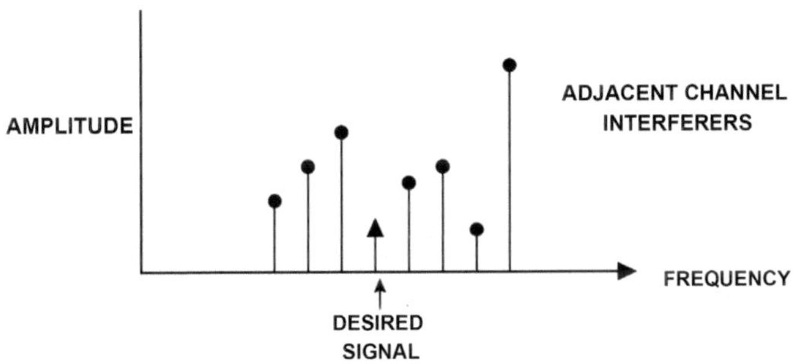

Figure 10.6. High-performance A/D converters transform amplified signals into digital forms, where they can be manipulated by various algorithms for further analysis.

Source: https://www.analog.com/en/_/media/analog/en/landing-pages/technical-articles/high-speed-converters-an-overview-of-what-why-and-how/figure2.jpg?la=en&rev=20ee01efa9ae469d8ffe14864815b9ef&sc_lang=en.

10.2.1. Overview of MEMS Technology

MEMS technology has transformed the area of signal conditioning and interface electronics. MEMS integrate mechanical structures, sensors, actuators, and electronics in a single chip, allowing for capabilities in real-time monitoring and control that are unmatched in many applications. Critical to their operation is their capability to translate physical phenomena like pressure, temperature, and motion into electrical signals. This is due to high-end signal filtering and amplification techniques that enhance data quality received from MEMS sensors. A digital representation of these signals through A/D conversion is crucial for further analysis and processing. These advancements demonstrate the importance of MEMS devices in facilitating signal conditioning and resulting in their expanding use in a variety of industrial applications. The use of MEMS technology, for instance, in areas such as health care and environmental monitoring offers significant challenges, primarily with respect to noise considerations in sensor interfacing. MEMS devices are subject to various sources of noise, which cause loss of signal integrity and accuracy. Therefore, robust interface electronics are inevitable to counter such effects and deliver reliable data acquisition. Various techniques like analog filtering and digital signal processing are employed to optimize the signal-to-noise ratio. The design also needs to consider the unique environmental conditions under which MEMS devices are likely to be used. For example, cardiovascular disease diagnosis and treatment smart stent device development focuses on achieving these noise specifications to provide high performance in extreme biophysical environments (Bussooa et al., 2018). Another important aspect of MEMS technology is power management, particularly as devices become increasingly complex and power-hungry. Proper power architecture is critical in ensuring the reliability and longevity of MEMS devices, particularly those requiring constant use such as environmental monitoring and wearable health products. Enhanced low-power design methodologies can lead to significant energy conservation, which is highly important for devices operating in hard-to-reach or remote areas. In addition, advances in energy harvesting technology can enable MEMS devices to function without the need for periodic battery replacement. This move in the direction of energy-efficient and self-sustaining MEMS devices is an indication of the growing search for green technology solutions and promises a promising future for the MEMS technology in many fields, including biologically inspired tactile sensing (Liñán et al., 2009).

10.2.2. Techniques for Amplifying MEMS Signals

Low-power and low-noise amplification is a key method employed with complementary metal-oxide-semiconductor (CMOS) integrated circuits to efficiently amplify signals generated by Micro-Electro-Mechanical Systems (MEMS) in many applications, most notably sensor performance and reliability. These circuits must be accommodated within the limited space of MEMS devices and must operate within power dissipation limits below thresholds that would otherwise impair their function—for example, the 1°C temperature increase limit described by Andersen et al. (2003). By means of efficient CMOS component organization and design, scientists are capable of establishing effective signal conditioning necessary to amplify exceedingly low-level signals from MEMS devices, creating a foundational bridge to the latter processing and data acquisition stages. Signal conditioning techniques such as filtering are significant mechanisms integrated into MEMS signal amplification. Advanced filtering methods are utilized to counteract noise, and particularly critical in high electromagnetic interference scenarios or when operating under severe conditions. For instance, chopper-stabilized topologies are utilized to boost signal integrity and also reduce offset variations and flicker noise. Implementation of programmable gains and adaptive filtering techniques allows for optimal performance within a sequence of sensor outputs, as observed in the modulator architectures described by Mor et al. (2005).

Figure 10.7. A critical technique involves the use of complementary metal-oxide-semiconductor (CMOS) integrated circuits that enable low-noise and power-efficient amplification.

Source: https://www.avaq.com/files/uploads/editor/b/20230616173549different-types-of-integrated-circuit.webp.

The sophisticated modulation techniques not only account for the improvement of transmission reliability of MEMS signals but also allow for smoother analog-to-digital conversion (A/D) processes, which further leads to precise data interpretation and utilization on various applications. Power management is a critical issue in MEMS signal amplification. In light of the power constraints that define MEMS devices, power optimization tactics become imperative to maximizing operational life and reliability. Low-power amplifier design is a must, particularly for continuous or long-term monitoring systems. CMOS technologies, as recently documented, allow for MEMS systems to be both extremely efficient and deliver unimaginable dynamic range without a loss of performance. Furthermore, placing signal amplification, filtering, and A/D conversion in one chip lowers the power consumption while amplifying function, with a compact solution for the intricacies of board-level MEMS signal conditioning and interface electronics. Such a strategy not only enhances the portability of devices but also provides opportunities for innovation in consumer electronics, biomedical implants, and automotive sensor systems, thereby emphasizing the critically crucial significance of efficient power management in MEMS applications (Figure 10.7).

10.2.3. Applications of MEMS Signal Amplification

The adaptability of micro-electromechanical systems (MEMS) in modern electronics significantly enhances the efficiency of signal amplification in many applications. The integration of MEMS devices with advanced signal conditioning techniques offers the precise conversion and processing of weak sensor signals into meaningful information. For instance, building a 4 x 4 array of POSFET devices entails sophisticated interface electronics such as multiplexers and voltage output buffers that make sensor connection to ancillary circuits easier. The design not only reduces interconnections but also increases the performance level of tactile sensing applications. Demonstrated sensitivity and response enhancement show how MEMS signal amplification can allow for uninterrupted data acquisition in environments demanding high accuracy, opening the door to CMOS implementation of complete on-chip tactile sensing systems with POSFETs (Adami et al., 2014). Yet another important application of MEMS signal amplification is through precision measurement instruments, such as gravimeters and microcontrollers for displacement sensing. Applicability of MEMS technology is particularly strong in gravitational wave detectors and other high-stability, precision measuring systems. An example is a microcontroller-based displacement sensor that was designed in particular for this application and exhibits a promising potential for portable, low-cost operation and excellent displacement sensitivity. With optical shadow sensing techniques augmented by digital lock-in amplification, such MEMS devices can be as sensitive as 10 nm/Hz on long measuring times. This innovation establishes the pivotal role of signal amplification as a determining influence in ensuring reliable data acquisition in applications where sensor precision is paramount, thereby enhancing functionality in scientific and industrial equipment.

Power management relevance is an important aspect in the case of MEMS signal amplification. These devices are often operated under stringent energy budgets, necessitating sensitive design for the purposes of minimizing power consumption while maximizing performance. Advanced power management methods in MEMS devices help extend operational lifetimes, particularly in battery-driven environments. For example, temperature sensing subsystems utilizing MEMS sensors typically require real-time accuracy in low-energy applications. In this scenario, proper conditioning circuits dealing with sensor signals not only maximize A/D and D/A conversion but also eliminate noise issues from consideration. In this regard, MEMS signal amplification offers robust interfacing properties that adapt to varying power limits, demonstrating its efficacy and versatility in practical use in multiple areas of study and industrial application.

10.3. FILTERING TECHNIQUES IN SIGNAL CONDITIONING

Signal conditioning is a significant factor in the enhancement of the performance of electronic devices, particularly in filtering. Filtering is employed to remove unwanted interference and noise from signals obtained from sensors to make the processed information reliable and accurate. For applications involving MEMS (Micro-Electro-Mechanical Systems), the devices are sensitive to the requirement of high-level filtering for the purpose of signal-to-noise ratio improvement. For example, advanced filtering technologies can be employed in MEMS signal amplifying circuits to improve output signal quality before A/D conversion. Efficient noise problems solved by such filtering play a major part towards the overall quality of the data acquisition process and therefore the resulting performance of interface electronics in their respective applications. Filter design and implementation are also closely linked with the conversion operations in signal conditioning. For A/D and D/A conversions, filtering is essential to prevent aliasing and permit only the desired frequency components to pass through. Low-pass, high-pass, band-pass, and notch filters are all common applications and serve a particular purpose depending on the requirement of the application. For instance, low-pass filters can filter out high-frequency noise, while band-pass filters allow a specified range of frequency to pass, and this is significant in communication systems or medical diagnostics. Efficiency of such filters as much as defines the validity of the signal conditioning process, highlighting the significance of employing an appropriate filtering technique based on the type of input signal and the specific application in interface electronics. Moreover, power management in MEMS components affects the efficacy of filtering techniques. As MEMS technology progresses, energy efficiency has emerged as a central concern, particularly in battery-powered or energy-harvesting applications. Signal-integrity preserving low-power filtering techniques are still extremely desirable. Adaptive filtering techniques that modify dynamically with changing signal conditions have the potential to enhance performance without excessively high-power consumption. These techniques can be particularly effective in systems like phased array systems, where efficient beamforming relies on properly filtered signals in order to drive large numbers of antenna elements. Techniques outlined in the signal conditioning category not only enhance signal quality but also allow sophisticated implementations while addressing power management challenges in MEMS and equivalent applications (Adami et al., 2010; Chan et al., 2025) (Figure 10.8).

Figure 10.8. Filtering is essential for removing unwanted noise and interference from signals captured by sensors, thereby ensuring that the data processed is both accurate and reliable.

Source: https://www.anypcba.com/upfile/blogs/articles/516/electronic-filter-circuit-board.webp.

10.3.1. Types of Filters Used

In interface electronics and signal conditioning, the intricate relationship between various filters is important in optimizing system performance as well as furnishing quality data acquisition. Of these, low-pass filters find extensive utilization to suppress high-frequency noise that can obscure desired signals, particularly in MEMS (Micro-Electro-Mechanical Systems) devices. By facilitating the flow of signals lower than a specified cutoff frequency while attenuating higher frequencies, these filters facilitate steeper analog-to-digital conversion. Low-pass filter design can vary widely, with uses employing passive RC networks and active implementations based on operational amplifiers to yield better figures of merit. With the increasing rate at which MEMS devices are being used in various applications, performance and efficiency of such filters need to be rigorously investigated for signal integrity optimization in dynamic conditions. While high-pass filters perform the key function of eliminating low-frequency noise and offset voltages that can interfere with accurate sensor readings. By way of filters, high-pass signals allow passage of signals above a given cutoff frequency, and therefore they are effective in those situations in which low-frequency drift might obscure useful data. In aerospace power systems, where the operating conditions should be as high-performance as possible, high-pass filtering is most commonly used to distinguish operational signals from those of non-ideal operation. The performance of these filters can be greatly enhanced through the proper selection of components and design, as in digital signal processing when feedback systems can be employed to reduce distortion effects. It is possible to tune these filters to provide improved system responsiveness and reliability, picking up on some of the considerations that are included in high-complexity project-based work. Band-pass filters are a mix of low-pass and high-pass filters, allowing only a specified range of frequencies to pass while reducing frequencies above this range. This class of filtering is critical for applications that necessitate the extraction of separate signals from within a larger spectrum, particularly in MEMS signal processing. Their applications typically involve sophisticated design methods in which both analog and digital approaches are fused to create specified filtering properties. Band-pass filter integration with advanced control systems like those listed in the ASDTIC control subsystem enables output control stability to be enhanced in difficult scenarios, ensuring reliable power supply performance under fluctuating input and output conditions (Lalli et al., 2025). Overall, the proper application of these various forms of filters is essential in the optimization of signal conditioning effectiveness, significantly impacting the overall quality of electronic system transmission data.

10.3.2. Design Considerations for Filters

In interface electronics and signal conditioning, the adequate design of filters is key to enforcing sensor signal integrity from the sensor chip, particularly MEMS devices, along the processing chain. Filters serve a critical role in shielding the signals of interest from noise and inhibiting the operation of subsequent stages, such as analog-to-digital conversion. In their design, cutoff frequency, response time, and order of the filter need to be well considered in order to meet the specific requirements of the application. A well-designed filter can well suppress the effect of varying noise levels and enhance the signal integrity to make the extracted data reliable and accurate, which subsequently contributes to the system reliability of the interface. Working with filters integrated into MEMS signal processing and amplification systems involves dealing with different kinds of challenges in terms of noise factors and power consumption. MEMS devices are typically utilized in environments where there are multiple electromagnetic interferences, hence the need for filters to work over a wide range of frequencies. Apart from traditional passive filtering, advanced techniques such as active filters using operational amplifiers can be employed to set the frequency response more precisely.

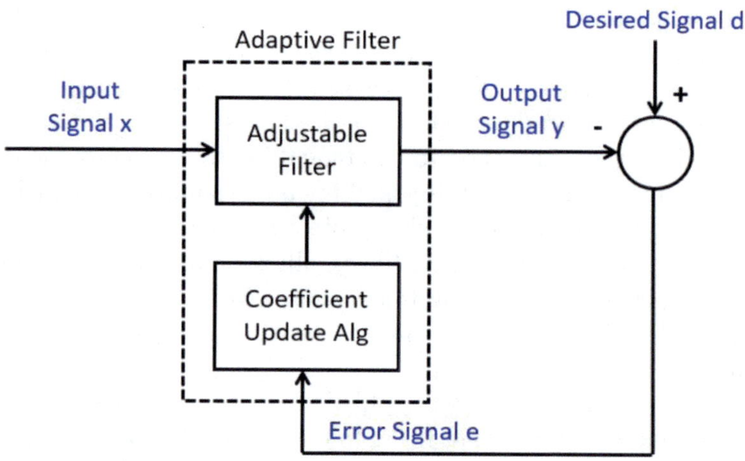

Figure 10.9. Filters serve a crucial role in distinguishing the relevant signals from unwanted noise, hindering the performance of subsequent stages, such as analog-to-digital conversion.

Source: https://schaumont.dyn.wpi.edu/ece4703b20/_images/adaptive.jpg.

Furthermore, power management care is necessary; filters add more elements that have a tendency to consume more power, which affects the overall energy efficiency of the system. Therefore, the design must optimize filter performance while maintaining thermal and power dissipation attributes in order to maintain the system within intended operation parameters. The design aspects extend beyond functionality to influence the thermal management and heat dissipation of electronic control units (ECUs) in various applications. For instance, as highlighted in recent studies, the use of high-temperature switches as well as their position in actuator driving topologies can have a pivotal influence on the thermal profile of ECUs. Choice of filter topography as well as the materials used also help minimize heat generation that results from power conditioning operations. Filter structure optimization enables one to achieve not only minimal dissipation of heat but also reliability enhancement of the ECU. As emphasized in solenoid driver research, striking an optimum balance between thermal management and the performance of the filter can yield phenomenal benefits in terms of performance metrics as well as system life-cycle cost (Beattie et al., 2025) (Figure 10.9).

10.3.3. Impact of Filtering on Signal Quality

In the area of signal conditioning, the efficacy of filtering mechanisms has a significant bearing on the quality of the output signal that is obtained. Filtering is a critical method for noise reduction, which has the propensity to desecrate precision and consistency in measurement in MEMS devices. By effective removal of undesired frequency components, filters purify sensor-generated signals to enhance signal quality. For instance, applying low-pass filtering allows signal-useful low-frequency signals to pass while eliminating high-frequency noise. These measures are essential in systems requiring precise A/D and D/A conversion, where any intrusion can lead to erroneous readings and reduced performance. Thus, proper application of filtering mechanisms not only preserves signal integrity but also improves the overall robustness of the interface electronics utilized in various applications, particularly in MEMS technology. Furthermore, the selection of an appropriate filter structure is important when it comes to improving responsiveness with minimal noise. All types of filters, either analog or digital, have distinct characteristics that can affect the quality of a signal in various ways. Analog filters, for instance, can cause phase delay and affect transient response, whereas digital filters are more flexible in parameter modification and can be devised to meet exacting needs. It is critical to recognize these dynamics when dealing with signals within sensitive environments where external interference is likely to occur. In the case of MEMS sensors, the integration of filters designed for use in specific operating conditions helps to ensure that the sensor outputs are highly accurate and reliable at all times. Finally, the meticulous engineering of filtering products within signal conditioning is directly related to improved signal quality, which is the cornerstone of successful data acquisition systems. In addition to the technical requirements, filtering impact on signal quality must consider broader implications for MEMS device power management. Not only does proper filtering influence signal purity, but it also contributes to system-wide QoS of energy consumption. As MEMS devices find more use in applications that are decreasing in size but of vital importance when it comes to power efficiency, filter optimization is a top design concern. For instance, filtering complexity reduction can lower power requirements so devices will run longer on batteries or other finite energy sources. To this, signal fidelity versus power consumption sensitivity adds an overlay of significance to filtering design choices in actual applications. This communication reveals the necessity of possessing a comprehensive understanding of filtering techniques since they not only enhance signal integrity but also play an important role while designing sustainable, effective systems in the rapid-evolving world of signal conditioning and interface electronics.

10.4. A/D AND D/A CONVERSION

In order to enable communication between the analog world and digital systems, conversion from analog-to-digital (A/D) and digital-to-analog (D/A) is a significant role in signal conditioning and interface electronics. A/D conversion translates continuous analog signals to discrete digital signals so that information from the real world like temperature, light intensity, or pressure can be accessed by digital systems. This relies on sampling and quantization, where the analogue signal is sampled at periodic intervals and later assigned a numerical value according to a given resolution. Successful A/D converters are fundamental in presenting digital equivalents fast and accurate enough, particularly in high-precision reading applications involving MEMS sensors, which have a direct influence on system performance and accuracy. Also essential, D/A conversion complements A/D operations by reversing analog signals from digital. D/A conversion is essential in applications where digital systems provide signals that must connect with the analog world, such as driving actuators or sound outputs. The quality of the D/A conversion significantly affects the fidelity of the output signal, demanding hardware that will have high resolution and low distortion. A good D/A converter is able to utilize various architectures such as resistor ladders or sigma-delta modulators to execute desired performance features. D/A converter design also happens to be noise-driven, and good layout and filtering are normally necessary in order to maintain interference low enough so as not to degrade signal quality, particularly in sensitive MEMS uses where signal fidelity is paramount. The convergence of A/D and D/A conversion technologies plays a crucial role in enhancing the performance and reliability of systems in modern electronics. New research in the field indicates new paths for optimizing conversion efficiency, particularly where there are real-time response and processing required. For instance, advances in MEMS signal filtering and amplification technologies improve interfacing performance, thereby necessitating ongoing development and research on A/D and D/A technologies. Enhancements in operating processes of these converters can lead not only to improved accuracy but to more varied applications in a variety of sectors, including aerospace, automobile, and consumer electronics. By optimizing these steps of conversion, engineers are able to design systems that meet not only but exceed the growing demands for precision and efficiency in the modern digital environment, with relevant studies pointing to potential areas of high payoff for further investigation, such as outlined in the IDEA study (Bailey et al., 2025).

10.4.1. Importance of Conversion in Signal Processing

The intricacies of modern electronics are mostly the function of the effectiveness of signal conversion processes, particularly in the integration of sensors across diverse applications. In the process of signal conditioning, Analog-to-Digital (A/D) and Digital-to-Analog (D/A) is crucial in acting as a go-between between the continuum of analog signals and the discreteness of digital systems. This conversion is essential in maintaining accurate data representation since it allows continuous signals, like those from MEMS sensors, to be processed by digital processors. Also, efficient A/D conversion minimizes the impact of noise and other interferences that can taint signals, allowing the processed data to accurately feed decision-making systems in numerous applications, from automobile technology to aerospace innovations. Appropriate choice and utilization of applicable conversion techniques further affect the performance and dependability of electronic systems as a whole. For instance, availability of high-resolution A/D converters may result in more efficient sampling of the signal, a very important requirement in applications where there is a need for accuracy such as environmental monitoring or medical diagnostics. The ability to maintain the noise properly at the signal interfacing stage cannot be stressed enough; it has a direct impact on the quality of the signal and thus on the success of the data analysis which follows. Error correction and filtering during conversion contribute to the robustness of the system, allowing it to function optimally even with undesirable external conditions. Thus, the importance of signal conversion as a consideration in signal processing extends beyond mere technical specification; it has broader implications on system effectiveness and reliability. In addition, more recent technologies emphasize the inclusion of advanced power management methods alongside signal conversion functions, especially in MEMS devices. The addition of digital power management systems is responsible for the drastic improvements along efficiency and performance with lower energy consumption and longer life in operation for embedded and portable applications. According to studies such as those by Bailey et al. (2025), context-aware systems that leverage digital-to-electric conversions have demonstrated significant reductions in fuel consumption along with improved operational parameters, highlighting the vast scope for sophisticated signal-processing approaches. The interconnectedness of these components implies that efforts to maximize signal conversion must also address the issues of power as well as environmental factors that can impact sensor operation. Understanding the nuances of conversion in signal processing is thus central to the design of efficient signal conditioning and interface electronics (Figure 10.10).

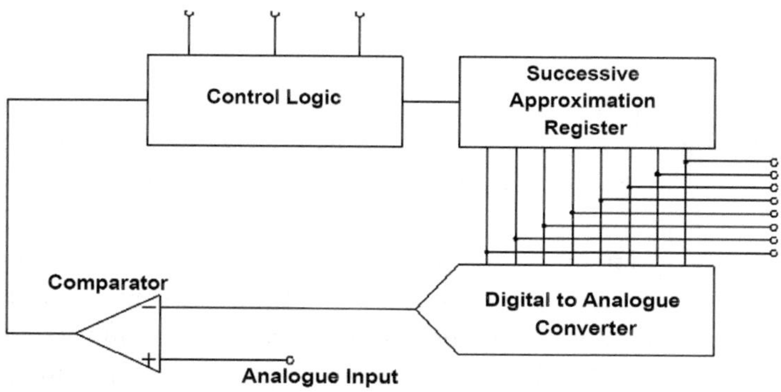

Figure 10.10. Analog-to-digital (A/D) and digital-to-analog (D/A) conversion plays a crucial role in bridging the gap between the continuum of analog signals and the discrete nature of digital systems.

Source: https://www.elprocus.com/wp-content/uploads/2014/10/4.jpg.

10.4.2. Types of A/D and D/A Converters

The effective conversion of analog signals to the digital format is crucial in modern electronics, particularly in MEMS (micro-electro-mechanical systems) applications. Among the various analog-to-digital converters (A/D converters), successive approximation register (SAR) converters offer an interesting trade-off between resolution and speed. The converters rely on a binary search algorithm to translate the analog input to a digital output with high effectiveness. This method reduces conversion time as well as power consumption—a very critical parameter for battery-operated MEMS devices. Additionally, the flash A/D converters provide high-speed conversion rates by utilizing numerous comparators, hence being suitable for high-speed applications, despite consuming more power and occupying a larger chip area. Thus, the choice of A/D converter not only affects performance specifications such as speed and accuracy but also affects power management significantly in MEMS applications. The digital-to-analog conversion (D/A conversion), however, is equally important to interface digital signals with analog circuits. Among the most well-known types of D/A converters, the binary-weighted resistor and R-2R ladder structures are ubiquitous due to their simplicity and ease of implementation. Although the binary-weighted resistor D/A converter, being suitable for low-resolution situations, might become unwieldy at high bit resolution due to the problem of resistor matching. On the other hand, R-2R ladder type is linear and scalable and therefore can be used where much accuracy is required. Moreover, advancements in signal conditioning techniques, such as the use of notch filters and resonant regulators, have enhanced overall stability and output impedance of D/A converters in most applications, particularly in advanced systems like dc microgrids, where harmonics can be critical performance concerns (Caldognetto et al., 2020). With increasing demands for strong and intelligent sensor interfaces, improvements in A/D and D/A converter technologies have been documented to optimize signal integrity. One of the most notable is the application of voltage-to-frequency converters in smart tactile sensors to handle the vast amount of data generated by sensitive skin-like sensor arrays. Through the initial signal processing at the sensor level, the converters effectively manage noise and reduce transmission errors, typically exaggerated by long cables. The techniques described demonstrate a clear path towards minimizing noise and maximizing efficiency along the signal chain. By synergy between A/D and D/A conversion technologies with MEMSs own signal filtering and amplification needs, engineers can design more accurate and sensitive interfaces in many advanced applications (Liñán et al., 2007).

10.4.3. Applications and Challenges in Conversion

The revolution towards renewable energy sources has propelled the need for efficient electronic systems that can transform different forms of energy to useful power. In counteraction to this, technologies such as piezoelectric energy harvesting emerged that utilize human kinetic power to power hand-held devices. The systems contain a wide range of electronic components from power conditioning circuits, amplifiers, and microcontrollers which function synergistically to enhance their performance and efficiency. As outlined by Mishra et al. (2025), the systems use elements like inverting amplifiers and buck converters to maintain voltage and current regulation and supply the desired voltage and current to drive mobile devices. The efficiency maxima, nevertheless, remain a challenge, with improvements required in materials, geometrical structures, and circuit configurations towards minimizing power losses while ensuring reliability. For this reason, overcoming such challenges is essential in ensuring large-scale adoption of energy harvesting technology within public spaces in the way towards a greener future. Apart from energy harvesting, signal conditioning and interface electronics are equally important in ensuring seamless A/D and D/A conversions in a range of applications. This is even more paramount in systems that require MEMS signal amplification and filtering, where accurate data extraction is of the highest priority. The complexity of noise factors in sensor interfacing cannot be underestimated because spurious interference can jeopardize the purity of the signals being processed. By using noise minimization methods and robust filtering algorithms, designers can improve data quality. Additionally, power management in MEMS devices adds another level of complexity because efficient utilization of energy enables device longevity with satisfactory performance. As discussed by Fang et al. (2021), the growing integration of power electronic converters imposes stability challenges on weak grids, and it requires innovative solutions to operate stably. This integration of signal conditioning and power management is the focus of ongoing research and development processes.

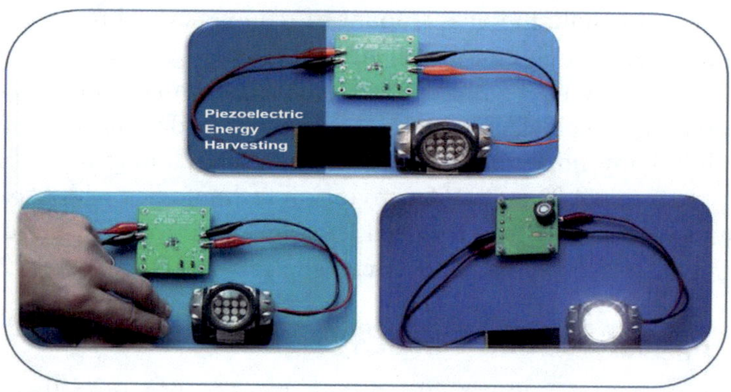

Figure 10.11. Applications such as piezoelectric energy harvesting have emerged, utilizing human kinetic energy for charging portable devices.

Source: https://cdnintech.com/media/chapter/80248/1512345123/media/F9.png.

The advancements in signal conditioning are accompanied by challenges in maintaining system stability, particularly at the interface with weak grid configurations. The impedance variation poses a special challenge that has vexed the power electronics community for more than a decade. Solutions are usually application-specific, tending to be less flexible across various operating conditions. The estimated solution proposed by Fang et al. (2021) places emphasis on the importance of precise grid information toward improving stability with a focus on voltage droop control as a likely solution. Implementation of such solutions is an interdisciplinary solution with the use of findings in control theory, electrical engineering, and real-time data analysis. Due to the growing need for renewable technologies, overcoming the obstacles related to conversion processes becomes crucial. In doing so, we can pave the way for innovations that not only improve efficiency but also accommodate global sustainable energy practices (Figure 10.11).

10.5. NOISE CONSIDERATIONS IN SENSOR INTERFACING

In the intricate context of sensor interfacing, the role of noise is a decisive factor in determining the performance and reliability of the entire system. Some sources of noise can significantly affect the accuracy of measurements by sensors, hence mandating detailed measures for their removal. For instance, electromagnetic interference (EMI) from external electronic components as well as thermal noise generated within the circuits can induce unstable fluctuations in the signal. Besides, the signal bandwidth should also be considered while choosing sensor components; a low bandwidth can lead to aliasing effects that pollute the signal outputs. So, a clear understanding of various noise considerations is required to achieve maximum signal integrity so that the output data is correct and meaningful to form a solid foundation for effective decision-making based on sensor readings. Signal conditioning is an important aspect of noise elimination in sensor interfacing, involving operations such as filtering and amplification. There are several techniques such as low-pass filtering, which are employed to remove high-frequency noise that might contaminate the signal so that only the desired frequency components are permitted to pass through for digitization. In addition, A/D and D/A conversion processes must also be well-developed to include noise reduction techniques. For example, the application of oversampling and averaging can strongly diminish random noise impact without compromising signal resolution. The novel three-axis teslameter, developed for high-precision measurements, demonstrates this method using cutting-edge analog signal conditioning to enable measurement fidelity and to resist noise. Thus, effective signal conditioning methods are essential in realizing legitimate sensor interfacing, which considerably enhances system performance under real-world conditions (Calvi et al., 2019). Another important facet of sensor interfacing is power management, particularly for MEMS devices. Power management approaches must be judiciously combined with noise reduction strategies to balance performance and power efficiency. For instance, low-power drive signals can reduce the possibility of electrical noise generation without sacrificing sensor response. Employing fiber optic communication structures can also be highly advantageous in this area. This approach naturally minimizes signal degradation associated with electrical conductors, ensuring a robust signal transmission with lower noise sensitivity. Such innovative designs with tighter layout constraints have been discovered to significantly enhance signal integrity in various applications such that the interfaced systems work at optimal efficiency and accuracy (Glasheen et al., 2025).

10.5.1. Sources of Noise in Signal Conditioning

The coupling between signal conditioning and environmental conditions plays an extremely dominant role in determining the efficiency of electronic devices, especially in sensor and data-acquisition systems. One of the major sources of disturbance is electromagnetic interference (EMI), which is prevalent in high electric and magnetic field environments, such as in the case of the ADITYA-U Tokamak fusion reactor. The signal conditioning electronics must be immune to this interference if they are to accurately measure and pass on reliable data. For instance, efficient noise rejection is needed in keeping the signal processing accurate from a photo multiplier tube. This is done through shielding, grounding methods, and careful signal trace routing to prevent unnecessary voltage variation. The systematic approach to addressing EMI not only enhances signal integrity but also optimizes the performance of the signal conditioning system, demonstrating how it is purely critical in developing fault-tolerant electronic interfaces in aggressive environments (Shah et al., 2024). Thermal noise is another extremely critical source of undesirable fluctuations in signal conditioning circuits in addition to EMI. Thermal noise or Johnson-Nyquist noise is a natural phenomenon in the resistive components and tends to cover small signals of importance in precise measurements in devices such as MEMS. Thermal fluctuation noise might be further amplified by variations in temperature, leading to uncontrolled variations in sensor performance. Effective filtering and amplification techniques can be used to alleviate such issues, enhancing the signal-to-noise ratio (SNR) of the measurement process. In addition to that, proper component selection that has low thermal noise figures is essential in high-performance signal conditioning systems design. With the high requirement for sensitive measurements continuing to increase, particularly in advanced sensing technologies, an understanding and controlling of thermal noise take precedence in engineering efficient signal interface solutions (Jacob et al., 2024). Furthermore, noise in digital signal processing introduces complexities that have a profound impact on data integrity. Quantization noise caused by the finite resolution of digitization processes is extremely vulnerable to analog-to-digital (A/D) and digital-to-analog (D/A) conversions. This type of noise becomes of outmost concern, especially in systems where high signal fidelity is needed, for instance, in environmental monitoring or medical applications. Sophisticated techniques, including oversampling and noise-shaping algorithms, may be used to improve the quality of the output of converters, thereby reducing the negative impacts of quantization. In signal conditioning, one must deal with these different sources of noise—whether from external conditions, thermal noise, or digital processing—to achieve effective and accurate electronic interfaces. With advances in technology, the use of good noise reduction methods remains essential in ensuring the best performance of signal conditioning systems across different applications (Shah et al., 2024) (Figure 10.12).

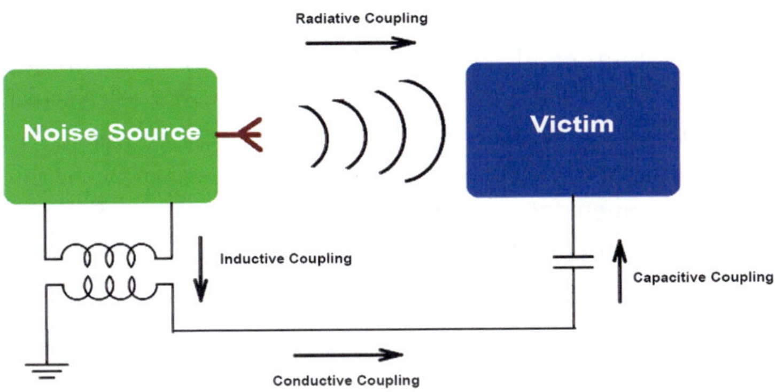

Figure 10.12. Major source of noise arises from electromagnetic interference (EMI), which is prevalent in environments with high electric and magnetic fields, such as those found in fusion reactors like the ADITYA-U Tokamak.

Source: https://magnoteknik.com/wp-content/uploads/2019/08/EMI-Coupling-Magno-Teknik-768x434.png.

10.5.2. Techniques for Noise Reduction

In modern signal conditioning, use of state-of-the-art techniques for reducing noise cannot be over-stressed. As sensors tend to generate microvolt-level signals, prone to contamination by environmental and electrical noise, stringent measures must be employed to ensure proper interpretation of the data. Techniques such as differential signaling and active filtering form the backbone in avoiding such unwanted interference. For instance, the use of differential amplifiers helps reject common-mode noise since it differentiates the signal and minimizes common interference. In addition, the incorporation of analog filtering before the conversion process significantly minimizes high-frequency noise, thus ensuring signal integrity. This strategic noise reduction is critical in uses ranging from MEMS for automotive sensors to sophisticated scientific instrumentation where precision is vital. Additionally, the design of analog-to-digital (A/D) and digital-to-analog (D/A) converters often utilizes noise-reduction techniques to provide higher fidelity in the converted signals. Oversampling is one such common process where the sampling of the signal is done at a rate much higher than the Nyquist rate. This improves the signal-to-noise ratio (SNR) as digital filters can now be applied that can effectively remove the quantization noise. Moreover, sigma-delta modulators that are commonly used in high-resolution A/D converters can dynamically quiet noise using their own noise shaping and oversampling capabilities. These features are most significantly useful in dealing with the high-frequency noise that is typical of MEMS devices since they ensure that the resulting digital signals more accurately reflect the physical phenomenon sensed. Thus, design choices work purposefully to prevent signal fidelity loss in electronic systems. Finally, power management of MEMS devices brings new challenges as well as opportunities to utilize noise reduction techniques. Power supply noise can couple into sensitive signal paths and lower the quality of retrieved data from MEMS sensors.

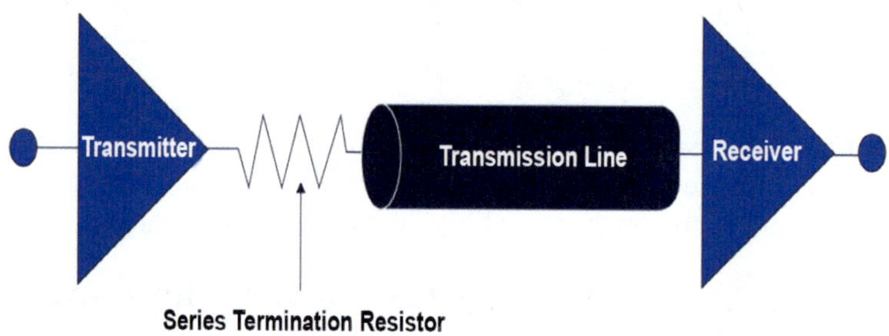

Figure 10.13. Techniques such as differential signaling and active filtering play crucial roles in mitigating these unwanted disturbances.

Source: https://media.monolithicpower.com/wysiwyg/Educational/Diagram_Recreation_-_Analog_vs._Digital_Control_Chapter_7_960x420_Fig_-26.png.

The employment of low-dropout regulators (LDOs) with low output noise is an effective technique in minimizing power supply contributions to overall noise levels. Also needed are proper grounding and shielding techniques to keep sensitive devices isolated from external electromagnetic interference. Guard traces located around sensitive analog lines and reduction of layout considerations that can cause noise coupling among power and signal lines are just a few of the techniques employed for this. By satisfying these attributes, designs can achieve a significant noise reduction, in that the critical characteristics of MEMS devices are preserved while accurate signal conditioning and reliable operational efficiency are maintained (Figure 10.13).

10.5.3. Impact of Noise on System Performance

In the broad universe of interface electronics and signal conditioning, achieving optimal system performance is greatly dependent on the function played by the presence of noise. Noise can be caused by electromagnetic interference, thermal noise, and inherent limitations in electronic devices. The harmful effect of noise is felt as accuracy and fidelity of processed signals, continuing to actual compromised performance. In the instance of experiments such as the EDGES High-Band, whose aim is to detect extremely feeble cosmic signals, low noise is absolutely imperative. Calibration of receivers is of greatest importance since it directly addresses the system's capability to discern between actual signals and noise, evidenced by calibration uncertainties that are capable of delivering up to 20 mK residuals even in areas of low foreground (Bowman et al., 2016). Reduction of such noise contributions is critical to offer the quality of experimental data and improve the overall system performance. Technological advancements, particularly in the field of MEMS devices, have indicated the imperative of proper amplification and filtering of signals to mitigate the impact of noise. With these systems functioning over various applications such as aerospace, medical, and consumer electronics, the requirement for precision amplifiers and stable filters cannot be overstated. Noise will corrupt sensor outputs, rendering them unusable to decipher or lead to system failure. In applications like digital magnetometers, for example, the application of noise-immune A/D and D/A conversion methods is critical. They offer more advanced signal processing that is necessary to guarantee sensitivity and accuracy, particularly in hostile environments (Drake et al., 2015). As such, proper design and implementation of signal conditioning elements are critical in reducing the impact of noise effects and ensuring maximum overall system reliability. Power management is also an integral part of noise performance in interfacing systems. The source of power can generate noise in the form of ripple and variations that compromise the signal integrity. Consequently, power management solutions for MEMS devices are well planned to shield them from power noise that would otherwise trigger interference in signal capture and processing. Engineers are required to trade parameters such as impedance matching and layout of the circuit in planning for minimal disruption of noise when coming up with an environment conducive to free signal propagation. By using high-quality parts and sophisticated circuit design methodologies, noise can be significantly minimized, which allows systems to uphold high-performance capabilities in spite of adverse conditions. Therefore, mitigation of noise at multiple stages—amplification, filtering, and power management—is still central to implementing effective signal conditioning and improving interface electronics (Figure 10.14).

Figure 10.14. Noise can stem from various sources, including electromagnetic interference, thermal fluctuations, and inherent limitations in electronic components.

Source: https://www.icdrex.com/wp-content/uploads/2024/12/board-4514999_1280.jpg.

10.6. POWER MANAGEMENT FOR MEMS DEVICES

The integration of microelectromechanical systems (MEMS) with advanced signal conditioning and interface electronics requires careful attention to power management. As MEMS devices demand strict size and power constraints, efficient power consumption is critical. These types of devices often require a balanced approach wherein sufficient power is provided to facilitate signal amplification, filtering, and conversion processes while minimizing heat production because the latter would otherwise detract from their performance. This is particularly important in application like neuroprosthetics, where temperature rise of less than 1°C from surrounding sensitive tissue is of key importance for safety and functionality (Andersen et al., 2003). Efficient power management strategies not only optimize performance but also optimize operational lifetime of MEMS devices, thus providing reliability in the intended applications. Power. MEMS device management also entails the development of multipurpose integrated circuits that can effectively carry out diverse functions such as signal amplification and analog-to-digital conversion. Such circuits must be made in such a way that they consume less power while possessing high-performance capabilities, especially in life-critical uses such as cardiovascular applications in which real-time monitoring and quick response are essential. Emerging advancements have led to the development of intelligent ultrasound stent technologies that use MEMS electrical properties to facilitate high-end diagnosis and treatment (Bussooa et al., 2018). This development shows, besides the significance of effective power management in providing integrity to merged ultrasound circuits, its critical role in opening up the opportunity to high-end biomedical devices that can significantly enhance patient outcomes. Additionally, problems with noise need to be addressed in power management design for MEMS devices. Noise, typically an undesirable byproduct of power fluctuations, can severely compromise the integrity of signal conditioning processes: a lower noise profile is crucial for converting analog signals into high-quality digital outputs. Low-noise amplifiers, adaptive filtering, and dynamic power scaling are critical in minimizing the effects. The design of front-end amplifiers, as observed in future neuroprosthetic devices, focuses on the provision of a stable power supply despite changes (Andersen et al., 2003). Hence, appropriate power management not only enhances the performance capability of MEMS devices but also maintains the integrity of the signal, thus promoting the greater vision of dependable signal conditioning across a broad spectrum of applications.

10.6.1. Importance of Power Management

Proper power management is a critical component of the discipline of interface electronics and signal conditioning when dealing with microelectromechanical systems (MEMS). Given that most of these devices find applications in applications where energy should be conserved, applying intelligent power management techniques guarantees performance and reliability. Through the integration of advanced power management systems, not only does wastage of energy avoid but performance of MEMS is also enhanced. Applied to signal filtering and amplification, for instance, dynamic power management enables devices to adjust power consumption in real time, a technique essential to extending battery life and peak operation efficiency. This adaptability makes more complex signal processing operations possible, of which more and more applications are being found in A/D and D/A conversion processes, thus demonstrating the inherent relationship between power management and system performance. Besides, power management is particularly important in considering the concerns with noise in sensor interfacing. High levels of noise would affect the integrity of signals processed by MEMS devices in such a way that there would be distorted output and poor performance. By applying effective power management techniques, the signal integrity would be maintained while reducing the power consumed by such devices. For instance, filter and sort algorithms may be used to sanitize signals for more clarity, and power management systems adjust the conditions under which the sensor functions dynamically based on real-time analysis of the data being received. This complementarity of noise reduction and power efficiency is critical as it directly contributes to the integrity of the data being utilized across applications from consumer products to sophisticated industrial systems. Furthermore, as already determined in investigations that have been investigating sophisticated technologies in aircraft systems, the merits of digital power management systems are still being valued in broader fields of engineering. The Enhanced Digital/Electric Aircraft Study identified how integrating digital systems with electric power distribution can yield monumental improvements in both fuel efficiency and operating expenses (Bailey et al., 2025). Similarly, efficient beamforming technologies utilizing digital technology prioritize power control in massive antenna array applications (Adami et al., 2010). The following are a few illustrations of how power management advantages are more than power conservation; they play a vital role in enhancing performance and reliability in sophisticated electronic systems, including those inherent in efficient use of MEMS devices in signal conditioning and interfacing applications (Figure 10.15).

Figure 10.15. Efficient power management is a crucial component in the realm of signal conditioning and interface electronics, particularly when dealing with micro-electromechanical systems (MEMS).

Source: https://images.surferseo.art/9cc1433b-9aa0-4786-a841-2a249e463d6c.png.

10.6.2. Techniques for Efficient Power Use

Technological innovations demand strong understanding of power management techniques, particularly when using Micro-Electro-Mechanical Systems

(MEMS). Efficient utilization of power in signal conditioning becomes necessary for achieving maximum performance of MEMS devices, which are likely to demand precise amplification and filtration of analog signals. Power consumption emerges as a significant factor that can decide the operational life and effectiveness of such systems. Innovative techniques such as adaptive power scaling enable us to reduce energy usage in idle periods, essentially getting devices to use only the power required to operate. Not only does this improve battery life, but it also helps towards making electronic systems more sustainable overall. As MEMS technology continues to advance, the integration of smart power management functionality with usage tracking will become increasingly important in the achievement of efficient use of power in a range of applications across signal processing and interface electronics.

The relationship between signal conditioning and power consumption is once more reflected in the case of analog-to-digital (A/D) and digital-to-analog (D/A) converters. The success of the conversions is most important because they directly affect signal integrity and energy consumption in electronic devices. The application of techniques such as multi-bit quantization and oversampling can enhance A/D converter resolution without compromising signal quality. Similarly, advanced D/A conversion methods allow for the delivery of higher fidelity output signals, thus power is not wasted in generating unwanted noise. These designs make way for optimization in throughput and data accuracy, essential in applications where data accuracy is a matter of life and death. Power consumption optimization is essential in this regime, as improved efficiency of operation automatically translates into greater device performance and reduced operation cost, making a good case for the inclusion of aggressive power management schemes in interface electronics. Noise factors need to be addressed of utmost priority while taking into account effective power usage in sensor interfacing because noise can potentially have an extreme influence on the quality of signals received. The emergence of filtering techniques such as active and passive filters helps to perform a vital function of minimizing interference while ensuring power consumption remains within acceptable ranges.

Additionally, utilizing low-power design techniques such as using low-noise amplifiers and circuit topologies optimizing is a way to minimize the unwanted signal disturbance while maintaining energy efficiency. As electronic systems become more sensor-rich for data acquisition, efficient noise management will continue to be more important. New materials and technologies such as nanomaterials are also set to enhance the performance of sensors, which can potentially be associated with autonomous signal processing capabilities. These advances in noise reduction and efficiency maximization pave the way for advancements in signal conditioning and interface electronics with a focus on the imperative position of power management strategies in achieving high-performance systems.

10.6.3. Challenges in Power Management for MEMS

Effective MEMS integration into applications is facilitated by an adequate knowledge of their power management challenges. With the increasing use of MEMS devices, such as sensors and actuators, in areas such as biomedical engineering and telecommunications, effective power management is a top concern. The natural limitations of small-size devices make it challenging to address issues of sustaining enough power levels while maintaining performance. For instance, the designs must save power in order to prolong battery life, something that is especially critical in applications like implanted medical devices or distant sensors. In addition, system designers also have the daunting task of ensuring power management solutions do not produce too much noise or disrupt the integrity of the signal of MEMS outputs, and this creates the need for creative power conditioning methods that enable maximum signal processing and noise mitigation. In addition, the twofold requirement of minimizing power consumption as well as enhancing device performance further makes the power management ecosystem for MEMS more difficult. Devices such as neuroprosthetics must adhere to strict power and size constraints in order to be safely implanted inside the body. For instance, MEMS electrode arrays integration with front-end CMOs circuits must necessitate there be a proper power dissipation consideration; any increase in temperature can adversely impact surrounding tissues and device performance as a whole. To address these challenges, the creation of low-noise, low-power operational amplifiers and high-efficiency voltage regulators is vital. These innovations allow for high-integrity signal amplification and filtering, critical for successful analog-to-digital and digital-to-analog conversion in MEMS technology. These power management methods not only have to satisfy the high-performance requirements but also the general goals of making the MEMS devices more durable and reliable. In pursuit of advanced power management methods for MEMS devices, it has become clearly evident that wireless power transmission technologies are of utmost significance. Where traditional wired power sources cannot be applied, such as in implanted cardiovascular devices, the development of high-tech wireless power transfer mechanisms has been an adequate alternative. Creating systems capable of efficient power harvesting from surrounding electromagnetic fields without causing too much signal distortion and loss during the process has remained the objective. The anticipated rollout of smart stent technology highlights as much the technical potential for wireless power as the technical challenge of providing communication integrity across varying operating conditions. As researchers create innovative solutions, the effective implementation of such MEMS devices will rely heavily on overcoming these power management challenges and funding match with the evolving environment of signal conditioning and interface electronics, which could, in turn, redefine patient treatment and care protocols (Andersen et al., 2003; Bussooa et al., 2018) (Figure 10.16).

Figure 10.16. Devices such as neuroprosthetics must adhere to strict size and power constraints to ensure safe implementation within the human body.

Source: https://media.springernature.com/full/springer-static/image/art%3A10.1038%2Fs41598-022-26673-2/MediaObjects/41598_2022_26673_Fig1_HTML.jpg.

10.7. CONCLUSION

The incorporation of advanced signal conditioning techniques is at the forefront of enhancing MEMS device performance across a wide spectrum of products from industrial sensors to wearable technology. Filtering and amplification of signals are critical to signal integrity and have a direct impact on data accuracy and reliability. The digital-to-analog (D/A) and analog-to-digital (A/D) conversion process is essential to fill the gap between MEMS sensor outputs' analog world and digital structures used in contemporary electronics. Minimizing noise issues in sensor interfacing allows designers to considerably improve the signal quality, which is essential for proper interpretation of data in such applications as environmental sensing and health monitoring. Thus, the facilitating role of signal conditioning is obvious, as it makes the interfacing possible and accurate readings in MEMS-based devices. Power management for MEMS devices is also very important, given the growing requirement for portable and low-power sensors. Effective power management strategies make MEMS devices usable without much energy consumption, which is essential for extending the battery life in wearable devices. The challenge, therefore, is how to reconcile demands for performance with energy efficiency, particularly in continuous monitoring devices. Advances in flexible electronics, as evidenced by studies on wearable sensors, illustrate how material technology can translate to more efficient devices with the capability to sustain longer operating times. With enhanced power management, MEMS applications in numerous fields, including personalized medicine, can become unprecedentedly effective, enhancing user experiences and enabling broader acceptance in health care settings. As signal conditioning and interface electronics evolve further, nanotechnology and novel materials will be needed to achieve beyond existing boundaries. The application of advanced nanomaterials at the cutting edge of technology in biosensing, as indicated by some recent research, presents staggering promise for enhancing sensor performance, including sensitivity and response rate (Naresh et al., 2021). Besides, the trend towards integrating flexible electronics with stable sensor technology highlights the scope for development and research in this direction (Gao et al., 2019). Future work must attempt to create more intelligent and responsive devices that continue to teach us more about health indices and environmental states. This ever-growing area of signal conditioning and interface electronics not only strives to improve the accuracy of measurements but also seeks new uses that will redefine technology's boundaries in everyday life.

10.7.1. Summary of Key Points

The intricate relationship between signal conditioning and interface electronics is the core to the successful adoption of Micro-electromechanical Systems (MEMS) into various applications. Specifically, the methods of MEMS amplification and filtering are core to the enhancement of the quality of sensor signals. A good interface can effectively remove undesirable noise so that the signals to be processed are as close as possible to the original measurements. Interestingly, A/D and D/A operations need to be optimized so that there can be proper digital representation of analog signals and vice versa. By utilizing complex filtering algorithms, interference can be minimized and signal-to-noise ratio maximized, which is crucial in the case of sensitive applications such as medical diagnosis and environmental monitoring. The use of these complex techniques finally makes MEMS devices efficient and reliable in actual application. In addition to signal filtering and amplification, power management is a critical concern which cannot be excluded in the context of MEMS devices. As these systems are becoming popular in industries, it is important that they are efficient in terms of operation while their power usage is controlled. Dynamic voltage scaling or sleep modes can significantly improve the battery life of handheld devices, depending on user experience and the overall system performance. Moreover, efficient power management assures robust operation under various conditions, which is highly effective for applications where sensors are installed in remote or inaccessible locations. Not only does an efficient power management scheme enhance the operation quality of MEMS devices, but it also serves as a central enabling factor for their application in diverse markets like consumer electronics and automotive systems. Last but not least, noise issues in sensor interfacing are a chronic problem that requires judicious attention. The intricate nature of signal paths must be assessed to identify how electromagnetic interference affects the sensor's performance adversely. Incorporating the appropriate methods of shielding and conservative signaling line routing can decrease susceptibility to noise, thus maintaining the integrity of data received. In Industry 4.0, with sophisticated monitoring and control equipment based on accurate sensor measurements, noise solution is a top priority for successful operation. As seen in recent publications, proper understanding of interactions between noise management, signal conditioning, and interface electronics will foster innovation and reliability in MEMS-based applications (Dóra et al., 2019; Yang et al., 2019). This information is critical for the creation of next-generation devices that will continue to expand the capabilities of MEMS in potential applications.

10.7.2. Future Trends in Signal Conditioning and Interface Electronics

MEMS technology is going to have a significant contribution to the upcoming interface electronics and signal conditioning trends. MEMS devices are especially renowned for their compactness and efficiency, and hence it becomes feasible for them to amplify the signal as well as filter it effectively at the sensor level.

This not only keeps the noise pertaining to the signal negligible but also leaves the overall digital data processing accuracy improved. As MEMS technology has advanced, signal conditioning circuits can be made more compact without sacrificing performance. This integration trend makes it possible to create more sophisticated systems required in applications like wearable tech and IoT devices, where space and power are limited. Continuing development of MEMS should lead to still greater advances in the precision and reliability of sensor data acquisition, enhancing a broad trend toward more efficient electronic systems throughout engineering and technology. As well as advances in MEMS, increasing application of analog-to-digital (A/D) and digital-to-analog (D/A) conversion technologies will set the future course for interface electronics. With more interconnected devices, high-fidelity signal conversion has become more in demand, and it takes advanced circuitry to efficiently handle high-speed signals. Highly efficient designs that provide greater power efficiency and reduced latency are required where applications call for processing data at high speeds, such as telecommunications and data acquisition systems. The future will see the coming together of A/D and D/A converters with advanced signal conditioning techniques, employing adaptive filtering algorithms that filter noise and enhance signal integrity. Such advancements not only make manufacturing easier but also optimize performance across a variety of digital applications requiring immediate and precise data processing (Sundberg et al., 2025). Another major to interface electronics and signal conditioning in the future is that there must be effective power management, particularly for MEMS devices. With these devices increasing in complexity and ubiquity, it is crucial that they are designed to last long without compromising power consumption to very low levels. Designing advanced power management methods that combine energy harvesting and low power design will be paramount in realizing maximum efficiency from MEMS devices. Incorporating feedback loops into signal conditioning functions may also be utilized to dynamically control power use with operational requirements. Resolving challenges faced by energy efficiency will not only contribute to making devices more efficient but also help in realizing sustainability goals in product design and production (Mohamad et al., 2022). As this industry evolves, the relationship between signal integrity, device miniaturization, and power efficiency will shine a light on significant innovation in signal conditioning and interface electronics.

10.7.3. Final Thoughts on the Importance of the Field

The intersection of signal-conditioning and interface electronic advanced technologies has revolutionized the face of modern sensing applications. As systems are increasingly more complicated, the need for efficient signal processing capabilities is more imperative than ever. The development of MEMS-based signal amplification and filtering techniques is a primary impetus for this advancement, enabling sensors to deliver reliable and precise information. For instance, the advancement in the design of Piezoelectric Oxide Semiconductor Field Effect Transistors (POSFET) has significantly enhanced tactile sensing capability and its performance is excellent with sensitivity of 102.4 mV/N without amplification. Such advancements highlight the need for interface electronics optimized to facilitate sensor efficiency so that the sensed signals reflect true environmental conditions. The trouble-free integration of these systems makes possible innovations in fields as diverse as robotics to medicine, underlining the critical contribution of this specialist field to technology innovation (Adami et al., 2014). In a consideration of the implications more generally of signal conditioning and interface electronics, the significance of noise considerations when interfacing sensors is clear. As devices detect and transform real-world signals, environmental noise can critically degrade the data quality, for which effective noise control methods must be used. Having integrated design methodologies considering noise from conceptualization itself can lead to more durable sensor systems. Besides, power control development of MEMS devices is also needed because such systems are likely to be employed in energy-constrained environments. Efficient use of energy not only prolongs equipment life but also optimizes performance, making the technology valid and applicable in a broad array of uses. Accordingly, with further technological improvements, it will be critical to grasp and resist these challenges in order to realize better sensor performance and integrity (Barlow et al., 2007). Finally, the general applicability of signal conditioning and interface electronics lies in their potential for cross-disciplinary applications. The ability to simulate a wide variety of sensor systems provides for a customized solution that accommodates a variety of technologies, thus promoting innovation in fields such as automotive engineering, environmental monitoring, and consumer electronics. With an acceptance of a more generic modeling platform, as suggested for infrared waveband sensors, researchers and engineers are able to acquire meaningful information on system dynamics that can be applied generically. This adaptable approach enables rapid generation of new technology, ultimately tackling complex challenges more effectively. With the growing demand for intelligent sensor systems, signal conditioning and interface electronics will play a significant role, leading to the next generation of innovative and efficient technological solutions (Barlow et al., 2007).

REFERENCES

1. Adami, A., Collini, C., Dahiya, R., et al. (2014). Tactile sensing chips with POSFET array and integrated interface electronics. https://core.ac.uk/download/42353920.pdf (accessed on 14th October 2025).
2. Adami, K. Z., de Vaate, J. G. B., Faulkner, A. J., et al. (2010). Beamforming techniques for large-N aperture arrays. http://arxiv.org/abs/1008.4047 (accessed on 14th October 2025).
3. Adamski, M., Węgrzyn, M., Węgrzyn, A., Barkalov, A., Titarenko, L., et al. (2007). Measurements, models, systems and design. https://core.ac.uk/download/pdf/15312083.pdf (accessed on 14th October 2025).
4. Ananda, C. M. (2009). Civil aircraft advanced avionics architectures–An insight into SARAS avionics, present and future perspective. https://core.ac.uk/download/11874849.pdf (accessed on 14th October 2025).
5. Andersen, R., Binkley, D., Blalock, B., et al. (2003). A miniaturized neuroprosthesis suitable for implantation into the brain. https://core.ac.uk/download/211392335.pdf (accessed on 14th October 2025).
6. Arias Drake, A., Carranza González, L., Ceballos Cáceres, J. F., et al. (2015). A front-end ASIC for a 3-D magnetometer for space applications by using anisotropic magneto resistors. https://core.ac.uk/download/158964962.pdf (accessed on 14th October 2025).
7. Bailey, A. R., Irish, L. A., & Tagge, G. E. (2025). Systems study for an Integrated Digital-Electric Aircraft (IDEA). https://core.ac.uk/download/pdf/42846809.pdf (accessed on 14th October 2025).
8. Balakumar, M., Kumar, N., Swathi, A., Maurya, S. P., & Maurya, P. K. (2024). Bio-electronics interface between electronics and biological systems for healthcare applications. *Power System Technology*. https://www.semanticscholar.org/paper/e881ee4dccc6db76787af311665f8aba6db82077 (accessed on 14th October 2025).
9. Barlow, M., Barron, D., et al. (2007). Mathematical and computer modeling of electro-optic systems using a generic modeling approach. https://core.ac.uk/download/9399574.pdf (accessed on 14th October 2025).
10. Beattie, E. C., & Zickwolf, H. C. (2025). Preliminary study, analysis and design for a power switch for digital engine actuators. https://core.ac.uk/download/pdf/42856790.pdf (accessed on 14th October 2025).
11. Berkery, E. A., & Krausz, A. (2025). Multi-kW DC power distribution system study program. https://core.ac.uk/download/pdf/42894219.pdf (accessed on 14th October 2025).

12. Bernardini Bressan, F., Degrave, J., Leman, M., et al. (2017). PLXTRM: Prediction-led eXtended-guitar tool for real-time music applications and live performance. https://core.ac.uk/download/84045349.pdf (accessed on 14th October 2025).
13. Bojoi, I. R., Griva, G. B., Limongi, L., et al. (2009). Digital current-control schemes. https://core.ac.uk/download/pdf/11404689.pdf (accessed on 14th October 2025).
14. Bowman, J. D., Monsalve, R. A., Mozdzen, T. J., et al. (2016). Calibration of the EDGES high-band receiver to observe the global 21-cm signature from the epoch of reionization. https://core.ac.uk/download/83235302.pdf (accessed on 14th October 2025).
15. Bussooa, A., Mercer, J. R., & Neale, S. (2018). Future of smart cardiovascular implants. https://core.ac.uk/download/161513601.pdf (accessed on 14th October 2025).
16. Caldognetto, T., Liu, G., Magnone, P., et al. (2020). Suppression of second-order harmonic current for droop-controlled distributed energy resource converters in DC microgrids. https://core.ac.uk/download/211056088.pdf (accessed on 14th October 2025).
17. Calvi, M., Cassar, J., Dimitrijevic, S., et al. (2019). Design and development of a reduced form-factor high accuracy three-axis telemeter. https://core.ac.uk/download/195724621.pdf (accessed on 14th October 2025).
18. Carlson, J. N., Kacholia, K. K., Marshall, R. D., et al. (2025). Spacecraft nitrogen generation. https://core.ac.uk/download/pdf/42890598.pdf (accessed on 14th October 2025).
19. Castellanos Ramos, J., Navas-Gonzalez, R. J., Oballe-Peinado, O., et al. (2001). Three realizations and comparison of hardware for piezoresistive tactile sensors. https://core.ac.uk/download/62896728.pdf (accessed on 14th October 2025).
20. Chan, A., Chin, B., Gregory, W., et al. (2025). High accuracy fuel flowmeter, Phase 1. https://core.ac.uk/download/pdf/42853063.pdf (accessed on 14th October 2025).
21. Eikenberry, S. S., Moon, D. S., & Pirger, B. E. (2001). A next generation high-speed data acquisition system for multi-channel infrared and optical photometry. https://core.ac.uk/download/25305902.pdf (accessed on 14th October 2025).
22. Escalera Morón, S., Guerra Vinuesa, O., Medeiro Hidalgo, F., et al. (2005). A 0.35 μm CMOS 17-bit@40-kS/s cascade 2-1 ΣΔ modulator with programmable gain and programmable chopper stabilization. https://core.ac.uk/download/286565025.pdf (accessed on 14th October 2025).

23. Fang, J., Yu, J., Zhang, Y., & Goetz, S. (2021). An estimation-based solution to weak-grid-induced small-signal stability problems of power converters. *IEEE Journal of Emerging and Selected Topics in Power Electronics, 9*, 4558–4572. https://www.semanticscholar.org/paper/88426bd7ed057a795 40864a1a967c9b7489360f4 (accessed on 14th October 2025).

24. Francisco, D. R. (2025). Electrical design of payload G-534: The Pool Boiling Experiment. https://core.ac.uk/download/pdf/42810594.pdf (accessed on 14th October 2025).

25. Gao, W., Ota, H., Kiriya, D., Takei, K., & Javey, A. (2019). Flexible electronics toward wearable sensing. *Accounts of Chemical Research, 52*, 523–533. https://doi.org/10.1021/acs.accounts.8b00500 (accessed on 14th October 2025).

26. Glasheen, W. M., & Poppel, G. L. (2025). Electro-optic architecture for servicing sensors and actuators in advanced aircraft propulsion systems. https://core.ac.uk/download/pdf/42810387.pdf (accessed on 14th October 2025).

27. Harrington, T. M., & Metzger, A. E. (2025). Addition of a gamma ray spectrometer to the alpha scattering experiment as designed for the Surveyor mission. https://core.ac.uk/download/pdf/85239718.pdf (accessed on 14th October 2025).

28. Horváth, D., & Szabó, R. Z. (2019). Driving forces and barriers of Industry 4.0: Do multinational and small and medium-sized companies have equal opportunities? *Technological Forecasting and Social Change, 146*, 119–132. https://doi.org/10.1016/j.techfore.2019.05.021 (accessed on 14th October 2025).

29. Lalli, V. R., & Schoenfeld, A. D. (2025). ASDTIC: A feedback control innovation. https://core.ac.uk/download/pdf/42899897.pdf (accessed on 14th October 2025).

30. Liñán Cembrano, G., Maldonado López, R., Rodríguez Vázquez, Á. B., et al. (2007). Integrated circuit interface for artificial skins. https://core.ac.uk/download/286564168.pdf (accessed on 14th October 2025).

31. Liñán Cembrano, G., Maldonado López, R., Rodríguez Vázquez, Á. B., et al. (2009). Integrated circuitry to detect slippage inspired by human skin and artificial retinas. https://core.ac.uk/download/286564015.pdf (accessed on 14th October 2025).

32. Luo, Y., Abidian, M. R., Ahn, J. H., Akinwande, D., Andrews, A. M., Antonietti, M., Bao, Z., et al. (2023). Technology roadmap for flexible sensors. *ACS Nano, 17*, 5211–5295. https://doi.org/10.1021/acsnano.2c12606 (accessed on 14th October 2025).

33. Martin, J. S., Downey, A., Baalousha, M., & Won, S. (2024). Rapid measurement of magnetic particle concentrations in wildland–urban interface fire ashes and runoff using compact NMR. *IEEE Sensors Journal, 24*, 7355–7363. https://www.semanticscholar.org/paper/3ee85ae3d7a1e7e553acd34aca4899865ad54dc4 (accessed on 14th October 2025).

34. Mishra, S., Sarangi, S. R., Bohidar, S. N., Mishra, D., & Parija, S. (2025). Footstep powered energy harvesting system. *2025 International Conference on Emerging Systems and Intelligent Computing (ESIC)*. https://www.semanticscholar.org/paper/28f0107fac1293a16ac8eca47e4b79bff5ced263 (accessed on 14th October 2025).

35. Mohamad, M. M., & Unadam, F. A. (2022). Industrial training: Challenges among students in construction industry. https://core.ac.uk/download/42955569.pdf (accessed on 14th October 2025).

36. Naresh, V., & Lee, N. (2021). A review on biosensors and recent development of nanostructured materials-enabled biosensors. *Sensors, 21*, 1109–1109. https://doi.org/10.3390/s21041109 (accessed on 14th October 2025).

37. Shah, M., Rajpal, R., Chowdhury, M. B., & Ramaiya, N. (2024). Design up-gradation of electronics for PMT sensor-based systems in ADITYA-U Tokamak. *2024 International Conference on Communication, Control, and Intelligent Systems (CCIS)*. https://www.semanticscholar.org/paper/5af660b9a01d8404ce052f706e9690e461cff757 (accessed on 14th October 2025).

38. Sundberg, G. R. (2025). Advances in solid state switchgear technology for large space power systems. https://core.ac.uk/download/pdf/42849258.pdf (accessed on 14th October 2025).

39. Yang, J. C., Mun, J., Kwon, S. Y., Park, S., Bao, Z., & Park, S. (2019). Electronic skin: Recent progress and future prospects for skin-attachable devices for health monitoring, robotics, and prosthetics. *Advanced Materials, 31*. https://doi.org/10.1002/adma.201904765 (accessed on 14th October 2025).

40. Zhao, Z., Nie, C., Jiang, C. T., Cao, S. H., Tian, K., Yu, S., & Gu, J. (2023). Modulating brain activity with invasive brain–computer interface: A narrative review. *Brain Sciences, 13*. https://www.semanticscholar.org/paper/79b760bebfe817d84ef5c5673a05b2f29d37dd8e (accessed on 14th October 2025).

INDEX

A

B

C

O

P

Q

R

S